C000066893

CODING FOR WIRELESS CHANNELS

Information Technology: Transmission, Processing, and Storage

Series Editors: Robert Gallager
                    Massachusetts Institute of Technology
                    Cambridge, Massachusetts

                    Jack Keil Wolf
                    University of California at San Diego
                    La Jolla, California

CODING FOR WIRELESS CHANNELS

Ezio Biglieri

 Springer

Library of Congress Cataloging-in-Publication Data

Biglieri, Ezio.
    Coding for wireless channels / Ezio Biglieri.
        p. cm. -- (Information technology---transmission, processing, and storage)
    Includes bibliographical references and index.
    ISBN 1-4020-8083-2 (alk. paper) -- ISBN 1-4020-8084-0 (e-book)
    1. Coding theory. 2. Wireless communication systems. I. Title. II. Series.

    TK5102.92 B57 2005
    621.3845'6--cd22
                                                                    2005049014

© 2005 Springer Science+Business Media, Inc.
All rights reserved. This work may not be translated or copied in whole or in part
without the written permission of the publisher (Springer Science+Business Media,
Inc., 233 Spring Street, New York, NY 10013, USA), except for brief excerpts in
connection with reviews or scholarly analysis. Use in connection with any form of
information storage and retrieval, electronic adaptation, computer software, or by
similar or dissimilar methodology now known or hereafter developed is forbidden.
The use in this publication of trade names, trademarks, service marks and similar
terms, even if they are not identified as such, is not to be taken as an expression of
opinion as to whether or not they are subject to proprietary rights.

Printed in the United States of America.

9 8 7 6 5 4 3 2 1   SPIN 11054627

springeronline.com

Contents

Preface

Dios te libre, lector, de prólogos largos
Francisco de Quevedo Villegas, El mundo por de dentro.

There are, so it is alleged, many ways to skin a cat. There are also many ways to teach coding theory. My feeling is that, contrary to other disciplines, coding theory was never a fully unified theory. To describe it, one can paraphrase what has been written about the Enlightenment: "It was less a determined swift river than a lacework of deltaic streams working their way along twisted channels" (E. O. Wilson, *Consilience*, 1999).

The seed of this book was sown in 2000, when I was invited to teach a course on coded modulation at Princeton University. A substantial portion of students enrolled in the course had little or no background in algebraic coding theory, nor did the time available for the course allow me to cover the basics of the discipline. My choice was to start directly with coding in the signal space, with only a marginal treatment of the indispensable aspects of "classical" algebraic coding theory. The selection of topics covered in this book, intended to serve as a textbook for a first-level graduate course, reflects that original choice. Subsequently, I had the occasion to refine the material now collected in this book while teaching Master courses at Politecnico di Torino and at the Institute for Communications Engineering of the Technical University of Munich.

While describing what can be found in this book, let me explain what cannot be found. I wanted to avoid generating an *omnium-gatherum,* and to keep the book length at a reasonable size, resisting encyclopedic temptations ($\mu\acute{\epsilon}\gamma\alpha$ $\beta\iota\beta\lambda\acute{\iota}o\nu$ $\mu\acute{\epsilon}\gamma\alpha$ $\kappa\alpha\kappa\acute{o}\nu$). The leitmotiv here is soft-decodable codes described through graphical structures (trellises and factor graphs). I focus on the basic principles underlying code design, rather than providing a handbook of code design. While an earlier exposure to coding principles would be useful, the material here only assumes that the reader has a firm grasp of the concepts usually presented in senior-lever courses on digital communications, on information theory, and on random processes.

Each chapter contains a topic that can be expatiated upon at book length. To include all facts deserving attention in this tumultuous discipline, and then to clarify their finer aspects, would require a full-dress textbook. Thus, many parts should be viewed akin to movie trailers, which show the most immediate and memorable scenes as a stimulus to see the whole movie.

As the mathematician Mark Kac puts it, a proof is that which convinces a reasonable reader; a rigorous proof is that which convinces an unreasonable reader. I assume here that my readers are reasonable, and hence try to avoid excessive rigor at the price of looking sketchy at times, with many treatments that should be taken modulo mathematical refinements.

The reader will observe the relatively large number of epexegetic figures, justified by the fact that engineers are visual animals. In addition, the curious reader may want to know the origin of the short sentences appearing at the beginning of each chapter. These come from one of the few literary works that was cited by C. E. Shannon in his technical writings. With subtle irony, in his citation he misspelled the work's title, thus proving the power of redundancy in error correction.

Some sections are marked ★. This means that the section's contents are crucial to the developments of this book, and the reader is urged to become comfortable with them before continuing.

Some of the material of this book, including a few proofs and occasional examples, reflects previous treatments of the subject I especially like: for these I am particularly indebted to sets of lecture notes developed by David Forney and by Robert Calderbank.

I hope that the readers of this book will appreciate its organization and contents; nonetheless, I am confident that Pliny the Elder is right when he claims that "there is no book so bad that it is not profitable in some part."

Many thanks are due to colleagues and students who read parts of this book and let me have their comments and corrections. Among them, a special debt of gratitude goes to the anonymous reviewers. I am also grateful to my colleagues Joseph Boutros, Marc Fossorier, Umberto Mengali, Alessandro Nordio, and Giorgio Taricco, and to my students Daniel de Medeiros and Van Thanh Vu. Needless to say, whatever is flawed is nobody's responsibility but mine. Thus, I would appreciate it if the readers who spot any mistake or inaccuracy would write to me at e.biglieri@ieee.org. An errata file will be sent to anyone interested.

Qu'on ne dise pas que je n'ai rien dit de nouveau:
la disposition des matières est nouvelle.
Blaise Pascal, Pensées, 65.

1

one gob, one gap, one gulp and gorger of all!

Tour d'horizon

In this chapter we introduce the basic concepts that will be dealt with in the balance of the book and provide a short summary of major results. We first present coding in the signal space, and the techniques used for decoding. Next, we highlight the basic differences between the additive white Gaussian noise channel and different models of fading channels. The performance bounds following Shannon's results are described, along with the historical development of coding theory.

1.1   Introduction and motivations

This book deals with coding in the signal space and with "soft" decoding. Consider a finite set $S = \{\mathbf{x}\}$ of information-carrying vectors (or *signals*) in the Euclidean N-dimensional space $\mathbb{R}^N$, to be used for transmission over a noisy channel. The output of the channel, denoted $\mathbf{y}$, is observed, and used to decode, i.e., to generate an estimate $\widehat{\mathbf{x}}$ of the transmitted signal. Knowledge of the channel is reflected by the knowledge of the conditional probability distribution $p(\mathbf{y} \mid \mathbf{x})$ of the observable $\mathbf{y}$, given that $\mathbf{x}$ was transmitted. In general, as in the case of fading channels (Chapters 4, 10), $p(\mathbf{y} \mid \mathbf{x})$ depends on some random parameters whose values may or may not be available at the transmitter and the receiver.

The decoder chooses $\widehat{\mathbf{x}}$ by optimizing a predetermined cost function, usually related to the error probability $P(e)$, i.e., the probability that $\widehat{\mathbf{x}} \neq \mathbf{x}$ when $\mathbf{x}$ is transmitted. A popular choice consists of using the maximum-likelihood (ML) rule, which consists of maximizing, over $\mathbf{x} \in S$, the function $p(\mathbf{y} \mid \mathbf{x})$. This rule minimizes the word error probability under the assumption that all code words are equally likely. If the latter assumption is removed, word error probability is minimized if we use the maximum a posteriori (MAP) rule, which consists of maximizing the function

$$p(\mathbf{x} \mid \mathbf{y}) = \frac{p(\mathbf{y} \mid \mathbf{x})p(\mathbf{x})}{p(\mathbf{y})} \propto p(\mathbf{y} \mid \mathbf{x})p(\mathbf{x}) \tag{1.1}$$

(here and in the following, the notation $\propto$ indicates proportionality, with a proportionality factor irrelevant to the decision procedure). To prove the above statements, denote by $\mathcal{R}(\mathbf{x})$ the decision region associated with the transmitted signal $\mathbf{x}$ (that is, the receiver chooses $\mathbf{x}$ if and only if $\mathbf{y} \in \mathcal{R}(\mathbf{x})$). Then

$$
\begin{aligned}
P(e) &\triangleq \sum_{\mathbf{x}} \mathbb{P}(\mathbf{y} \notin \mathcal{R}(\mathbf{x}), \mathbf{x}) \\
&= 1 - \sum_{\mathbf{x}} \int_{\mathcal{R}(\mathbf{x})} p(\mathbf{y}, \mathbf{x}) \, d\mathbf{y} \\
&= 1 - \sum_{\mathbf{x}} \int_{\mathcal{R}(\mathbf{x})} p(\mathbf{x} \mid \mathbf{y}) p(\mathbf{y}) \, d\mathbf{y}
\end{aligned}
$$

$P(e)$ is minimized by independently maximizing each term in the sum, which is obtained by choosing $\mathcal{R}(\mathbf{x})$ as the region where $p(\mathbf{x} \mid \mathbf{y})$ is a maximum over $\mathbf{x}$: thus, the MAP rule yields the minimum $P(e)$. If $p(\mathbf{x})$ does not depend on $\mathbf{x}$, i.e., $p(\mathbf{x})$ is the same for all $\mathbf{x} \in S$, then the $\mathbf{x}$ that maximizes $p(\mathbf{x} \mid \mathbf{y})$ also maximizes $p(\mathbf{y} \mid \mathbf{x})$, and the MAP and ML rules are equivalent.

Selection of S consists of finding practical ways of communicating discrete messages reliably on a real-world channel: this may involve satellite communications, data transmission over twisted-pair telephone wires or shielded cable-TV wires, data storage, digital audio/video transmission, mobile communication, terrestrial radio, deep-space radio, indoor radio, or file transfer. The channel may involve several sources of degradation, such as attenuation, thermal noise, intersymbol interference, multiple-access interference, multipath propagation, and power limitations.

The most general statement about the selection of S is that it should make the best possible use of the resources available for transmission, viz., bandwidth, power, and complexity, in order to achieve the quality of service (QoS) required. In summary, the selection should be based on four factors: error probability, bandwidth efficiency, the signal-to-noise ratio necessary to achieve the required QoS, and the complexity of the transmit/receive scheme. The first factor tells us how reliable the transmission is, the second measures the efficiency in bandwidth expenditure, the third measures how efficiently the transmission scheme makes use of the available power, and the fourth measures the cost of the equipment.

Here we are confronted with a crossroads. As discussed in Chapter 3, we should decide whether the main limit imposed on transmission is the bandwidth- or the power-limitation of the channel.

To clarify this point, let us define two basic parameters. The first one is the spectral (or bandwidth) efficiency R_b/W, which tells us how many bits per second (R_b) can be transmitted in a given bandwidth (W). The second parameter is the *asymptotic power efficiency* γ of a signal set. This parameter is defined as follows. Over the additive white Gaussian noise channel with a high signal-to-noise ratio (SNR), the error probability can be closely approximated by a complementary error function, whose argument is proportional to the ratio between the energy per transmitted information bit $\mathcal{E}_b$ and twice the noise power spectral density of the noise N_0. The proportionality factor γ expresses how efficiently a modulation scheme makes use of the available signal energy to generate a given error probability. Thus, we may say that, at least for high SNR, a signal set is better than another if its asymptotic power efficiency is greater (at low SNR the situation is much more complicated, but the asymptotic power efficiency still plays some role). Some pairs of values of R_b/W and γ that can be achieved by simple choices of S (called *elementary constellations*) are summarized in Table 1.1.

The fundamental trade-off is that, for a given QoS requirement, increased spectral efficiency can be reliably achieved only with a corresponding increase in the minimum required SNR. Conversely, the minimum required SNR can be reduced only by decreasing the spectral efficiency of the system. Roughly, we may say

$\mathcal{S}$	R_b/W	γ
PAM	$2\log_2 M$	$\dfrac{3\log_2 M}{M^2-1}$
PSK	$\log_2 M$	$\sin^2\dfrac{\pi}{M}\cdot\log_2 M$
QAM	$\log_2 M$	$\dfrac{3}{2}\dfrac{\log_2 M}{M-1}$
FSK	$2\dfrac{\log_2 M}{M}$	$\dfrac{1}{2}\log_2 M$

Table 1.1: *Maximum bandwidth- and power-efficiency of some M-ary modulation schemes: PAM, PSK, QAM, and orthogonal FSK.*

that we work in a bandwidth-limited regime if the channel constraints force us to work with a ratio R_b/W much higher than 2, and in a power-limited regime if the opposite occurs. These regimes will be discussed in Chapter 3.

1.2   Coding and decoding

In general, the optimal decision on the transmitted code word may involve a large receiver complexity, especially if the dimensionality of $\mathcal{S}$ is large. For easier decisions it is useful to introduce some structure in $\mathcal{S}$. This process consists of choosing a set $\mathcal{X}$ of *elementary* signals, typically one- or two-dimensional, and generating the elements of $\mathcal{S}$ as vectors whose components are chosen from $\mathcal{X}$: thus, the elements of $\mathcal{S}$ have the form $\mathbf{x} = (x_1, x_2, \ldots, x_n)$ with $x_i \in \mathcal{X}$. The collection of such $\mathbf{x}$ will be referred to as a *code in the signal space*, and $\mathbf{x}$ as a *code word*. In some cases it is also convenient to endow $\mathcal{S}$ with an algebraic structure: we do this by defining a set $\mathcal{C}$ where operations are defined (for example, $\mathcal{C} = \{0, 1\}$ with mod-2 addition and multiplication), and a one-to-one correspondence between elements of $\mathcal{S}$ and $\mathcal{C}$ (in the example above, we may choose $\mathcal{S} = \{+\sqrt{\mathcal{E}}, -\sqrt{\mathcal{E}}\}$, where $\mathcal{E}$ is the average energy of $\mathcal{S}$, and the correspondence $\mathcal{C} \to \mathcal{S}$ obtained by setting $0 \to +\sqrt{\mathcal{E}}, 1 \to -\sqrt{\mathcal{E}}$).

The structure in $\mathcal{S}$ may be described algebraically (we shall deal briefly with this choice in Chapter 3) or by a graphical structure on which the decoding process may be performed in a simple way. The graphical structures we describe in this book are *trellises* (Chapters 5, 6, and 7) and *factor graphs* (Chapters 8 and 9).

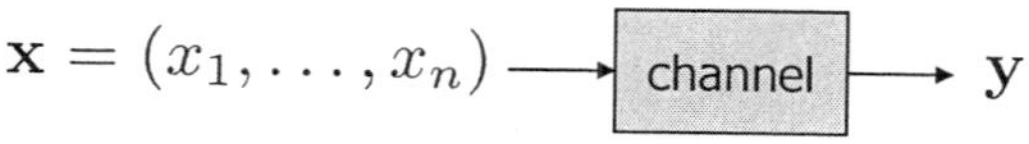

$$\mathbf{x} = (x_1, \ldots, x_n) \longrightarrow \boxed{\text{channel}} \longrightarrow \mathbf{y}$$

Figure 1.1: *Observing a channel output when* $\mathbf{x}$ *is transmitted.*

We shall examine, in particular, how a given code can be described by a graphical structure and how a code can be directly designed, once its graphical structure has been chosen. Trellises used for convolutional codes (Chapter 6) are still the most popular graphical models: the celebrated Viterbi decoding algorithm can be viewed as a way to find the shortest path through one such trellis. Factor graphs (Chapter 8) were introduced more recently. When a code can be represented by a cycle-free factor graph, then the structure of the factor graph of a code lends itself naturally to the specification of a finite algorithm (the sum-product, or the max-sum algorithm) for optimum decoding. If cycles are present, then the decoder proceeds iteratively (Chapter 9), in agreement with a recent trend in decoding, and in general in signal processing, that favors iterative (also known as *turbo*) algorithms.

1.2.1 Algebraic vs. soft decoding

Consider transmission of the n-tuple $\mathbf{x} = (x_1, \ldots, x_n)$ of symbols chosen from $\mathcal{X}$. At the output of the transmission channel, the vector $\mathbf{y} = (y_1, \ldots, y_n)$ is observed (Figure 1.1).

In *algebraic decoding*, a time-honored yet suboptimal decoding method, "hard" decisions are separately made on each component of the received signal $\mathbf{y}$, and then the vector $\widetilde{\mathbf{x}} \triangleq (\hat{x}_1, \ldots, \hat{x}_n)$ is formed. This procedure is called *demodulation* of the elementary constellation. If $\widetilde{\mathbf{x}}$ is an element of $\mathcal{S}$, then the decoder selects $\widehat{\mathbf{x}} = \widetilde{\mathbf{x}}$. Otherwise, it claims that $\widetilde{\mathbf{x}}$ "contains errors," and the structure of $\mathcal{S}$ (usually an algebraic one, hence the name of this decoding technique) is exploited to "correct" them, i.e., to change some components of $\widetilde{\mathbf{x}}$ so as to make $\widetilde{\mathbf{x}}$ an element of $\mathcal{S}$. The channel is blamed for making these errors, which are in reality made by the demodulator.

A substantial improvement in decoding practice occurs by substituting algebraic decoders with soft decoders. In the first version that we shall consider (*soft block decoding*), an ML or a MAP decision is made on the entire code word, rather than symbol by symbol, by maximizing, over $\mathbf{x} \in \mathcal{S}$, the function $p(\mathbf{y} \mid \mathbf{x})$ or $p(\mathbf{x} \mid \mathbf{y})$, respectively. Notice the difference: in soft decoding, the demodulator does not make mistakes that the decoder is expected to correct. Demodulator and decoder are not separate entities of the receiver, but rather a single block: this makes it

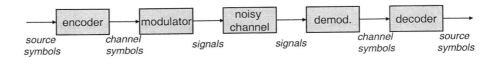

Figure 1.2: *Illustrating error-correction coding theory.*

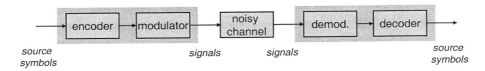

Figure 1.3: *Illustrating error-control coding theory.*

more appropriate to talk about *error-control* rather than *error-correcting* codes. The situation is schematized in Figures 1.2 and 1.3. Soft decoding can be viewed as an application of the general principle [1.11]

> *Never discard information prematurely that may be useful in making a decision until after all decisions related to that information have been completed,*

and often provides a considerable improvement in performance. An often-quoted ballpark figure for the SNR advantage of soft decoders versus algebraic is 2 dB.

Example 1.1

Consider transmission of binary information over the additive white Gaussian channel using the following signal set (a *repetition code*). When the source emits a 0, then three equal signals with positive polarity and unit energy are transmitted; when the source emits a 1, then three equal signals with negative polarity are transmitted. Algebraic decoding consists of individually demodulating the three signals received at the channel output, then choosing a 0 if the majority of demodulated signals exhibits a positive polarity, and choosing a 1 otherwise. The second strategy (soft decoding) consists of demodulating the entire block of three signals, by choosing, between $+++$ and $---$, the one with the smaller Euclidean distance from the received signal.

Assume for example that the signal transmitted is $\mathbf{x} = (+1, +1, +1)$, and that the signal received is $\mathbf{y} = (0.8, -0.1, -0.2)$. Individual demodulation of these signals yields a majority of negative polarities, and hence the (wrong) decision that a 1 was transmitted. On the other hand, the squared Euclidean distances between the received and transmitted signals are

$$d_E^2[(0.8, -0.1, -0.2), (+1, +1, +1)] = (0.8-1)^2 + (-0.1-1)^2 + (-0.2-1)^2 = 2.69$$

and

$$d_E^2[(0.8, -0.1, -0.2), (-1, -1, -1)] = (0.8+1)^2 + (-0.1+1)^2 + (-0.2+1)^2 = 4.69$$

which leads to the (correct) decision that a 0 was transmitted. We observe that in this example the hard decoder fails because it decides without taking into account the fact that demodulation of the second and third received samples is unreliable, as they are relatively close to the zero value. The soft decoder combines this reliability information in the single parameter of Euclidean distance.

The probability of error obtained by using both decoding methods can be easily evaluated. Algebraic decoding fails when there are two or three demodulation errors. Denoting by p the probability of one demodulation error, we have for hard decoding the error probability

$$P_A(e) = 3p^2(1-p) + p^3$$

where $p = Q(\sqrt{2/N_0})$, $N_0/2$ the power spectral density of the Gaussian noise, and $Q(\cdot)$ the Gaussian tail function. For small-enough error probabilities, we have $p \approx \exp(-1/N_0)$, and hence

$$P_A(e) \approx 3p^2 \approx 3\exp(-2/N_0)$$

For soft decoding, $P(e)$ is the same as for transmission of binary antipodal signals with energy 3 [1.1]:

$$P_S(e) = Q\left(\sqrt{\frac{6}{N_0}}\right) \approx \exp(-3/N_0)$$

This result shows that soft decoding of this code can achieve (even disregarding the factor of 3) the same error performance of algebraic decoding with a signal-to-noise ratio smaller by a factor of 3/2, corresponding to 1.76 dB. $\qquad\square$

In Chapters 5, 6, and 7, we shall see how trellis structures and the Viterbi algorithm can be used for soft block decoding.

Symbol-by-symbol decoding

Symbol-by-symbol soft decoders may also be defined. They minimize *symbol error probabilities*, rather than *word error probabilities*, and work, in contrast to algebraic decoding, by supplying, rather than "hard" tentative decisions for the

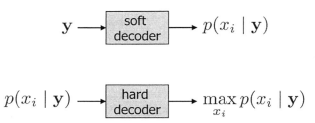

Figure 1.4: *MAP decoding: soft and hard decoder.*

various symbols, the so-called *soft decisions*. A soft decision for x_i is the a posteriori probability distribution of x_i given $\mathbf{y}$, denoted $p(x_i|\mathbf{y})$. A *hard decision* for x_i is a probability distribution such that $p(x_i|\mathbf{y})$ is equal either to 0 or to 1. The combination of a soft decoder and a hard decoder (the task of the former usually being much harder that the latter's) yields symbol-by-symbol maximum a posteriori (MAP) decoding (Figure 1.4). We can observe that the task of the hard decoder, which maximizes a function of a discrete variable (usually taking a small number of values) is far simpler than that of the soft decoder, which must *marginalize* a function of several variables. Chapter 8 will discuss how this marginalization can be done, once the code is given a suitable graphical description.

1.3   The Shannon challenge

In 1948, Claude E. Shannon demonstrated that, for any transmission rate less than or equal to a parameter called *channel capacity*, there exists a coding scheme that achieves an arbitrarily small probability of error, and hence can make transmission over the channel perfectly reliable. Shannon's proof of his capacity theorem was nonconstructive, and hence gave no guidance as to how to find an actual coding scheme achieving the ultimate performance *with limited complexity*. The cornerstone of the proof was the fact that if we pick a long code *at random*, then its average probability of error will be satisfactorily low; moreover, there exists at least one code whose performance is at least as good as the average. Direct implementation of random coding, however, leads to a decoding complexity that prevents its actual use, as there is no practical encoding or decoding algorithm. The general decoding problem (find the maximum-likelihood vector $\mathbf{x} \in \mathcal{S}$ upon observation of $\mathbf{y} = \mathbf{x} + \mathbf{z}$) is NP-complete [1.2].

Figure 1.5 summarizes some of Shannon's finding on the limits of transmission at a given rate ρ (in bits per dimension) allowed on the additive white Gaus-

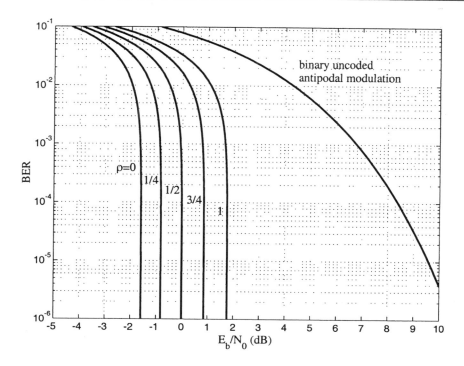

Figure 1.5: *Admissible region for the pair BER,* $\mathcal{E}_b/N_0$. *For a given code rate* ρ, *only the region above the curve labeled* ρ *is admissible. The BER curve corresponding to uncoded binary antipodal modulation is also shown for comparison.*

sian noise channel with a given bit-error rate (BER). This figure shows that the ratio $\mathcal{E}_b/N_0$, where $\mathcal{E}_b$ is the energy spent for transmitting one bit of information at a given BER over an additive white Gaussian noise channel and $N_0/2$ is the power spectral density of the channel noise, must exceed a certain quantity. In addition, a code exists whose performance approaches that shown in the Figure. For example, for small-BER transmission at rate $\rho = 1/2$, Shannon's limits dictate $\mathcal{E}_b/N_0 > 0$ dB, while for a vanishingly small rate one must guarantee $\mathcal{E}_b/N_0 > -1.6$ dB. Performance limits of coded systems when the channel input is restricted to a certain elementary constellation could also be derived. For example, for $\rho = 1/2$, if we restrict the input to be binary we must have $\mathcal{E}_b/N_0 > 0.187$ dB.

Since 1948, communication engineers have been trying hard to develop practically implementable coding schemes in an attempt to approach ideal performance, and hence channel capacity. In spite of some pessimism (for a long while the motto of coding theorists was "good codes are messy") the problem was eventually solved in the early 1990s, at least for an important special case, the additive white Gaussian channel. Among the most important steps towards this solution, we may recall Gallager's low-density parity-check (LDPC) codes with iterative decoding (discovered in 1962 [1.9] and rediscovered much later: see Chapter 9); binary convolutional codes, which in the 1960s were considered a practical solution for operating about 3 dB away from Shannon's limit; and Forney's concatenated codes (a convolutional code concatenated with a Reed–Solomon code can approach Shannon's limit by 2.3 dB at a BER of 10^{-5}). In 1993, a new class of codes called *turbo codes* was disclosed, which could approach Shannon's bound by 0.5 dB. Turbo codes are still among the very best codes known: they combine a random-like behavior (which is attractive in the light of Shannon's coding theorem) with a relatively simple structure, obtained by concatenating low-complexity compound codes. They can be decoded by separately soft-decoding their component codes in an iterative process that uses partial information available from all others. This discovery kindled a considerable amount of new research, which in turn led to the rediscovery, 40 years later, of the power and efficiency of LDPC codes as capacity-approaching codes. Further research has led to the recognition of the *turbo principle* as a key to decoding capacity-approaching codes, and to the belief that almost any simple code interconnected by a large pseudorandom interleaver and iteratively decoded will yield near-Shannon performance [1.7]. In recent years, code designs have been exhibited which progressively chip away at the small gap separating their performance from Shannon's limit. In 2001, Chung, Forney, Richardson, and Urbanke [1.5] showed that a certain class of LDPC codes with iterative decoding could approach that limit within 0.0045 dB.

1.3.1 Bandwidth- and power-limited regime

Binary error-control codes can be used in the power-limited (i.e., wide-bandwidth, low-SNR) regime to increase the power efficiency by adding redundant symbols to the transmitted symbol sequence. This solution requires the modulator to operate at a higher data rate and, hence, requires a larger bandwidth. In a bandwidth-limited environment, increased efficiency in power utilization can be obtained by choosing solutions whereby higher-order elementary constellations (e.g., 8-PSK instead of 2-PSK) are combined with high-rate coding schemes. An early solution consisted of employing uncoded multilevel modulation; in the mid-1970s the invention of trellis-coded modulation (TCM) showed a different way [1.10]. The TCM solution (described in Chapter 7) combines the choice of a modulation scheme with that of a convolutional code, while the receiver does soft decoding. The redundancy necessary to power savings is obtained by a factor-of-2 expansion of the size of the elementary-signal constellation $\mathfrak{X}$. Table 1.2 summarizes some of the energy savings ("coding gains") in dB that can be obtained by doubling the constellation size and using TCM. These refer to coded 8-PSK (relative to uncoded 4-PSK) and to coded 16-QAM (relative to uncoded 8-PSK). These gains can actually be achieved only for high SNRs, and they decrease as the latter decrease. The complexity of the resulting decoder is proportional to the number of states of the trellis describing the TCM scheme.

Number of states	coding gain (8-PSK)	coding gain (16-QAM)
4	3.0	4.4
8	3.6	5.3
16	4.1	6.1
32	4.6	6.1
64	4.8	6.8
128	5.0	7.4
256	5.4	7.4

Table 1.2: *Asymptotic coding gains of TCM (in dB).*

1.4 The wireless channel

Coding choices are strongly affected by the channel model. We examine first the Gaussian channel, because it has shaped the coding discipline. Among the many other important channel models, some arise in digital wireless transmission. The consideration of wireless channels, where nonlinearities, Doppler shifts, fading, shadowing, and interference from other users make the simple AWGN channel model far from realistic, forces one to revisit the Gaussian-channel paradigms described in Chapter 3. Over wireless channels, due to fading and interference the signal-to-disturbance ratio becomes a random variable, which brings into play a number of new issues, among them optimum power allocation. This consists of choosing, based on channel measurements, the minimum transmit power that can compensate for the channel effects and hence guarantee a given QoS.

Among the most common wireless channel models (Chapters 2, 4), we recall the flat independent fading channel (where the signal attenuation is constant over one symbol interval, and changes independently from symbol to symbol), the block-fading channel (where the signal attenuation is constant over an N-symbol block, and changes independently from block to block), and a channel operating in an interference-limited mode. This last model takes into consideration the fact that in a multiuser environment a central concern is overcoming interference, which may limit the transmission reliability more than noise.

1.4.1   The flat fading channel

This simplest fading channel model assumes that the duration of a signal is much greater than the delay spread caused by multipath propagation. If this is true, then all frequency components in the transmitted signal are affected by the same random attenuation and phase shift, and the channel is frequency-flat. If in addition the channel varies very slowly with respect to the elementary-signal duration, then the fading level remains approximately constant during the transmission of one signal (if this does not occur, the fading process is called *fast.*)

The assumption of a frequency-flat fading allows it to be modeled as a process affecting the transmitted signal in a multiplicative form. The additional assumption of slow fading reduces this process to a sequence of random variables, each modeling an attenuation that remains constant during each elementary-signal interval. In conclusion, if x denotes the transmitted elementary signal, then the signal received at the output of a channel affected by slow, flat fading, and additive white Gaussian noise, and demodulated coherently, can be expressed in the form

$$y = Rx + z \qquad (1.2)$$

where z is a complex Gaussian noise and R is a Gaussian random variable, having a Rice or Rayleigh pdf.

It should be immediately apparent that, with this simple model of fading channel, the only difference with respect to an AWGN channel, described by the input–output relationship

$$y = x + z \qquad (1.3)$$

resides in the fact that R, instead of being a constant attenuation, is now a random variable whose value affects the amplitude, and hence the power, of the received signal. A key role here is played by the channel state information (CSI), i.e., the fade level, which may be known at the transmitter, at the receiver, or both. Knowledge of CSI allows the transmitter to use power control, i.e., to adapt to the fade level the energy associated with x, and the receiver to adapt its detection strategy.

Figure 4.2 compares the error probability over the Gaussian channel with that over the Rayleigh fading channel without power control (a binary, equal-energy uncoded modulation scheme is assumed, which makes CSI at the receiver irrelevant). This simple example shows how considerable the loss in energy efficiency is. Moreover, in the power-limited environment typical of wireless channels, the simple device of increasing the transmitted energy to compensate for the effect of fading is not directly applicable. A solution is consequently the use of coding, which can compensate for a substantial portion of this loss.

Coding for the slow, flat Rayleigh fading channel

Analysis of coding for the slow, flat Rayleigh fading channel proves that Hamming distance (also called *code diversity* in this context) plays the central role here. Assume transmission of a coded sequence $\mathbf{x} = (x_1, x_2, \ldots, x_n)$, where the components of $\mathbf{x}$ are signals selected from an elementary constellation. We do not distinguish here among block or convolutional codes (with soft decoding), or block- or trellis-coded modulation. We also assume that, thanks to perfect (i.e., infinite-depth) interleaving, the fading random variables affecting the various signals x_k are independent. Finally, it is assumed that the detection is coherent, i.e., that the phase shift due to fading can be estimated and hence removed.

We can calculate the probability that the receiver prefers the candidate code word $\widehat{\mathbf{x}}$ to the transmitted code word $\mathbf{x}$ (this is called the *pairwise error probability* and is the basic building block of any error probability evaluation). This probability is approximately inversely proportional to the *product* of the squared Euclidean distances between the components of $\mathbf{x}, \widehat{\mathbf{x}}$ that differ, and, to a more relevant extent, to a power of the signal-to-noise ratio whose exponent is the Hamming distance between $\mathbf{x}$ and $\widehat{\mathbf{x}}$, called the *code diversity*. This result holds under the assumption that perfect CSI is available at the receiver.

Robustness

From the previous discussion, it is accepted that coding schemes optimum for this channel should maximize the Hamming distance between code words. Now, if the channel model is uncertain or is not stationary enough to design a coding scheme closely matched to it, then the best proposition may be that of a "robust" solution, that is, a solution that provides suboptimum (but close to optimum) performance on a wide variety of channel models. The use of antenna diversity with maximal-ratio combining (Section 4.4.1) provides good performance on a wide variety of fading environments. The simplest approach to understanding receive-antenna diversity is based on the fact that, since antennas generate multiple transmission channels, the probability that the signal will be simultaneously faded on all channels can be made small, and hence the detector performance improves. Another perspective is based upon the observation that, under fairly general conditions, a channel affected by fading can be turned into an additive white Gaussian noise (AWGN) channel by increasing the number of antenna-diversity branches and using maximum-ratio combining (which requires knowledge of CSI at the receiver). Consequently, it can be expected (and verified by analyses and simulations) that a coded modulation scheme designed to be optimal for the AWGN channel will perform asymptotically well also on a fading channel with diversity, at the cost only of an increased receiver complexity.

We may also think of space or time or frequency diversity as a special case of coding. In fact, the various diversity schemes may be seen as implementations of the simple repetition code, whose Hamming distance turns out to be equal to the number of diversity branches. Another robust solution is offered by bit-interleaved coded modulation, which consists of separating encoder and modulator with a bit interleaver, as described in Section 7.9.

1.5 Using multiple antennas

Multiple receive antennas can be used as an alternative to coding, or in conjunction with it, to provide rate and diversity gain. Assume that t transmit and r receive antennas are used. Then, a multiplicity of transmit antennas creates a set of parallel channels that can be used to potentially increase the data rate up to a factor of $\min\{t, r\}$ (with respect to single-antenna transmission) and hence generate a *rate gain*. The other gain is due to the number of independent paths traversed by each signal, which has a maximum value rt. There is a fundamental trade-off between these two gains: for example, maximum rate gain, obtained by simultaneously sending independent signals, entails no diversity gain, while maximum diversity gain, obtained by sending the same signal from all antennas, generates no rate gain. This point is addressed in Section 10.14.

Recent work has explored the ultimate performance limits in a fading environment of systems in which multiple antennas are used at both transmitter and receiver side. It has been shown that, in a system with t transmit and r receive antennas and a slow fading channel modeled by an $t \times r$ matrix with random i.i.d. complex Gaussian entries (the *independent Rayleigh fading* assumption), the average channel capacity with perfect CSI at the receiver is about $m \triangleq \min\{t, r\}$ times larger than that of a single-antenna system for the same transmitted power and bandwidth. The capacity increases by about m bit/s/Hz for every 3-dB increase in signal-to-noise ratio (SNR). A further performance improvement can be achieved under the assumption that CSI is available at the transmitter as well. Obtaining transmitter CSI from multiple transmitting antennas is particularly challenging because the transmitter should achieve instantaneous information about the fading channel. On the other hand, if transmit CSI is missing, the transmission scheme employed should guarantee good performance with the majority of possible channel realizations. Codes specifically designed for a multiple-antenna system use degrees of freedom in both space and time and are called *space–time codes*.

1.6 Some issues not covered in this book

1.6.1 Adaptive coding and modulation techniques

Since wireless channels exhibit a time-varying response, adaptive transmission strategies look attractive to prevent insufficient utilization of the channel capacity. The basic idea behind adaptive transmission consists of allocating power and rate to take advantage of favorable channel conditions by transmitting at high speeds, while at the same time counteracting bad conditions by reducing the throughput. For an assigned QoS, the goal is to increase the average spectral efficiency by taking advantage of the transmitter having knowledge of the CSI. The amount of performance improvement provided by such knowledge can be evaluated in principle by computing the Shannon capacity of a given channel with and without it. However, it should be kept in mind that capacity results refer to a situation in which complexity and delay are not constrained. Thus, for example, for a Rayleigh fading channel with independently faded elementary signals, the capacity with channel state information (CSI) at the transmitter and the receiver is only marginally larger than for a situation in which only the receiver has CSI. This result implies that if very powerful and complex codes are used, then CSI at the transmitter can buy little. However, in a delay- and complexity-constrained environment, a considerable gain can be achieved. Adaptive techniques are based on two steps: (a) measurement of the parameters of the transmission channel and (b) selection of one or more transmission parameters based on the optimization of a preassigned cost function. A basic assumption here is that the channel does not vary too rapidly; otherwise, the parameters selected might be badly matched to the channel. Thus, adaptive techniques can only be beneficial in a situation where the Doppler spread is not too wide. This conclusion makes adaptive techniques especially attractive in an indoors environment, where propagation delays are small and the relative speed between transmitter and receiver is typically low. In these conditions, adaptive techniques can work on a frame-by-frame basis.

1.6.2   Unequal error protection

In some analog source coding applications, like speech or video compression, the sensitivity of the source decoder to errors in the coded symbols is typically not uniform: the quality of the reconstructed analog signal is rather insensitive to errors affecting certain classes of bits, while it degrades sharply when errors affect other classes. This happens, for example, when analog source coding is based on some form of hierarchical coding, where a relatively small number of bits carry the "fundamental information" and a larger number of bits carries the "details," like in the case of MPEG standards.

If we assume that the source encoder produces frames of binary coded symbols, each frame can be partitioned into classes of symbols of different "importance" (i.e., of different sensitivity). Then, it is apparent that the best coding strategy aims at achieving lower BER levels for the important classes while admitting higher BER levels for the unimportant ones. This feature is referred to as *unequal error protection*.

A conceptually similar solution to the problem of avoiding degradations of the channel having a catastrophic effect on the transmission quality is *multiresolution modulation*. This process generates a hierarchical protection scheme by using a signal constellation consisting of clusters of points spaced at different distances. The minimum distance between two clusters is higher than the minimum distance within a cluster. The most significant bits are assigned to clusters, and the least significant bits to signals in a cluster.

1.7   Bibliographical notes

Comprehensive reviews of coding-theory development and applications can be found in [1.4, 1.6, 1.8]. Ref. [1.3] gives an overview of the most relevant information-theoretic aspects of fading channels.

References

[1.1] S. Benedetto and E. Biglieri, *Digital Transmission Principles with Wireless Applications*. New York: Kluwer/Plenum, 1999.

[1.2] E. R. Berlekamp, R. J. McEliece, and H. C. A. van Tilborg, "On the intractability of certain coding problems," *IEEE Trans. Inform. Theory*, Vol. 24, pp. 384–386, May 1978.

[1.3] E. Biglieri, J. Proakis, and S. Shamai (Shitz), "Fading channels: Information-theoretic aspects," *IEEE Trans. Inform. Theory*, Vol. 44, No. 6, pp. 2169–2692, October 1998.

[1.4] A. R. Calderbank, "The art of signaling: Fifty years of coding theory," *IEEE Trans. Inform. Theory*, Vol. 44, No. 6, pp. 2561–2595, October 1998.

[1.5] S.-Y. Chung, G. D. Forney, Jr., T. J. Richardson, and R. Urbanke, "On the design of low-density parity-check codes within 0.0045 dB of the Shannon limit," *IEEE Commun. Letters*, Vol. 5, No. 2, pp. 58–60, February 2001.

[1.6] D. J. Costello, Jr., J. Hagenauer, H. Imai, and S. B. Wicker, "Applications of error-control coding," *IEEE Trans. Inform. Theory*, Vol. 44, No. 6, pp. 2531–2560, October 1998.

[1.7] G. D. Forney, Jr., "Codes on graphs: News and views," *Proc. 2nd Int. Symp. Turbo Codes and Related Topics*, Brest, France, pp. 9-16, September 4-7, 2000.

[1.8] G. D. Forney, Jr., and G. Ungerboeck, "Modulation and coding for linear Gaussian channels," *IEEE Trans. Inform. Theory*, Vol. 44, No. 6, pp. 2384–2415, October 1998.

[1.9] R. G. Gallager, *Low-density Parity-check Codes*. Cambridge, MA: MIT Press, 1963.

[1.10] G. Ungerboeck, "Channel coding with multilevel/phase signals," *IEEE Trans. Inform. Theory*, Vol. 28, No. 1, pp. 55–67, January 1982.

[1.11] A. J. Viterbi, "Wireless digital communication: A view based on three lessons learned," *IEEE Communications Magazine*, Vol. 29, No. 9, pp. 33–36, September 1991.

2

Two. Else there is danger of. Solitude.

Channel models for digital transmission

The aim of this chapter is to examine in some detail the problem of constructing models for the channels to be used for digital transmission. The emphasis here is on wireless channels, and special attention will be paid to modeling of fading. The impact of the model on the performance assessment of a coded transmission system is also discussed.

2.1   Time- and frequency-selectivity

We work with baseband-equivalent channel models, both continuous time and discrete time. In this Chapter we use the following notations: in continuous time, $s(t)$, $y(t)$, and $w(t)$ denote the transmitted signal, the received signal, and the additive noise, respectively. In discrete time we use the notations $s(n)$, $y(n)$, and $w(n)$, with n the discrete time. We consider only *linear* channels here. The most general model is

$$y(t) = \int h(t;\tau)s(t-\tau)\,d\tau + w(t) \qquad y(n) = \sum_k h(n;k)s(n-k) + w(n) \quad (2.1)$$

where $h(t;\tau)$ is the channel response at time t to a unit impulse $\delta(\,\cdot\,)$ transmitted at time $t - \tau$. Similarly, $h(n;k)$ is the channel impulse response at time n to a unit impulse $\delta(n)$ transmitted at time $n - k$. This channel is said to be *time selective and frequency selective*, where time selectivity refers to the presence of a time-invariant impulse response and frequency selectivity to an input–output relationship described by a convolution between input and impulse response. By assuming that the sum in (2.1) includes $L + 1$ terms, we can represent the discrete channel by using the convenient block diagram of Figure 2.1, where z^{-1} denotes unit delay.

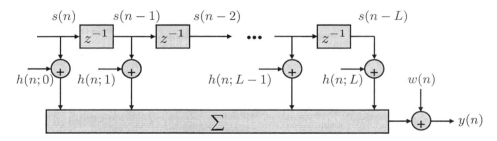

Figure 2.1: *Block diagram of a discrete time-selective, frequency-selective channel.*

If the channel is time invariant, then $h(t;\tau)$ is a constant function of t. We write $h(\tau) \triangleq h(0;\tau)$ for the (time-invariant) response of the channel to a unit impulse transmitted at time 0, and we have the following model of a *non-time-selective, frequency-selective* channel:

$$y(t) = \int h(\tau)s(t-\tau)\,d\tau + w(t) \qquad y(n) = \sum_k h(k)s(n-k) + w(n) \quad (2.2)$$

The block diagram of Figure 2.1 is still valid for this channel, provided that we write $h(k)$ in lieu of $h(n;k)$.

The model of a *time-selective, non-frequency-selective* channel is obtained by assuming that $h(t;\tau) = h(t)\delta(\tau)$ (or, for discrete channels, $h(n;k) = h(n)\delta(k)$). Then we have

$$
\begin{aligned}
y(t) &= \int h(t;\tau)s(t-\tau)\,d\tau + w(t) \\
&= \int h(t)\delta(\tau)s(t-\tau)\,d\tau + w(t) \\
&= h(t)s(t) + w(t)
\end{aligned}
\tag{2.3}
$$

and

$$
\begin{aligned}
y(n) &= \sum_k h(n;k)s(n-k) + w(n) \\
&= \sum_k h(n)\delta(k)s(n-k) + w(n) \\
&= h(n)s(n) + w(n)
\end{aligned}
\tag{2.4}
$$

We observe that in (2.3) and (2.4) the channel impulse response affects the transmitted signal multiplicatively, rather than through a convolution.

Finally, a *non-time-selective, non-frequency-selective* channel model is obtained by assuming that, in (2.3), $h(t;\tau)$ does not depend on t; if it has the form $h(t;\tau) = h\delta(\tau)$ (or, for discrete channels, $h(n;k) = h\delta(n)$), we obtain

$$
y(t) = hs(t) + w(t) \qquad y(n) = hs(n) + w(n)
\tag{2.5}
$$

The simplest situation here occurs when h is a deterministic constant (later on we shall examine the case of h being a random variable). If in addition $w(t)$ is white Gaussian noise, the resulting channel model is called an *additive white Gaussian noise* (AWGN) channel. Typically, it is assumed that $h = 1$ so that the only parameter needed to characterize this channel is the power spectral density of $w(t)$.

2.2 Multipath propagation and Doppler effect

The received power in a radio channel is affected by attenuations that are conveniently characterized as a combination of three effects, as follows:

(a) The *path loss* is the signal attenuation due to the fact that the power received by an antenna at distance D from the transmitter decreases as D increases.

Empirically, the power attenuation is proportional to D^α, with α an exponent whose typical values range from 2 to 4. In a mobile environment, D varies with time, and consequently so does the path loss. This variation is the slowest among the three attenuation effects we are examining here.

(b) The *shadowing* loss is due to the absorption of the radiated signal by scattering structures. It is typically modeled by a random variable with log-normal distribution.

(c) The *fading loss* occurs as a combination of two phenomena, whose combination generates random fluctuations of the received power. These phenomena are *multipath propagation* and *Doppler frequency shift*. In the following we shall focus our attention on these two phenomena, and on mathematical models of the fading they generate.

Multipath propagation occurs when the electromagnetic field carrying the information signal propagates along more than one "path" connecting the transmitter to the receiver. This simple picture of assuming that the propagation medium includes several paths along which the electromagnetic energy propagates, although not very accurate from a theoretical point of view, is nonetheless useful to understand and to analyze propagation situations that include reflection, refraction, and scattering of radio waves. Such situations occur, for example, in indoor propagation, where the electromagnetic waves are perturbed by structures inside the building, and in terrestrial mobile radio, where multipath is caused by large fixed or moving objects (buildings, hills, cars, etc.).

Example 2.1 (Two-path propagation)

Assume that the transmitter and the receiver are fixed and that two propagation paths exist. This is a useful model for the propagation in terrestrial microwave radio links. The received signal can be written in the form

$$y(t) = x(t) + b\,x(t - \tau) \tag{2.6}$$

where b and τ denote the relative amplitude and the differential delay of the reflected signal, respectively (in other words, it is assumed that the direct path has attenuation 1 and delay 0). Equation (2.6) models a static multipath situation in which the propagation paths remain fixed in their characteristics and can be identified individually. The channel is linear and time invariant. Its transfer function

$$H(f) = 1 + b\,e^{-j2\pi f\tau}$$

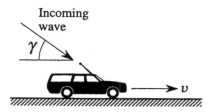

Incoming
wave

γ

v

Figure 2.2: *Effect of movement: Doppler effect.*

in which the term $b \exp(-j2\pi f\tau)$ describes the multipath component, has magnitude

$$|H(f)| = \sqrt{(1 + b\cos 2\pi f\tau)^2 + b^2 \sin^2 2\pi f\tau}$$
$$= \sqrt{1 + b^2 + 2b\cos 2\pi f\tau}$$

For certain delays and frequencies, the two paths are essentially in phase alignment, so $\cos 2\pi f\tau \approx 1$, which produces a large value of $|H(f)|$. For some other values, the paths nearly cancel each other, so $\cos 2\pi f\tau \approx -1$, which produces a minimum of $|H(f)|$ usually referred to as a *notch*. $\qquad\square$

When the receiver and the transmitter are in relative motion with constant radial speed, the received signal is subject to a constant frequency shift (the *Doppler shift*) proportional to this speed and to the carrier frequency. Consider the situation depicted in Figure 2.2. Here the receiver is in relative motion with respect to the transmitter. The latter transmits an unmodulated carrier with frequency f_0. Let v denote the speed of the vehicle (assumed constant), and γ the angle between the direction of propagation of the electromagnetic plane wave and the direction of motion. The Doppler effect causes the received signal to be a tone whose frequency is displaced (decreased) by an amount

$$f_D = f_0 \frac{v}{c} \cos\gamma \tag{2.7}$$

(the *Doppler frequency shift*), where c is the speed of propagation of the electromagnetic field in the medium. Notice that the Doppler frequency shift is either greater or lower than 0, depending on whether the transmitter is moving toward the receiver or away from it (this is reflected by the sign of $\cos\gamma$).

By disregarding for the moment the attenuation and the phase shift affecting the received signal, we can write it in the form

$$y(t) = A\exp[j2\pi(f_0 - f_D)t] \tag{2.8}$$

Notice that we have assumed a constant vehicle speed, and hence a constant f_D. Variations of v would cause a time-varying f_D in (2.8).

More generally, consider now the transmission of a bandpass signal $x(t)$, and take attenuation $\alpha(t)$ and delay $\tau(t)$ into account. The complex envelope of the received signal is

$$\tilde{y}(t) = \alpha(t)e^{-j\theta(t)}\tilde{x}[t - \tau(t)]$$

where

$$\theta(t) = 2\pi\left[(f_0 + f_D)\tau(t) - f_D t\right]$$

This channel can be modeled as a time-varying linear system with low-pass equivalent impulse response

$$h(t; \tau) = 2\alpha(t)\,e^{-j\theta(t)}\,\delta[t - \tau(t)]$$

2.3 Fading

In general, the term *fading* describes the variations with time of the received signal strength. Fading, due to the combined effects of multipath propagation and of relative motion between transmitter and receiver, generates time-varying attenuations and delays that may significantly degrade the performance of a communication system.

With multipath and motion, the signal components arriving from the various paths with different delays combine to produce a distorted version of the transmitted signal. A simple example will illustrate this fact.

Example 2.2 (A simple example of fading)

Consider now the more complex situation represented in Figure 2.3. A vehicle moves at constant speed v along a direction that we take as the reference for angles. The transmitted signal is again an unmodulated carrier at frequency f_0. It propagates along two paths, which for simplicity we assume to have the same delay (zero) and the same attenuation. Let the angles under which the two paths are received be 0 and γ. Due to the Doppler effect, the received signal is

$$y(t) = A\exp\left[j2\pi f_0\left(1 - \frac{v}{c}\right)t\right] + A\exp\left[j2\pi f_0\left(1 - \frac{v}{c}\cos\gamma\right)t\right] \qquad (2.9)$$

We observe from the above equation that the transmitted sinusoid is received as a pair of tones: this effect can be viewed as a spreading of the transmitted signal frequency, and hence as a special case of frequency dispersion caused by the channel and due to the combined effects of Doppler shift and multipath propagation.

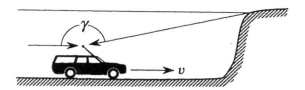

Figure 2.3: *Effect of a two-path propagation and movement.*

Equation (2.9) can be rewritten in the form

$$y(t) = A \left[\exp\left(-j2\pi f_0 \frac{v}{c} t\right) + \exp\left(-j2\pi f_0 \frac{v}{c} \cos\gamma t\right) \right] e^{j2\pi f_0 t} \qquad (2.10)$$

The magnitude of the term in square brackets provides the instantaneous envelope of the received signal:

$$R(t) = 2A \left| \cos\left[2\pi \frac{v}{c} f_0 \frac{1 - \cos\gamma}{2} t \right] \right|$$

The last equation shows an important effect: the envelope of the received signal exhibits a sinusoidal variation with time, occurring with frequency

$$\frac{v}{c} f_0 \frac{1 - \cos\gamma}{2}$$

The resulting channel has a time-varying response. We have time-selective fading, and, as observed before, also frequency dispersion. $\qquad\square$

A more complex situation, occurring when the transmission environment includes several reflecting obstacles, is described in the example that follows.

Example 2.3 (Multipath propagation and the effect of movement)

Assume that the transmitted signal (an unmodulated carrier as before) is received through N paths. The situation is depicted in Figure 2.4. Let the receiver be in motion with velocity v, and let A_i, θ_i, and γ_i denote the amplitude, the phase, and the angle of incidence of the ray from the ith path, respectively. The received signal contains contributions with a variety of Doppler shifts: in the ith path the carrier frequency f_0 is shifted by

$$f_i \triangleq f_0 \frac{v}{c} \cos\gamma_i, \qquad i = 1, 2, \ldots, N$$

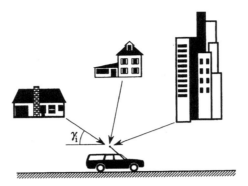

Figure 2.4: *Effect of N-path propagation and movement.*

Thus, the (analytic) received signal can be written in the form

$$y(t) = \sum_{i=1}^{N} A_i \exp j[2\pi(f_0 - f_i)t + \theta_i] \qquad (2.11)$$

The complex envelope of the received signal turns out to be

$$R(t)e^{j\Theta(t)} = \sum_{i=1}^{N} A_i e^{-j(2\pi f_i t - \theta_i)}$$

□

2.3.1 Statistical models for fading channels

As we can observe from the previous examples, our ability to model the channel is connected to the possibility of deriving the relevant propagation parameters. Clearly, this is increasingly difficult and becomes quickly impractical as the number of parameters increases. A way out of this impasse, and one that leads to models that are at the same time accurate and easily applicable, is found in the use of the central limit theorem whenever the propagation parameters can be modeled as random variables (RV) and their number is large enough. To be specific, let us refer to the situation of Example 2.3. For a large number N of paths, we may assume that the attenuations A_i and the phases $2\pi f_i t - \theta_i$ in (2.11) are random variables that can be reasonably assumed to be independent of each other. Then, invoking the central limit theorem, we obtain that at any instant, as the number of contributing paths become large, the sum in (2.11) approaches a Gaussian RV. The complex envelope of the received signal becomes a lowpass Gaussian process whose real and imaginary parts are independent and have mean zero and the same variance σ^2. In these conditions, $R(t)$ and $\Theta(t)$ turn out to be independent processes, with $\Theta(t)$ being uniformly distributed in $(0, 2\pi)$ and $R(t)$ having a Rayleigh probability density function (pdf), viz.,

$$p_R(r) = \begin{cases} \dfrac{r}{\sigma^2}e^{-r^2/2\sigma^2}, & 0 \leq r < \infty \\ 0, & r < 0 \end{cases} \tag{2.12}$$

Here the average power of the envelope is given by

$$\mathbb{E}[R^2] = 2\sigma^2 \tag{2.13}$$

A channel whose envelope pdf is (2.12) is called a *Rayleigh fading channel*. The Rayleigh pdf is often used in its "normalized" form, obtained by choosing $\mathbb{E}[R^2] = 1$:

$$p_R(r) = 2re^{-r^2} \tag{2.14}$$

An alternative channel model can be obtained by assuming that, as often occurs in practice, the propagation medium has, in addition to the N weaker "scatter" paths, one major strong fixed path (often called a *specular* path) whose magnitude is known. Thus, we may write the received-signal complex envelope in the form

$$R(t)e^{j\Theta(t)} = u(t)e^{j\alpha(t)} + v(t)e^{j\beta(t)}$$

where, as before, $u(t)$ is Rayleigh distributed, $\alpha(t)$ is uniform in $(0, 2\pi)$, and $v(t)$ and $\beta(t)$ are deterministic signals. With this model, $R(t)$ has the *Rice* pdf

$$p_R(r) = \frac{r}{\sigma^2}\exp\left\{-\frac{r^2 + v^2}{2\sigma^2}\right\}I_0\left(\frac{rv}{\sigma^2}\right) \tag{2.15}$$

for $r \geq 0$. ($I_0(\cdot)$ denotes the zeroth-order modified Bessel function of the first kind.) Its mean square is $\mathbb{E}[R^2] = v^2 + 2\sigma^2$. This pdf is plotted in Figure 2.5 for some values of v and $\sigma^2 = 1$.

Here $R(t)$ and $\Theta(t)$ are not independent, unless we further assume a certain amount of randomness in the fixed-path signal. Specifically, assume that the phase β of the fixed path changes randomly and that we can model it as a RV uniformly distributed in $(0, 2\pi)$. As a result of this assumption, $R(t)$ and $\Theta(t)$ become independent processes, with Θ uniformly distributed in $(0, 2\pi)$ and $R(t)$ still a Rice random variable.

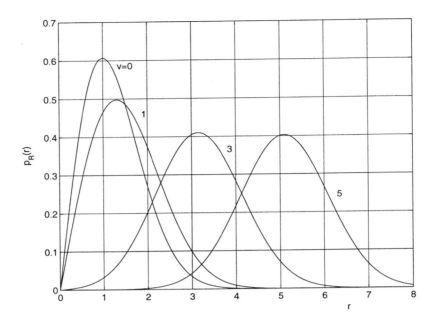

Figure 2.5: *Rice pdf with $\sigma^2 = 1$.*

Notice that, in (2.15), v denotes the envelope of the fixed-path component of the received signal, while $2\sigma^2$ is the power of the Rayleigh component (see (2.13) above). Thus, the "Rice factor"

$$K = \frac{v^2}{2\sigma^2}$$

denotes the ratio between the power of the fixed-path component and the power of the Rayleigh component. Sometimes the Rice pdf is written in a normalized form, obtained by assuming $\mathbb{E}[R^2] = v^2 + 2\sigma^2 = 1$ and exhibiting the Rice factor explicitly:

$$p_R(r) = 2r(1 + K) \exp\left\{-(1 + K)r^2 - K\right\} I_0\left(2r\sqrt{K(1 + K)}\right) \qquad (2.16)$$

for $r \geq 0$.

As $K \to 0$—i.e., as the fixed path reduces its power—since $I_0(0) = 1$, the Rice pdf becomes a Rayleigh pdf. On the other hand, if $K \to \infty$, i.e., the fixed-path power is considerably higher than the power in the random paths, then the Gaussian pdf is a good approximation for the Rice density.

Yet another statistical model for the envelope R of the fading is the Nakagami-m distribution. The probability density function of R is

$$p_R(r) = \frac{2}{\Gamma(m)} \left(\frac{m}{\Omega}\right)^m r^{2m-1} e^{-mr^2/\Omega}, \qquad r \geq 0 \qquad (2.17)$$

which has $\mathbb{E}[R^2] = \Omega$. The parameter m, called *fading figure*, is a ratio of moments:

$$m \triangleq \frac{\Omega^2}{\mathbb{V}[R^2]} \geq \frac{1}{2} \qquad (2.18)$$

For integer values of m, (2.17) is the pdf of the RV

$$Y \triangleq \sqrt{\sum_{i=1}^{m} X_i^2} \qquad (2.19)$$

where $X_1, \ldots, X_m$ are independent, Rayleigh-distributed RVs. As special cases, the choice $m = 1$ yields the Rayleigh distribution, while $m = 1/2$ yields a single-sided Gaussian distribution.

We observe that the Nakagami-m distribution is characterized by *two* parameters, and consequently it provides some extra flexibility if the mathematical model of the fading must be matched to experimental data.

2.4 Delay spread and Doppler-frequency spread

A simple yet useful classification of fading channels can be set up on the basis of the definition of two quantities called *coherence time* and *coherence bandwidth* of the physical channel.

Multipath fading occurs because different paths are received, each with a different Doppler shift: when the receiver and the transmitter are in relative motion with constant radial speed, the Doppler effect, in conjunction with multipath propagation, causes *time-* and *frequency-selective fading*. Consider these propagation paths, each characterized by a delay and attenuation, and examine how they change with time to generate a time-varying channel response. First, observe that significant changes in the attenuations of different paths occur at a rate much lower than significant changes in their phases. If $\tau_i(t)$ denotes the delay in the ith path, the corresponding phase is $2\pi f_0(t - \tau_i(t))$, which changes by 2π when $\tau_i(t)$ changes by $1/f_0$, or, equivalently, when the path length changes by c/f_0. Now, if the path length changes at velocity v_i, this change occurs in a time $c/(f_0 v_i)$, the inverse of the Doppler shift in the ith path. Consequently, significant changes in the channel occur in a time T_c whose order of magnitude is the inverse of the maximum Doppler shift B_D among the various paths, called the *Doppler spread* of the channel. The time T_c is called the *coherence time* of the channel, and we have

$$T_c \triangleq \frac{1}{B_D} \qquad (2.20)$$

The significance of T_c is as follows. Let T_x denote the duration of a transmitted signal.[1] If it is so short that during transmission the channel does not change appreciably in its features, then the signal will be received undistorted. Its distortion becomes noticeable when T_x is above T_c, which can be interpreted as the delay between two time components of the signal beyond which their attenuations become independent. We say the channel is *time selective* if $T_x \gtrsim T_c$.

The coherence time shows how rapidly a fading channel changes with time. Similarly, the quantity dual to it, called *coherence bandwidth*, shows how rapidly the channel changes in frequency. Consider paths i and j and the phase difference between them, i.e., $2\pi f(\tau_i(t) - \tau_j(t))$. This changes significantly when f changes by an amount proportional to the inverse of the difference $\tau_i(t) - \tau_j(t)$. If T_d, called the *delay spread* of the channel, denotes the maximum among these differences, a significant change occurs when the frequency change exceeds the inverse of T_d. We define the *coherence bandwidth* of the channel as

$$B_c \triangleq \frac{1}{T_D} \qquad (2.21)$$

This measures the signal bandwidth beyond which the frequency distortion of the transmitted signal becomes relevant. In other words, the coherence bandwidth is the frequency separation at which two frequency components of the signal undergo independent attenuations. A signal with $B_x \gtrsim B_c$ is subject to frequency-selective fading. More precisely, the envelope and phase of two unmodulated carriers at different frequencies will be markedly different if their frequency spacing exceeds B_c so that the cross-correlation of the fading fluctuations of the two tones decreases toward zero. The term *frequency-selective fading* expresses this lack of correlation among different frequency components of the transmitted signal.

In addition to coherence time and bandwidth, it is sometimes useful to define the *coherence distance* of a channel in which multiple antennas are used (see especially Chapter 10). This is the maximum spatial separation of two antennas over which

---

[1] Since we shall be considering *coded* signal for most of this work, from now on we may think of T_x as the duration of a code word.

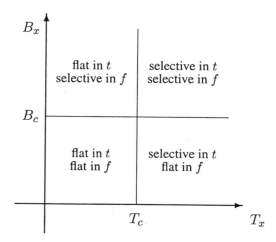

Figure 2.6: *Radio-channel classification.*

the channel response can be assumed constant: specifically, we say that the channel is *space selective* if the separation between antennas is larger than the coherence distance.

2.4.1 Fading-channel classification

From the previous discussion we have two quantities B_c and T_c describing how the channel behaves for the transmitted signal. Specifically,

(a) If $B_x \ll B_c$, there is no frequency-selective fading and hence no time dispersion. The channel transfer function looks constant, and the channel is called *flat* (or *nonselective*) in frequency. The fading affects the transmitted signal multiplicatively, by a factor which varies with time.

(b) If $T_x \ll T_c$, there is no time-selective fading, and the channel is called *flat* (or *nonselective*) in time.

Qualitatively, the situation appears as shown in Figure 2.6. The channel flat in t and f is not subject to fading either in time or in frequency. The channel flat in time and selective in frequency is called an *intersymbol-interference channel*. The channel flat in frequency is a good model for several terrestrial mobile radio channels. The channel selective both in time and in frequency is not a good model for terrestrial mobile radio channels, but it can be useful for avionic communications, in which high speeds (and hence short coherence times) combine with long delays due to earth reflections (and hence narrow coherence bandwidths).

The product $T_d B_D = 1/T_c B_c$ is called the *spread factor* of the channel. If $T_d B_D < 1$, the channel is said to be *underspread*; otherwise, it is *overspread*. Generally, if the spread factor $T_d B_D \ll 1$, the channel impulse response can be easily measured, and that measurement can be used by the receiver in the demodulation of the received signal and by the transmitter to optimize the transmitted signal. Measurement of the channel impulse response of an overspread channel is extremely difficult and unreliable, if not impossible. Since, in general, signal bandwidth and signal duration are such that $B_x T_x \gg 1$ (as otherwise there would be no hope for reliable communication, even in a nonfaded time-invariant channel, as, for example, the AWGN channel), it follows that a slowly fading, frequency nonselective channel is underspread.

Finally, we say that the channel is *ergodic* if the signal (i.e., the code word) is long enough to experience essentially all the states of the channel. This situation occurs when $T_x \gg T_c$. Thus, we discriminate between slow and fast fading and ergodic and nonergodic channels according to the variability of the fading process in terms of the whole code word transmission duration.

The preceding discussion is summarized in Table 2.1. (See [2.2] for further details.)

$B_x \ll B_c$	frequency-flat fading
$B_x \gtrsim B_c$	frequency-selective channel
$T_x \ll T_c$	time-flat (slow) fading
$T_x \gtrsim T_c$	time-selective (fast) channel
$T_c B_c > 1$	underspread channel
$T_c B_c \ll 1$	overspread channel
$T_x \ll T_c$	nonergodic channel
$T_x \gg T_c$	ergodic channel

Table 2.1: Classification of fading channels.

2.5   Estimating the channel

As we shall see in subsequent chapters, the performance of a transmission system over a fading channel may be greatly improved if the value taken on by the fading random variable affecting the propagation is known, at the receiver only or at both transmitter and receiver. Here we examine a technique for measuring a channel described as in Figure 2.1. We use "probing signals," to be transmitted in addition to information-bearing signals each time the channel changes significantly (and hence at least once every T_c).

A good set of probing signals is generated by a *pseudonoise (PN) sequence* $u(1), \ldots, u(N)$; it has the property that its autocorrelation $c(m)$ is approximately an ideal impulse. For simplicity we assume here that the channel is real, that the sequence is binary ($u(j) = \pm A$ for $1 \leq j \leq N$), and that we have exactly

$$c(m) \triangleq \sum_{j=1}^{N} u(j)u(j+m) = \begin{cases} A^2 N, & m = 0 \\ 0 & m \neq 0 \end{cases} \tag{2.22}$$

where we take $u(j) = 0$ whenever $j < 1$ or $j > N$. Without noise, the channel response to the PN sequence is the convolution

$$r'(n) = \sum_{k=0}^{L} h(n; k)u(n - k) \tag{2.23}$$

This response can be nonzero only from time $n = 1$ to time $n = N + L$: in this period we assume that the channel, albeit random, remains constant, so that we can rewrite (2.23) as

$$r'(n) = \sum_{k=0}^{L} h(k)u(n - k) \tag{2.24}$$

with $h(k)$, $k = 0, \ldots, L$, a sequence of complex random variables. Now, correlate the noiseless channel output $r'(n)$ with the PN sequence. Using (2.22) we obtain

$$\begin{aligned} \rho'(-m) &\triangleq \sum_{n=m+1}^{m+N} r'(n)u(n - m) \\ &= \sum_{n=m+1}^{m+N} \sum_{k=0}^{L} h(k)u(n - k)u(n - m) \\ &= \sum_{k=0}^{L} h(k) \sum_{j=1}^{N} u(j + m - k)u(j) \\ &= \sum_{k=0}^{L} h(k)c(m - k) \\ &= A^2 N h(m) \end{aligned} \tag{2.25}$$

which is proportional to the mth sample of the channel impulse response.

Consider now the effect of an additive white Gaussian noise $w(n)$ with variance σ^2. The noisy-channel response to the PN sequence is

$$y(n) = y'(n) + w(n) \qquad (2.26)$$

Correlating the channel output $y(n)$ with the PN sequence, we obtain

$$\rho(-m) = \rho'(-m) + \sum_{n=M+1}^{m+N} w(n)u(n-m) \qquad (2.27)$$

where the additional term is again a Gaussian RV with mean zero and variance

$$\sum_{n=m+1}^{m+N} \sum_{n'=m+1}^{m+N} \mathbb{E}[w(n)w^*(n')]u(n-m)u(n'-m) = \sigma^2 \sum_{n=m+1}^{m+N} u^2(n-m)$$
$$= \sigma^2 N A^2 \qquad (2.28)$$

In conclusion, we observe a correlation $\rho(-m)$ which is the sum of two terms: one is proportional to N times the impulse-response sample that we wish to estimate, while the other is a noise term whose variance is proportional to the PN sequence length N. The resulting signal-to-noise ratio is proportional to N: thus, by increasing the sequence length (and hence the measurement length) we can make the channel measure arbitrarily good. Notice, however, that making N very long leads to an accurate estimate but decreases the data-transmission rate. Two techniques, used, for example, in the GSM standard of digital cellular telephony, allow one to increase the ratio between the information symbols and the probe symbols: the first one consists of placing the probe symbols in the middle of a data frame, the second one of interpolating between the previous and the next channel measurement.

2.6   Bibliographical notes

Ref. [2.2] contains an extensive review of the information-thoretical and communications aspects of fading channels. Engineering aspects of wireless channels and modeling problems are treated, for example, in [2.3–2.5].

References

[2.1] S. Benedetto and E. Biglieri, *Digital Transmission Principles with Wireless Applications*. New York: Kluwer/Plenum, 1999.

[2.2] E. Biglieri, J. Proakis, and S. Shamai (Shitz), "Fading channels: Information-theoretic aspects," *IEEE Trans. Inform. Theory*, Vol. 44, No. 6, pp. 2169–2692, October 1998.

[2.3] W. Jakes, *Microwave Mobile Communications*. New York: J. Wiley & Sons, 1974.

[2.4] V. Veeravalli and A. Sayeed, *Wideband Wireless Channels: Statistical Modeling, Analysis and Simulation*. New York: J. Wiley & Sons, *in press*.

[2.5] M. D. Yacoub, *Foundations of Mobile Radio Engineering*. Boca Raton, FL: CRC Press, 1993.

3

and Horssmayres Prosession tyghting up under the threes.

Coding in a signal space

In this chapter we introduce signal constellations as sets $\mathcal{S} = \{\mathbf{x}\}$ of vectors in an n-dimensional space. Codes in signal spaces are defined as constellations whose elements are n-tuples of "elementary signals" chosen in a set $\mathcal{X}$. We evaluate the error probability obtained when a code in a signal space is used for transmission over the additive white Gaussian noise channel. Constellations are then compared on the basis of their bandwidth and power efficiencies. Capacity theorems yield ultimate bounds on the achievable performance. Next, we examine some symmetry properties of signal sets. A class of codes having a special algebraic structure, viz., linear binary codes, is introduced and described in some depth.

3.1   Signal constellations

Consider a finite signal constellation, i.e., a set $\mathcal{S} = \{\mathbf{x}\}$ of vectors in the Euclidean N-dimensional space $\mathbb{R}^N$ (also called *points*, or *signals*), to be used for transmission over a noisy channel. The squared norm $\|\mathbf{x}\|^2$ will be referred to as the *energy* of $\mathbf{x}$.

Let $M \triangleq |\mathcal{S}|$ denote the number of elements of $\mathcal{S}$, i.e., the number of available signals. Then, the maximum amount of information carried by $\mathcal{S}$ is $\log |\mathcal{S}|$ bits; this maximum information is the entropy of the constellation corresponding to equally likely signals (see Appendix A). We assume for simplicity that $|\mathcal{S}|$ is a power of 2, that is,

$$M = 2^m \tag{3.1}$$

A one-to-one map, called a *labeling*, can be defined, which associates with every element $\mathbf{x} \in \mathcal{S}$ an m-tuple of binary digits. The association of this m-tuple with $\mathbf{x}$ is called *modulation*. Notice that choice of a labeling affects also the design of the device (the *demodulator*) that transforms the received signal back into a sequence of binary digits to be delivered to the end user.

★ **Measuring the information transmitted: the rate.**   We list here various possible definitions of the *rate R* at which information is transmitted over the channel. The maximum amount of information transmitted by using the constellation $\mathcal{S}$ can be measured by $m \triangleq \log M$, the *number of bits per signal*. Since every signal has N dimensions, we may define the transmission rate in *bits per dimension* as $\rho \triangleq \log M / N$. If the speed of information transmission R_b (in bits per second) is of concern, and an N-dimensional signal is transmitted every T seconds (that is, we have T^{-1} signals transmitted every second), we transmit at a rate N/T dimensions per second, or

$$R_b = \frac{\log M}{N} \frac{\text{bit}}{\text{dimension}} \times \frac{N}{T} \frac{\text{dimension}}{\text{s}} = \frac{\log M}{T} \frac{\text{bit}}{\text{s}} \tag{3.2}$$

If, in addition, the bandwidth need of the signals selected is β Hz per dimension per second, then the bandwidth requirement for the transmission will be $\beta N/T$ Hz. Notice that, although the actual value with real-life waveforms may vary, it is often assumed for simplicity that $\beta = 1/2$: in fact, the dimensionality of a set of signals with duration T and bandwidth W is approximately $N = 2WT$ (see [3.2, p. 80], [3.11], and *infra*, Section 3.4.1).

In some cases, rather than using the number of bits carried by one signal, it may be preferable to use the number of bits *per channel use*: this option should be chosen whenever the definition of a channel entails the transmission of more than one signal, as in the case (to be examined later in Chapter 10) of multiple-antenna transmitters.

The demodulation problem.   If we denote by $\mathbf{y}$ the vector observed at the output of the channel (in general, the dimension of $\mathbf{y}$ may differ from the dimension of $\mathbf{x}$), the *demodulation* problem consists of selecting in $\mathcal{S}$ an estimate of $\mathbf{x}$, which we denote $\widehat{\mathbf{x}}$, which is optimum under some performance criterion.

The signal design problem.   This is the problem of selecting, under a suitable set of constraints (for example, the average transmitter power), the set $\mathcal{S}$ so that, under a prescribed demodulation criterion, performance is maximized.

The *additive white Gaussian noise* (AWGN) channel is defined as the channel that transforms $\mathbf{x}$ into

$$\mathbf{y} = \mathbf{x} + \mathbf{z} \tag{3.3}$$

that is, $\mathbf{x}$ is perturbed additively by a noise vector $\mathbf{z}$ independent of $\mathbf{x}$ and whose components are independent, zero-mean Gaussian random variables with common variance $N_0/2$ (this parameter is referred to as the *power spectral density* of the white noise).

We assume that the blocks of m binary digits output by the information source are equally likely and statistically independent, so the signals transmitted over the channel are also equally likely and independent. Assume also that the channel is stationary memoryless (see Section A.1 in refAppendix A)), i.e., that it processes every signal independently and irrespectively of its transmission time. In this situation, the behavior of the channel is described by the conditional probability density function (pdf) $p(\mathbf{y} \mid \mathbf{x})$, i.e., the pdf of $\mathbf{y}$ given that $\mathbf{x}$ was transmitted. (With an additive channel described by (3.3), $p(\mathbf{y} \mid \mathbf{x})$ is simply the pdf of the noise $\mathbf{z}$ with mean $\mathbf{x}$.) Moreover, it makes sense to choose, as the performance criterion for the demodulator, the minimization of the signal error probability, i.e., the probability that when $\mathbf{x}$ is transmitted the demodulator selects $\widehat{\mathbf{x}} \neq \mathbf{x}$:

$$P(e) \triangleq \frac{1}{M} \sum_{\mathbf{x}} \mathbb{P}(\widehat{\mathbf{x}} \neq \mathbf{x} \mid \mathbf{x}) \tag{3.4}$$

or the *bit error probability*, that is, the probability that a binary digit output by the source and mapped into $\mathbf{x}$ is transformed into a different binary digit.

3.2 Coding in the signal space

For easier modulation, demodulation, and labeling of the constellation S, it is convenient to introduce some structure in it. This may consist of choosing a set X of *elementary* signals, typically one- or two-dimensional, and $S \subset X^n$, that is, generating the vectors of S as having n components in X. This way, the elements of S have the form $\mathbf{x} = (x_1, x_2, \ldots, x_n)$, with $x_i \in X$. The collection of such $\mathbf{x}$ will be referred to as a *block code in the signal space*, and $\mathbf{x}$ as a *code word*.

Example 3.1

Choose $X = \{+1, -1\}$ and the following constellation of 4 signals in $\mathbb{R}^3$: $S = \{(+1,+1,+1), (+1,-1,-1), (-1,+1,-1), (-1,-1,+1)\}$. The components of each element in S are chosen so that the third one equals the product of the first two: consequently, even when the source emits independent pairs of binary digits, the elementary signals transmitted over the channel are not independent. Observe also that in this case we may define a channel use as the transmission either of a single ± 1 or of a triplet. Finally, observe that a simple labeling of the four signals can be obtained as follows: let a binary "0" correspond to the elementary signal "+1," and a binary "1" to the elementary signal "−1." This generates the first two components of $\mathbf{x}$, while the third one is obtained as their product. $\qquad\square$

A special case of S is the "uncoded constellation" $S = X^n$, the set of all n-tuples of elementary signals. Here $|S| = |X|^n$, and S has $n \times D$ dimensions, where D is the dimensionality of X.

3.2.1 Distances

A way of characterizing the quality of a code in the signal space used for transmission over a noisy channel is through the distinguishability of its elements, which in turn leads to the definition of a distance $d(\mathbf{x}, \mathbf{x}')$ between pairs of words $\mathbf{x}, \mathbf{x}'$.

Euclidean distance

This is the quantity

$$d_{\mathrm{E}}(\mathbf{x}, \mathbf{x}') \triangleq \|\mathbf{x} - \mathbf{x}'\| \tag{3.5}$$

It can be seen that for a code in the signal space we have

$$d_{\mathrm{E}}^2(\mathbf{x}, \mathbf{x}') = \sum_{i=1}^{n} \|x_i - x_i'\|^2 \tag{3.6}$$

The Euclidean distance is especially useful with AWGN channels with low noise spectral density. In particular, the minimum Euclidean distance between any two vectors in S

$$d_{E,\min} \triangleq \min_{x \neq x'} d_E(x, x') \tag{3.7}$$

is a parameter that, as we shall see in the following, often plays a central role in code design, in the sense that codes with larger minimum Euclidean distance exhibit a lower error probability at high SNR. Notice, however, that the above statement may be deceiving if the SNR at which the constellation is used is not large enough (see Problem 2 in the Problem section of this chapter, and our discussion of turbo codes in Chapter 9).

Hamming distance

The Hamming distance $d_H(x, x')$ between x and x' is defined as the number of components in which the two vectors differ. The minimum Hamming distance

$$d_{H,\min} \triangleq \min_{x \neq x'} d_H(x, x') \tag{3.8}$$

also plays a central role in code design, as we shall see later on in the context of coding for low-noise independent Rayleigh fading channels (Chapter 4).

Bhattacharyya distance

Another useful distinguishability measure is provided by the Bhattacharyya distance, which depends explicitly on the channel on which the transmission takes place. This is defined as

$$d_B(x, x') \triangleq -\ln \int_{\mathbb{R}^N} \sqrt{p(y \mid x)p(y \mid x')} \, dy \tag{3.9}$$

The average Bhattacharyya distance over the constellation:

$$\overline{B} \triangleq \frac{1}{M^2} \sum_{x, x'} d_B(x, x') \tag{3.10}$$

yields sometimes a suitable measure for signal selection.

Example 3.2

> Given two signals $\mathbf{x}$ and $\mathbf{x}'$ with equal energies, that is $\|\mathbf{x}\|^2 = \|\mathbf{x}'\|^2 = \mathcal{E}$ and scalar product $(\mathbf{x}, \mathbf{x}') = \mu \mathcal{E}$, their Bhattacharyya distance over the AWGN channel is proportional to the signal-to-noise ratio $\mathcal{E}/4N_0$ and to $(1 - \mu)$: thus, it is maximized when $\mu = -1$, i.e., when the two signals are antipodal.   □

★ Relation between Hamming and Euclidean distance

For general constellations, there is no immediate relation between d_H and d_E. However, consider the special case of the binary elementary constellation $\mathcal{X} = \{x, -x\}$. The squared Euclidean distance between two code words $\mathbf{x}$, $\mathbf{x}'$ differing in $d_\mathrm{H}(\mathbf{x}, \mathbf{x}')$ places is given by

$$d_\mathrm{E}^2(\mathbf{x}, \mathbf{x}') = 4\mathcal{E}d_\mathrm{H}(\mathbf{x}, \mathbf{x}') \tag{3.11}$$

where $\mathcal{E} \triangleq \|x\|^2$. This result shows in particular that, for a code with a binary elementary constellation, maximizing the Hamming distance is tantamount to maximizing the Euclidean distance and consequently getting a good performance over the low-noise AWGN channel. This observation prompts us to examine in special detail those codes based on a binary elementary constellation. Specifically, if we represent the signals $\pm x$ through two elements in the binary Galois field $\mathbb{F}_2$, we can introduce two operations (mod-2 sum and product) that allow the code to be endowed with nice algebraic properties that facilitate the study of these "binary" codes. Later in the Chapter we shall delve into this.

Example 3.3

> Another example of a constellation in which Euclidean and Hamming distances are proportional is shown in Figure 3.1, where the 4-PSK constellation is labeled in such a way that the Euclidean distance is related to Hamming distance by
>
> $$d_\mathrm{E}^2(\mathbf{x}, \mathbf{x}') = 2\mathcal{E}d_\mathrm{H}(\mathbf{x}, \mathbf{x}') \tag{3.12}$$
>
> □

3.3 Performance evaluation: Error probabilities

Consider now demodulation, under the minimum-$P(e)$ criterion, of a constellation used for transmission over the AWGN channel. It is known from detection theory

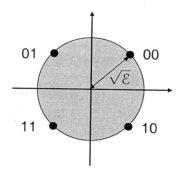

Figure 3.1: *In Gray-coded 4-PSK the squared Euclidean distance is proportional to Hamming distance.*

that optimum demodulation consists of selecting as $\widehat{\mathbf{x}}$ the vector with the minimum Euclidean distance from the observed vector $\mathbf{y}$. It is convenient here to define the *Voronoi region* (or *decision region*) associated with $\mathbf{x} \in \mathcal{S}$ as the set of vectors in $\mathbb{R}^N$ that are closer to $\mathbf{x}$ than to any other element of $\mathcal{S}$:

$$\mathcal{R}(\mathbf{x}) = \{\mathbf{y} \in \mathbb{R}^n \mid \|\mathbf{y} - \mathbf{x}\| = \min_{\mathbf{x}' \in \mathcal{S}} \|\mathbf{y} - \mathbf{x}'\|\}$$

The probability of an erroneous decoding when $\mathbf{x}$ is the transmitted signal is then given by

$$
\begin{aligned}
P(e \mid \mathbf{x}) &= 1 - \mathbb{P}[\mathbf{y} \in \mathcal{R}(\mathbf{x}) \mid \mathbf{x}] \\
&= 1 - \mathbb{P}[\mathbf{x} + \mathbf{z} \in \mathcal{R}(\mathbf{x})] \\
&= 1 - \int_{\mathcal{R}(\mathbf{x})} p(\mathbf{z} - \mathbf{x}) \, d\mathbf{z}
\end{aligned}
\tag{3.13}
$$

where $p(\mathbf{z})$ is the pdf of the noise, so that $p(\mathbf{z} - \mathbf{x})$ is the pdf of $\mathbf{x} + \mathbf{z}$.

Since in most cases the exact expression (3.13) is too hard to compute, it is often expedient to resort to a simple upper bound. Let $P(\mathbf{x} \to \widehat{\mathbf{x}})$ denote the probability that, when $\mathbf{x}$ is transmitted, $\mathbf{y}$ is closer to $\widehat{\mathbf{x}}$ than to $\mathbf{x}$. This is called the *pairwise error probability* (PEP) because if the transmission system uses only two signals, viz., $\mathbf{x}$ and $\widehat{\mathbf{x}}$, then $P(e \mid \mathbf{x}) = P(\mathbf{x} \to \widehat{\mathbf{x}})$. Observe that $P(e \mid \mathbf{x})$ can be expressed as the probability that at least one $\widehat{\mathbf{x}} \neq \mathbf{x}$ is closer than $\mathbf{x}$ to $\mathbf{y}$. Using the upper bound to the probability of a union of events, we can write

$$P(e \mid \mathbf{x}) \leq \sum_{\widehat{\mathbf{x}} \neq \mathbf{x}} P(\mathbf{x} \to \widehat{\mathbf{x}}) \tag{3.14}$$

Finally, the error probability $P(e)$ is given by

$$P(e) = \frac{1}{M} \sum_{\mathbf{x} \in \mathcal{S}} P(e \mid \mathbf{x}) \leq \frac{1}{M} \sum_{\mathbf{x} \in \mathcal{S}} \sum_{\widehat{\mathbf{x}} \neq \mathbf{x}} P(\mathbf{x} \to \widehat{\mathbf{x}}) \tag{3.15}$$

Now, the PEP can be easily computed in closed form as follows:

$$\begin{aligned} P(\mathbf{x} \to \widehat{\mathbf{x}}) &= \mathbb{P}(\|\mathbf{y} - \widehat{\mathbf{x}}\|^2 < \|\mathbf{y} - \mathbf{x}\|^2 \mid \mathbf{x}) \\ &= \mathbb{P}(\|(\mathbf{x} + \mathbf{z}) - \widehat{\mathbf{x}}\|^2 < \|(\mathbf{x} + \mathbf{z}) - \mathbf{x}\|^2) \\ &= \mathbb{P}(\|\mathbf{z} + (\mathbf{x} - \widehat{\mathbf{x}})\|^2 < \|\mathbf{z}\|^2) \\ &= \mathbb{P}((\mathbf{z}, \widehat{\mathbf{x}} - \mathbf{x}) > \|\mathbf{x} - \widehat{\mathbf{x}}\|^2/2) \end{aligned} \tag{3.16}$$

Observe that the scalar product $(\mathbf{z}, \widehat{\mathbf{x}} - \mathbf{x})$, being a linear transformation of the Gaussian vector $\mathbf{z}$, is itself a Gaussian RV, with mean 0 and variance $N_0 \|\mathbf{x} - \widehat{\mathbf{x}}\|^2/2$. Thus, since for a Gaussian RV X with mean 0 and variance σ^2 we have

$$\mathbb{P}(X > x) = Q\left(\frac{x}{\sigma}\right) \tag{3.17}$$

we obtain

$$P(\mathbf{x} \to \widehat{\mathbf{x}}) = Q\left(\frac{\|\mathbf{x} - \widehat{\mathbf{x}}\|}{\sqrt{2N_0}}\right) \tag{3.18}$$

where $Q(\cdot)$ denotes the Gaussian tail function:

$$Q(x) \triangleq \frac{1}{\sqrt{2\pi}} \int_x^\infty e^{-z^2/2}\, dz$$

This function is related to the often-used *complementary error function* erfc$(\cdot)$ by

$$Q(x) = \frac{1}{2} \operatorname{erfc}\left(\frac{x}{\sqrt{2}}\right)$$

A simpler approximation (the *Bhattacharyya bound*) is based on the exponential bound (which is equivalent to the Chernoff bound derived in the Problem section)

$$Q(x) \leq e^{-x^2/2}, \quad x \geq 0 \tag{3.19}$$

and yields

$$P(\mathbf{x} \to \widehat{\mathbf{x}}) \leq \exp\left\{-\|\mathbf{x} - \widehat{\mathbf{x}}\|^2/4N_0\right\} \tag{3.20}$$

Example 3.4

With uncoded binary signaling ($M = 2$) we have $P(e) = P(\mathbf{x} \to \hat{\mathbf{x}})$, and consequently, by choosing the antipodal signal set $\mathcal{S} = \{+\sqrt{\mathcal{E}}, -\sqrt{\mathcal{E}}\}$, we obtain from (3.18):

$$P(e) = Q\left(\sqrt{\frac{2\mathcal{E}}{N_0}}\right) \qquad (3.21)$$

$\square$

3.3.1 Asymptotics

As $N_0 \to 0$, since $Q(\cdot)$ is a decreasing function, the right-hand side of (3.15) will be dominated by the pairs of signals at minimum Euclidean distance. By retaining only these dominant terms, we can write

$$P(e) \overset{\sim}{\leq} \alpha\, P(\mathbf{x} \to \hat{\mathbf{x}})|_{\|\mathbf{x}-\hat{\mathbf{x}}\|=d_{\mathrm{E,min}}} = \alpha Q\left(\frac{d_{\mathrm{E,min}}}{\sqrt{2N_0}}\right) \qquad (3.22)$$

where the notation $\overset{\sim}{\leq}$ indicates that the inequality holds only approximately, unless N_0 is vanishingly small, and where α is a constant (see Problem 1 for its interpretation as the average number of nearest neighbors in the constellation).

The above approximation (3.22) shows that for low enough N_0 we may choose, as a criterion for signal selection, and hence for code selection, the maximization of the minimum Euclidean distance.

3.3.2 Bit error probabilities

The above calculations were based on symbol error probability. To allow comparisons among modulation schemes with different values of M and hence whose signals carry different numbers of bits, a better performance measure is the *bit error probability* $P_b(e)$, often referred to as *bit-error rate* (BER). This is the probability that a binary digit emitted by the source will be received erroneously by the user.

In general, it can be said that the calculation of $P(e)$ is a far simpler task than the calculation of $P_b(e)$. Moreover, the latter depends also on the mapping of the source bits onto the signals in the modulator's constellation. A simple bound on $P_b(e)$ can be derived by observing that, since each signal carries $\log M$ bits, one symbol error produces at least one bit error and at most $\log M$ bit errors. Therefore,

$$\frac{P(e)}{\log M} \leq P_b(e) \leq P(e) \qquad (3.23)$$

Since (3.23) is valid in general, we should try to keep $P_b(e)$ as close as possible to its lower bound $P(e)/\log M$ (this is sometimes referred to as *error probability per bit*). One way of achieving this goal is by choosing the labeling in such a way that, whenever a symbol error occurs, the signal erroneously chosen by the demodulator differs from the transmitted one by the least number of bits. Since for high signal-to-noise ratios we may expect that errors occur when a signal is mistaken for one of its nearest neighbors, then a reasonable choice is a labeling such that neighboring signal points correspond to binary sequences that differ by only one digit. When this choice is made, we say that the signals are *Gray labeled*, and we approximate $P_b(e)$ by its lower bound in (3.23). Figure 3.1 shows an example of Gray labeling (see also our discussion of bit-interleaved coded modulation in Section 7.9).

3.4   Choosing a coding/modulation scheme

We now discuss some criteria that should guide the choice of a coding/modulation scheme over the AWGN channel. These are bandwidth efficiency, power efficiency, and error probability.

3.4.1   ★ Bandwidth occupancy

Since in the following we shall be interested in a comparison among coding/modulation schemes that leaves out of consideration the actual waveform shapes and focuses instead on the geometric features of the signal constellations, it is convenient to use the following definition of bandwidth. The *2WT-theorem* [3.2, p. 80], [3.11], states that, for large T and W, the dimensionality of a set of signals with duration T and bandwidth occupancy W is approximately $N = 2WT$. This motivates our definition of the *Shannon bandwidth* of a signal set with N dimensions as

$$W = \frac{N}{2T} \tag{3.24}$$

This bandwidth can of course be expressed in Hz, but it may be appropriate in several instances to express it in *dimension pairs per second*. The Shannon bandwidth is the minimum amount of bandwidth that the signal *needs*, in contrast to the (several possible) definitions of *Fourier bandwidth* of the modulated signal. The latter expresses the amount of bandwidth that the signal actually *uses*. In most cases, Shannon bandwidth and Fourier bandwidth differ little: however, there are examples of modulated signals (spread-spectrum signals) whose Fourier bandwidth is much larger than their Shannon bandwidth.

Example 3.5 (PSK)

An M-PAM has 1 dimension, so its Shannon bandwidth is $W = 1/2T$. An M-PSK signal has 2 dimensions, and hence $W = 1/T$.   □

Note that in general, for any sensible definition of the bandwidth W, we have $W = \alpha/T$, which reflects the fundamental fact from Fourier theory that the time duration of a signal is inversely proportional to its bandwidth occupancy. The actual value of α depends on the definition of bandwidth and on the actual shapes of the waveforms used by the modulator.

3.4.2 ★ Signal-to-noise ratio

Recall that the information rate of the source, R_b, is related to the number of waveforms used by the modulator, M, and to the duration of these waveforms, T, by the equality

$$R_b = \frac{\log M}{T} \tag{3.25}$$

This is the rate, in bit/s, that can be accepted by the modulator. The average power expended by the modulator is

$$\mathcal{P} = \frac{\mathcal{E}}{T}$$

where $\mathcal{E}$ is the average energy of the modulator signals, i.e.,

$$\mathcal{E} \triangleq \frac{1}{|\mathcal{S}|} \sum_{x \in \mathcal{S}} \|x\|^2 \tag{3.26}$$

Each signal carries $\log M$ information bits. Thus, defining $\mathcal{E}_b$ as the average energy expended by the modulator to transmit one bit, so that $\mathcal{E} = \mathcal{E}_b \log M$, we have

$$\mathcal{P} = \mathcal{E}_b \frac{\log M}{T} = \mathcal{E}_b R_b \tag{3.27}$$

We define the *signal-to-noise ratio* as the ratio between the average signal power and the average noise power. The latter equals $(N_0/2) \cdot 2W = N_0 W$, where we assume conventionally that the equivalent noise bandwidth of the receiving filter is the Shannon bandwidth. We have

$$\mathrm{SNR} \triangleq \frac{\mathcal{P}}{N_0 W} = \frac{\mathcal{E}_b}{N_0} \frac{R_b}{W} \tag{3.28}$$

3.4.3 ★ Bandwidth efficiency and asymptotic power efficiency

Expression (3.28) shows that the signal-to-noise ratio is the product of two quantities, viz., $\mathcal{E}_b/N_0$, the energy per bit divided by twice the power spectral density, and R_b/W, the *bandwidth* (or *spectral*) *efficiency* of a modulation scheme. In fact the latter, measured in bit/s/Hz, tells us how many bits per second are transmitted in a given bandwidth W. The higher the bandwidth efficiency, the more efficient is the use of the available bandwidth made by the modulation scheme.

We also observe that R_b/W may be conveniently related to the number of bits per dimension. We have

$$\frac{R_b}{W} = 2\,\frac{\log M}{N} = 2\rho \qquad (3.29)$$

Since ρ is the number of bits transmitted per dimension, we can interpret the above equality by saying that the spectral efficiency R_b/W represents the number of bits transmitted *per dimension pair* (this interpretation is especially useful when we use two-dimensional elementary constellations, which is often the case).

We now define the *asymptotic power efficiency* γ. We have seen that, for high signal-to-noise ratios, the error probability is approximated by a Gaussian tail function whose argument is $d_{\mathrm{E,min}}/\sqrt{2N_0}$. Define γ as the quantity satisfying

$$\sqrt{\gamma\frac{2\mathcal{E}_b}{N_0}} = \frac{d_{\mathrm{E,min}}}{\sqrt{2N_0}}$$

that is,

$$\gamma \triangleq \frac{d_{\mathrm{E,min}}^2}{4\mathcal{E}_b} \qquad (3.30)$$

so that

$$P(e) \approx \alpha\,Q\left(\sqrt{\gamma\frac{2\mathcal{E}_b}{N_0}}\right) \qquad (3.31)$$

In words, γ expresses how efficiently a constellation makes use of the available signal energy to generate a given minimum distance. Thus, we may say that, at least for high signal-to-noise ratios, a constellation is better than another (having a comparable average number α of nearest neighbors) if its asymptotic power efficiency is greater.

For example, the antipodal constellation [3.2] $\mathcal{S} = \{\pm\sqrt{\mathcal{E}}\}$ has $\sqrt{\mathcal{E}_b} = \sqrt{\mathcal{E}}$ and $d_{\mathrm{E,min}} = 2\sqrt{\mathcal{E}}$, so $\gamma = 1$. This may serve as a baseline figure. Other values of γ and R_b/W are shown in Table 1.1 and in Figure 3.2.

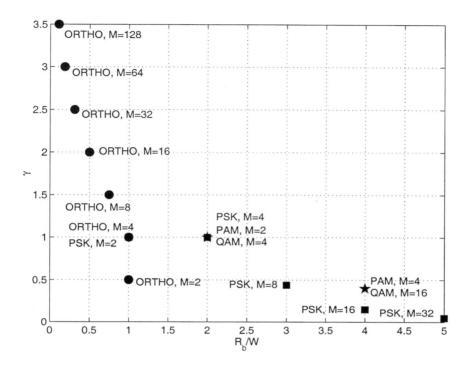

Figure 3.2: *Power efficiency vs. bandwidth efficiency of some constellations (OR-THO stands for orthogonal constellations). Observe how, with orthogonal signaling, increasing the constellation size M yields an increase of γ and a decrease of R_b/W. The opposite occurs with PAM, QAM, and PSK.*

3.4.4 Tradeoffs in the selection of a constellation

In summary, the evaluation of a constellation may be based on the following three parameters: the error probability $P(e)$, the signal-to-noise ratio $\mathcal{E}_b/N_0$ necessary to achieve $P(e)$, and the bandwidth efficiency R_b/W. The first tells us about the reliability of the transmission, the second measures the efficiency in power expenditure, and the third measures how efficiently the modulation scheme makes use of the bandwidth. For low error probabilities, we may simply consider the asymptotic power efficiency γ and the bandwidth efficiency.

The ideal system achieves a small $P(e)$ with a low $\mathcal{E}_b/N_0$ and a high R_b/W: now, we shall evaluate bounds on the values of these parameters that can be achieved by any modulation scheme. In addition, complexity considerations force us to move further apart from the theoretical limits. Consequently, complexity should also be introduced among the parameters that force a tradeoff in the selection of a modulation scheme.

3.5 Capacity of the AWGN channel

We now evaluate the capacity of the AWGN channel; specifically, we examine the Gaussian channel and a code built out of one-dimensional (i.e., real) elementary signals. For every channel use, the input is x and the output is the real random variable $y = x + z$. Assume initially that no constraint is put on the input and output alphabets $\mathcal{X}$ and $\mathcal{Y}$, except for a constraint on the energy of the input signal, which has the form $\mathbb{E}\, x^2$. Since $z \perp\!\!\!\perp x$, we have (see Appendix A for the relevant definitions):

$$\mathrm{H}(y \mid x) = \mathrm{H}(x + z \mid x) = \mathrm{H}(z \mid x) = \mathrm{H}(z) \tag{3.32}$$

and hence

$$
\begin{aligned}
\mathrm{I}(x; y) &= \mathrm{H}(y) - \mathrm{H}(y \mid x) \\
&= \mathrm{H}(y) - \mathrm{H}(z)
\end{aligned}
\tag{3.33}
$$

Now (Theorem A.3.1, Appendix A),

$$\mathrm{H}(z) = \frac{1}{2} \log 2\pi e\, \mathbb{E}\, z^2 \tag{3.34}$$

and, since $\mathbb{E}\, z = 0$,

$$\mathbb{E}\, y^2 = \mathbb{E}\,(x + z)^2 = \mathbb{E}\, x^2 + \mathbb{E}\, z^2 \tag{3.35}$$

Thus, the entropy of $\mathcal{Y}$ is bounded above by $\frac{1}{2} \log 2\pi e(\mathbb{E}\, x^2 + \mathbb{E}\, z^2)$, and in conclusion

$$
\begin{aligned}
\mathrm{I}(x; y) &\leq \frac{1}{2} \log 2\pi e(\mathbb{E}\, x^2 + \mathbb{E}\, z^2) - \frac{1}{2} \log 2\pi e\, \mathbb{E}\, z^2 \\
&= \frac{1}{2} \log \left(1 + \frac{\mathbb{E}\, x^2}{\mathbb{E}\, z^2}\right)
\end{aligned}
\tag{3.36}
$$

and the maximum of $\mathrm{I}(\mathcal{X}; \mathcal{Y})$ is attained when x is a Gaussian random vector with zero mean and variance $\mathbb{E}\, x^2$. This maximum value is the information capacity of the Gaussian channel:

$$C = \frac{1}{2} \log\left(1 + \mathrm{SNR}\right) \quad \text{bit/dimension} \tag{3.37}$$

where

$$\mathrm{SNR} \triangleq \frac{\mathbb{E}\, x^2}{\mathbb{E}\, z^2} \tag{3.38}$$

Observation 3.5.1   If x and z are complex, then the maximum value of the mutual information is achieved for x Gaussian, with zero mean, variance $\mathbb{E}|x|^2$, and independent real and imaginary parts. Moreover, it is convenient to express C in bit/dimension pair:

$$C = \log(1 + \text{SNR}) \quad \text{bit/dimension pair} \tag{3.39}$$

where now $\text{SNR} \triangleq \mathbb{E}|x|^2/\mathbb{E}|z|^2$.

Observation 3.5.2   The SNR (3.38) can be given different expressions as follows. Assume x to be N-dimensional. The signal variance is $\mathcal{E}$, while the noise variance is $NN_0/2$. Since the Shannon bandwidth of a signal is $W = N/2T$, we may write

$$\text{SNR} = \frac{\mathcal{E}}{NN_0/2} = \frac{\mathcal{E}/T}{N_0 W} = \frac{\mathcal{P}}{N_0 W} \tag{3.40}$$

Recalling (3.28), we can also express the SNR in the form

$$\text{SNR} = \frac{\mathcal{E}_b R_b}{N_0 W} \tag{3.41}$$

Since $\text{SNR} = 2\mathcal{E}/(NN_0)$, we see that, as $N \to \infty$, if $\mathcal{E}/N_0$ remains constant then the number of bits per dimension expressed by C tends to zero, because $\text{SNR} \to 0$. The number NC of bits that can be reliably transmitted over N dimensions tends to the constant limit $\log(e)\mathcal{E}/N_0$. We shall return on this in Section 3.5.1.

Sketch of the proof of the capacity theorem

The capacity (3.37) is also the maximum achievable rate for the channel. A fundamental theorem of Information Theory (Appendix A) shows that there exists a sequence of codes with rate C and block length n such that, as $n \to \infty$, the error probability tends to 0. Here we provide a qualitative summary of the proof, in the form originally given by Shannon.

As we are considering one-dimensional elementary signals and code words with length n, the dimensionality of the signal constellation is n. Observe now that the volume of a n-dimensional sphere Σ_n with radius r is proportional to r^n; thus, the volume of the shell between $r - \epsilon$ (with $0 < \epsilon < r$) and r is proportional to $r^n - (r - \epsilon)^n$. The ratio between the volume of the shell and the volume of the sphere is

$$\frac{r^n - (r - \epsilon)^n}{r^n} = 1 - \left(1 - \frac{\epsilon}{r}\right)^n$$

and tends to 1 as $n \to \infty$, no matter what the thickness ϵ of the shell is. This phenomenon, called *sphere hardening*, is summarized by saying that the volume of a n-dimensional sphere tends to concentrate near its surface as $n \to \infty$.

Next, consider a set of code words $\mathbf{x}$ whose components are subject to the energy constraint $\mathbb{E}x^2 \leq \mathcal{E}$, and let the received vector be $\mathbf{y} = \mathbf{x} + \mathbf{z}$. Let us first apply the sphere-hardening concept to the noise vector $\mathbf{z}$. As n grows to infinity, due to the law of large numbers, the squared length of vector $\mathbf{z}$ tends to a constant value:

$$\sum_{i=1}^{n} z_i^2 \approx n\mathbb{E}z^2 = nN_0/2$$

where z_i are the independent, equally distributed random components of $\mathbf{z}$. Sphere hardening assures that, while fluctuations of the length of $\mathbf{z}$ are possible, they tend to vanish as $n \to \infty$, so that $\mathbf{x} + \mathbf{z}$ lies on the surface of the sphere $\Sigma_n(\mathbf{x})$, centered at $\mathbf{x}$ and with radius $\sqrt{nN_0/2}$. Thus, signals differing by a Euclidean distance less than $\sqrt{nN_0/2}$ cannot be detected without ambiguity. Conversely, $\mathbf{x}$ can be detected with vanishingly small ambiguity if $\Sigma_n(\mathbf{x})$ is disjoint from the spheres associated with the other code words: in fact, $\Sigma_n(\mathbf{x})$ is contained in the Voronoi region of $\mathbf{x}$.

Further, consider the received vector $\mathbf{y}$. Its squared length tends to

$$\sum_{i=1}^{n} y_i^2 \approx n\mathbb{E}y^2 = n(\mathbb{E}x^2 + \mathbb{E}z^2) \leq n(\mathcal{E} + N_0/2)$$

and consequently $\mathbf{y}$ lies within a sphere with radius $\sqrt{n(\mathcal{E} + N_0/2)}$. In these conditions, the maximum number of disjoint spheres $\Sigma_n(\mathbf{x})$ that can be accommodated inside the sphere with radius $\sqrt{n(\mathcal{E} + N_0/2)}$ is no more than the ratio of the volumes

$$\frac{[n(\mathcal{E} + N_0/2)]^{n/2}}{[n(N_0/2)]^{n/2}} = (1 + \text{SNR})^{n/2}$$

This is the number $|\mathcal{S}|$ of code words. Thus, the rate of the code is

$$C = \frac{\log |\mathcal{S}|}{n} = \frac{1}{2}\log(1 + \text{SNR}) \quad \text{bit/dimension} \tag{3.42}$$

This "sphere-packing" argument also shows that we cannot hope to send information at a rate greater than C with low probability of error.

3.5.1 The bandlimited Gaussian channel

Assume now that the transmission of signal $\mathbf{x}$ takes a time T. Assuming as usual that the dimensionality of the constellation is $N = 2WT$, with W its (Shannon-)

bandwidth occupancy, we transmit $2W$ dimensions per second. Thus, using (3.40), Equation (3.37) can be rewritten in the form

$$C = W \log \left(1 + \frac{\mathcal{P}}{N_0 W} \right) \quad \text{bit/s} \qquad (3.43)$$

which expresses the capacity of the bandlimited AWGN channel.

Notice that, as $W \to \infty$, we have

$$C \to \frac{\mathcal{P}}{N_0} \log e \quad \text{bit/s} \qquad (3.44)$$

which shows how capacity grows linearly with signal power, rather than logarithmically as in (3.37). Equation (3.44) also shows that, for a given $\mathcal{P}/N_0$, the capacity remains bounded even though W (and hence the number of signal dimensions) grows without bounds. This occurs because $\mathcal{P}$ is fixed, and hence the power per Hz tends to zero.

If (3.41) is used, we obtain

$$C = W \log \left(1 + \frac{\mathcal{E}_b R_b}{N_0 W} \right) \quad \text{bit/s}$$

Since for reliable transmission we must have $R_b < C$, we require that

$$\frac{R_b}{W} < \log \left(1 + \frac{\mathcal{E}_b R_b}{N_0 W} \right)$$

Solving this inequality for the minimum allowable $\mathcal{E}_b/N_0$, we obtain

$$\frac{\mathcal{E}_b}{N_0} > \frac{2^{R_b/W} - 1}{R_b/W}$$

as plotted in Figure 3.3. The curve in this figure demarcates the region in which arbitrary low $P(e)$ can be reached: for any given R_b/W there exists a minimum value of $\mathcal{E}_b/N_0$ that must be exceeded if arbitrarily high reliability must be achieved. Notice that, as W increases, the required $\mathcal{E}_b/N_0$ approaches the lower limit

$$\lim_{W \to \infty} \frac{2^{R_b/W} - 1}{R_b/W} = \ln 2 \Leftrightarrow -1.6 \text{ dB}$$

Moreover, as $R_b/W > 2$, that is, when bandwidth is constrained, the energy-to-noise ratio required for reliable transmission increases dramatically. The region

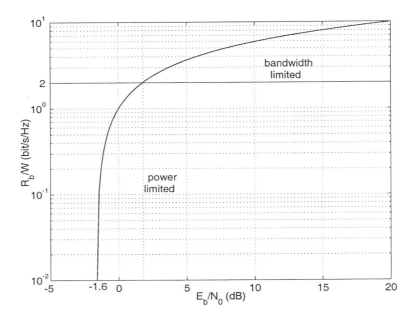

Figure 3.3: *Capacity limits for the bandlimited AWGN channel.*

where $R_b/W > 2$ (more than 2 bit/s/Hz, or equivalently more than 1 bit per dimension) is usually referred to as the *bandwidth-limited region*, and the region where $R_b/W < 2$ as the *power-limited region*. Figure 3.3 suggests that if the available power is severely limited, then we should compensate for this limitation by increasing the bandwidth occupancy, while the cost of a bandwidth limitation is an increase in the transmitted power.

Example 3.6

In this example we exhibit explicitly an M-ary signal constellation that, with no bandwidth constraint, has an error probability that tends to 0, as $M \to \infty$, provided that $\mathcal{E}_b/N_0 > \ln 2$, and hence shows the best possible behavior asymptotically. This is the set of M orthogonal, equal-energy signals defined by

$$(\mathbf{x}_i, \mathbf{x}_j) = \begin{cases} 0, & i \neq j \\ \mathcal{E}, & i = j \end{cases} \tag{3.45}$$

This signal set has dimensionality $N = M$. Due to the special symmetry of this signal set, the Voronoi regions of the signals are all congruent (more on this *infra*, in Section 3.6), and hence the error probability $P(e \mid \mathbf{x}_i)$ is the same for all transmitted signals x_i. Thus, we can assume with no loss of generality that x_1 was transmitted, and write $P(e) = P(e \mid x_1)$. The Voronoi region $\mathcal{R}(x_1)$ is bounded by the $M-1$ hyperplanes $y_1 = y_2, y_1 = y_3, \ldots, y_1 = y_M$. Hence, the probability of correct reception is given by

$$P(c) = P(c \mid x_1) = \mathbb{P}(y_1 > y_2, y_1 > y_3, \ldots, y_1 > y_M \mid x_1) \qquad (3.46)$$

Now, given that x_1 was transmitted, the RVs $y_1, \ldots, y_M$ are iid Gaussian, with equal variance $N_0/2$ and mean values

$$\mathbb{E}\left[y_i \mid x_1\right] = \begin{cases} \sqrt{\mathcal{E}}, & i = 1, \\ 0, & i \neq 1 \end{cases}$$

The events $\{y_1 > y_i \mid x_1\}$ are not independent; however, they are conditionally independent given y_1, so, using (3.17), we have

$$
\begin{aligned}
P(c) &= \mathbb{E}\left[\mathbb{P}\left(y_1 > y_2, y_1 > y_3, \ldots, y_1 > y_M \mid x_1, y_1\right)\right] \\
&= \mathbb{E}\left[1 - Q\left(\sqrt{\frac{2}{N_0}} y_1\right)\right]^{M-1}
\end{aligned}
$$

where the expectation is taken with respect to the RV y_1, whose conditional pdf is

$$p(y_1 \mid x_1) = (\pi N_0)^{-1/2} \exp[-(y_1 - \sqrt{\mathcal{E}})^2/N_0]$$

We now rewrite $P(c)$, after observing that $\mathcal{E} = \mathcal{E}_b \log M$, in the form

$$P_M(c) = \mathbb{E}\left[1 - Q(X + \sqrt{2 \log M \cdot \mathcal{E}_b/N_0})\right]^{M-1} \qquad (3.47)$$

where $X \sim \mathcal{N}(0,1)$, and examine the behavior of $P(c)$ as $M \to \infty$. Take the logarithm of the argument of the expectation in (3.47), and observe that

$$
\begin{aligned}
\lim_{M \to \infty} &\ln\left[1 - Q(X + \sqrt{2 \log M \cdot \mathcal{E}_b/N_0})\right]^{M-1} \\
&= \lim_{M \to \infty} \frac{\ln\left[1 - Q(X + \sqrt{2 \log M \cdot \mathcal{E}_b/N_0})\right]}{(M-1)^{-1}}
\end{aligned}
$$

Using l'Hôpital's rule, the above is equal to

$$\lim_{M \to \infty} -\frac{(M-1)^2}{M^{1+(\mathcal{E}_b/N_0)/\ln 2}} A(\log M) \qquad (3.48)$$

where

$$A(\log M) \triangleq \sqrt{\frac{\mathcal{E}_b/N_0}{4\pi \log M}} \frac{\exp\left(-X^2/2 - X\sqrt{\log(M)\mathcal{E}_b/N_0}\right)}{1 - Q(X + \sqrt{2\log(M)\mathcal{E}_b/N_0})}$$

Define now $\varepsilon \triangleq 1 - (\mathcal{E}_b/N_0)/\ln 2$. By observing that, as $M \to \infty$, we have $\sqrt{\ln M} = o(\ln M)$ and $\ln \ln M = o(\ln M)$, we can see that asymptotically

$$\ln((M-1)^2/M^{1+(\mathcal{E}_b/N_0)/\ln 2})A(M) \sim \varepsilon \ln M + o(\ln M) \to \mathrm{sgn}(\varepsilon) \cdot \infty$$

Thus,

$$\ln \left[1 - Q(X + \sqrt{2\log M \cdot \mathcal{E}_b/N_0})\right]^{M-1} \to \begin{cases} -\infty & \text{if } \varepsilon > 0 \\ 0 & \text{if } \varepsilon < 0 \end{cases}$$

and, in conclusion,

$$P(c) \to \begin{cases} 0 & \text{if } \mathcal{E}_b/N_0 < \ln 2 \\ 1 & \text{if } \mathcal{E}_b/N_0 > \ln 2 \end{cases}$$

which shows that the error probability tends to zero as $M \to \infty$, provided that $\mathcal{E}_b/N_0$ exceeds the threshold value $\ln 2$ given by the capacity formula. $\square$

3.5.2 ★ Constellation-constrained AWGN channel

The calculation of the channel capacity developed above involves no constraint on the use of an elementary constellation, except for the assumption of signals x with limited energy. We now evaluate the capacity of the AWGN channel when a specific elementary constellation is chosen as $\mathcal{X}$. To avoid a maximization of the mutual information over the a priori probabilities of the transmitted signals (as involved by the definition of channel capacity), we make the simplifying assumption that the elementary signals $x \in \mathcal{X}$ are transmitted with equal probabilities. Under our assumption of equally likely signals, the capacity of the channel is given by

$$C = \mathrm{H}(x) - \mathrm{H}(x \mid y) \tag{3.49}$$

where $H(x)$ is the maximum number of bits that can be carried by each elementary signal (this is $\log |\mathcal{X}|$). Using Bayes's rule, and observing that the a priori probabilities of x are equal, we obtain

$$\begin{aligned} H(x \mid y) &= \mathbb{E}_{x,y} \left[\log \frac{1}{p(x \mid y)}\right] \\ &= \mathbb{E}_{x,y} \left[\log \frac{\displaystyle\sum_{x' \in \mathcal{X}} p(y \mid x')}{p(y \mid x)}\right] \end{aligned} \tag{3.50}$$

so that

$$C = \log |\mathcal{X}| - \mathbb{E}_{x,y} \left[\log \frac{\displaystyle\sum_{x' \in \mathcal{X}} p(y \mid x')}{p(y \mid x)} \right] \tag{3.51}$$

This capacity can be conveniently evaluated by using Monte Carlo integration, by picking first a value of x, then taking the expectation with respect to y given x, and finally taking the expectation with respect to x. Under our assumptions of equally likely signals, the latter is computed as

$$\mathbb{E}_x[\cdot] = \frac{1}{|\mathcal{X}|} \sum_{x \in \mathcal{X}} [\cdot] \tag{3.52}$$

Notice that for the AWGN channel we have

$$p(y \mid x) = c e^{-\|y - x\|^2 / N_0} \tag{3.53}$$

where c is a normalization constant, which is irrelevant here. Figure 3.4 shows C for some two-dimensional constellations as a function of the signal-to-noise ratio

$$\mathrm{SNR} = \frac{\mathcal{E}_b R_b}{N_0 W} = \frac{\mathcal{E}_b}{N_0} \log M = \frac{\mathcal{E}}{N_0}$$

which follows from our assumption of two-dimensional signals, which entails $R_b/W = \log M$. The uppermost curve of Figure 3.4 describes the capacity of the AWGN channel with no constraint on the constellation. It is seen that, for low SNRs, the capacity loss due to the use of a specific constellation may be very small, while it increases at high SNR. From this we can infer that binary transmission is a reasonable proposition for small SNR, whereas we should choose a large constellation if the channel has a large signal-to-noise ratio.

3.5.3 ★ How much can we achieve from coding?

We now examine the amount of energy savings that the use of the code $\mathcal{S}$ allows one to achieve with respect to the transmission of the elementary uncoded constellation $\mathcal{X}$. If we plot the error probability achieved with and without coding as a function of $\mathcal{E}_b/N_0$, a typical behavior of the two error-probability curves emerges (see Figure 3.5. In particular, it is seen that, with coding, the same value of $P(e)$ may be achieved at a lower $\mathcal{E}_b/N_0$ than without coding. If this occurs, we say that the coding scheme yields a *coding gain*, usually measured in dB. Notice that

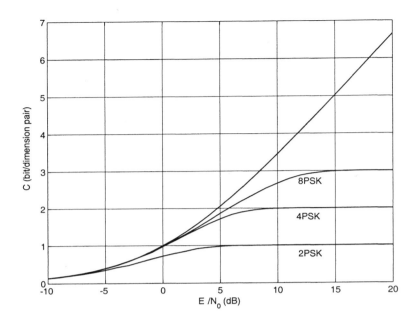

Figure 3.4: *Capacity of some two-dimensional constellations over the AWGN channel. The unconstrained capacity* $\log(1 + \mathcal{E}/N_0)$ *is also shown*

the coding gain might be negative if $P(e)$ is not low enough: this shows that coding may improve the quality of a channel only if the channel itself is not too bad. Also, the coding gain usually increases as $P(e)$ decreases: the limiting value as $P(e) \to 0$ is called *asymptotic coding gain*.

The asymptotic coding gain can be evaluated by simply taking the ratio between the values of the asymptotic power efficiency for coded and uncoded constellations.

Example 3.1 (continued)

> This constellation has minimum squared Euclidean distance 8 and $\mathcal{E}_b = 3/2$. Thus, its asymptotic power efficiency is $4/3$. The baseline γ (uncoded binary PAM) is 1, so the asymptotic coding gain of this constellation is $4/3$.   □

A way to evaluate the potential performance of an elementary constellation used with coding consists of computing its performance with reference to a code achiev-

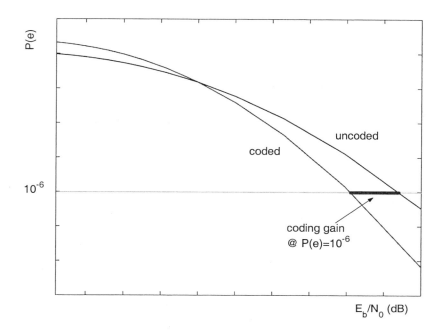

Figure 3.5: *Illustration of coding gain.*

ing capacity at a certain SNR. Define first the *normalized SNR*

$$\text{SNR}_o \triangleq \frac{\text{SNR}}{C^{-1}(\rho)} \tag{3.54}$$

Here C denotes the capacity of the channel, interpreted as a function of SNR, such that $C^{-1}(\rho)$ is the minimum value of SNR required to support the actual data rate ρ, in bit/dimension. Thus, SNR_o measures how much the SNR exceeds this minimal value. For a capacity-achieving coding scheme, $\rho = C$, and hence $\text{SNR}_o = 1$ (that is, 0 dB). A practical scheme (which has $\rho < C$) requires an SNR that is larger than SNR_o by some factor, which is precisely the normalized SNR. Thus, the value of the normalized signal-to-noise ratio signifies how far a system is operating from the capacity limit.

Recall that, with no constraint on the choice of the (one-dimensional) signal constellation $\mathcal{X}$, the channel capacity is given by

$$C = \frac{1}{2} \log(1 + \text{SNR}) \qquad \text{bit/dimension} \tag{3.55}$$

where SNR $= 2\mathcal{E}/N_0$. The capacity is actually achieved by a Gaussian-distributed set of elementary signals. Now, rewrite the capacity formula (3.55) as

$$\text{SNR} = 2^{2C} - 1 \tag{3.56}$$

which, in conjunction with (3.54), yields the definition of normalized SNR in the form

$$\text{SNR}_o \triangleq \frac{\text{SNR}}{2^{2\rho} - 1} \tag{3.57}$$

where ρ is the actual rate.

Example 3.7

Consider PAM as a baseline one-dimensional constellation; this has $|\mathcal{X}|$ points equally spaced on the real line, and centered at the origin. Let d denote the spacing between two adjacent signals; then the error probability over the AWGN channel is known to be

$$Q\left(\frac{d}{\sqrt{2N_0}}\right)$$

for the two outer points of the constellation, and

$$2Q\left(\frac{d}{\sqrt{2N_0}}\right)$$

for the $|\mathcal{X}| - 2$ inner points [3.2, Section 5.2]. Hence, by observing that the average energy of the constellation is

$$\mathcal{E} = (|\mathcal{X}|^2 - 1)\frac{d^2}{12}$$

and that SNR $= 2\mathcal{E}/N_0$, the average error probability is given by

$$P(e) = 2\left(1 - \frac{1}{|\mathcal{X}|}\right)Q\left(\sqrt{3\frac{\text{SNR}}{|\mathcal{X}|^2 - 1}}\right) \tag{3.58}$$

For uncoded PAM, $\rho = \log|\mathcal{X}|$ bit/dimension. From (3.57) we have

$$\text{SNR}_o = \frac{\text{SNR}}{|\mathcal{X}|^2 - 1} \tag{3.59}$$

so that the error probability for PAM can be written in the form

$$\begin{aligned}
P(e) &= 2\left(1 - \frac{1}{|\mathcal{X}|}\right)Q\left(\sqrt{3\,\text{SNR}_o}\right) \\
&\approx 2Q\left(\sqrt{3\,\text{SNR}_o}\right)
\end{aligned} \tag{3.60}$$

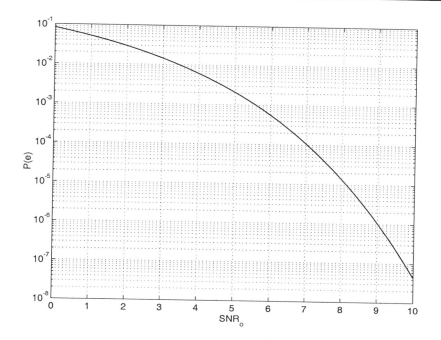

Figure 3.6: *Error probability of uncoded PAM vs. the normalized SNR.*

where the last approximation holds for large constellation size and makes $P(e)$ independent of $|\mathcal{X}|$. With $P(e)$ plotted versus SNR_o (Figure 3.6), it can be seen that, for $P(e) = 10^{-6}$, uncoded PAM is about 9 dB away from the capacity limit, which indicates that the use of coding can in principle buy that much. Observe also that the available coding gain decreases as the error probability increases. □

A dual way of describing the performance of a coding/modulation scheme is to plot its error probability versus the data rate for a fixed SNR level. The *normalized rate*

$$\rho_o \triangleq \frac{\rho}{C(\text{SNR})} \tag{3.61}$$

indicates again how far (in terms of rate) a system is operating from the capacity limit. For the AWGN channel we have

$$\rho_o = \frac{2\rho}{\log(1 + \text{SNR})} \tag{3.62}$$

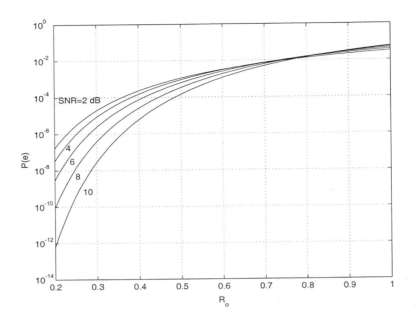

Figure 3.7: *Error probability of uncoded PAM vs. the normalized rate ρ_o.*

Example 3.7 (continued)

With uncoded PAM, we have, using (3.62) and (3.58),

$$|\mathfrak{X}| = 2^\rho = (1 + \text{SNR})^{\rho_o/2}$$

and

$$P(e) = \left(1 - (1 + \text{SNR})^{-\rho_o/2}\right) Q\left(\sqrt{3\frac{\text{SNR}}{(1 + \text{SNR})^{\rho_o} - 1}}\right)$$

which is plotted in Figure 3.7. We observe, for example, that, at an error probability of 10^{-4} and SNR$= 10$ dB, the rate achieved is about one half of the capacity. $\square$

3.6 Geometrically uniform constellations

Here we want to characterize the symmetries of the constellation $\mathcal{S}$ used for transmission over the AWGN channel. In particular, we want to develop tools useful to assess whether a given constellation has certain symmetries that are important in communications.

In general, the conditional error probability $P(e \mid \mathbf{x})$ depends on $\mathbf{x}$, i.e., different $\mathbf{x}$ may have different error probabilities. We are interested in finding constellations $\mathcal{S} = \{\mathbf{x}\}$ such that $P(e \mid \mathbf{x})$ is independent of $\mathbf{x}$. We have two options here:

(a) If the error probability is estimated via the union bound

$$P(e \mid \mathbf{x}) \leq \sum_{\widehat{\mathbf{x}} \neq \mathbf{x}} Q\left(\frac{\|\mathbf{x} - \widehat{\mathbf{x}}\|}{\sqrt{2N_0}}\right)$$

then this estimate of $P(e \mid \mathbf{x})$ does not depend on $\mathbf{x}$ if the set of Euclidean distances (including multiplicities) from $\mathbf{x}$ to any point of $\mathcal{S}$ do not depend on $\mathbf{x}$. In this case we say that $\mathcal{S}$ is *distance uniform*.

(b) If we use the exact expression of error probability:

$$P(e \mid \mathbf{x}) = 1 - \mathbb{P}[\mathbf{x} + \mathbf{n} \in \mathcal{R}(\mathbf{x})]$$

then $P(e \mid \mathbf{x})$ does not depend on $\mathbf{x}$ if the Voronoi regions are *all congruent*. We say that $\mathcal{S}$ is *Voronoi uniform* if all Voronoi regions are congruent.

The significance of these two properties is the following. Uniformity properties can be used to simplify the evaluation of error probability $P(e)$ for transmission of $\mathcal{S}$ over the AWGN channel, because the union bound on $P(e \mid \mathbf{x})$ does not depend on the transmitted vector $\mathbf{x}$ if $\mathcal{S}$ is distance-uniform. Similarly, the *exact* error probability does not depend on $\mathbf{x}$ if $\mathcal{S}$ is Voronoi uniform. Voronoi uniformity implies distance uniformity.

In practice, it is convenient to define a higher level of uniformity, called *geometric uniformity*, that implies both Voronoi and distance uniformity. Let us first recall some definitions.

Definition 3.6.1 *An* isometry *of* $\mathbb{R}^N$ *is a transformation* $u : \mathbb{R}^N \to \mathbb{R}^N$ *that preserves Euclidean distances:*

$$\|u(\mathbf{x}) - u(\mathbf{y})\| = \|\mathbf{x} - \mathbf{y}\|$$

Consider next a set $\mathcal{S}$ of points in $\mathbb{R}^N$.

Definition 3.6.2 *An isometry* u *that leaves* $\mathcal{S}$ *invariant, i.e., such that*

$$u(\mathcal{S}) = \mathcal{S}$$

is called a symmetry *of* $\mathcal{S}$. *The symmetries of* $\mathcal{S}$ *form a group under composition of isometries, called the* symmetry group *of* $\mathcal{S}$ *and denoted* $\Gamma(\mathcal{S})$.

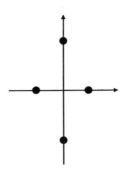

Figure 3.8: *A non-GU signal set.*

We are now ready to define *geometric uniformity.*

Definition 3.6.3 S *is geometrically uniform if, given any two points* $\mathbf{x}_i$, $\mathbf{x}_j$ *in* S, *there exists an isometry* $u_{i \to j}$ *that transforms* $\mathbf{x}_i$ *in* $\mathbf{x}_j$ *and leaves* S *invariant:*

$$u_{i \to j}(\mathbf{x}_i) = \mathbf{x}_j \qquad\qquad u_{i \to j}(S) = S$$

A geometrically uniform (GU) signal set can thus be generated by the action of a group of isometries on an "initial" vector.

Example 3.8

The constellation of Figure 3.8 is not geometrically uniform: in fact its symmetry group (which has four elements: the identity, the reflection along the horizontal axis, the reflection along the vertical axis, and the combination of two reflections) does not act transitively to generate the four signal points from an initial $\mathbf{x} \in S$. $\square$

Example 3.9

Take the initial vector $[1/\sqrt{2}, 1/\sqrt{2}]'$ and the rotation group R_4 represented by the four matrices

$$\begin{bmatrix} 1 & 0 \\ 0 & 1 \end{bmatrix} \quad \begin{bmatrix} 0 & 1 \\ -1 & 0 \end{bmatrix} \quad \begin{bmatrix} -1 & 0 \\ 0 & -1 \end{bmatrix} \quad \begin{bmatrix} 0 & -1 \\ 1 & 0 \end{bmatrix}.$$

This generates a four-point constellation (4-PSK). We learn from this example that it may happen that the symmetry group $\Gamma(S)$ of a geometrically uniform signal set is larger than necessary to generate S. In fact, 4-PSK has the 8-element dihedral group D_4 as its symmetry group. In addition to the rotation matrices shown above, D_4 has the matrices

$$\begin{bmatrix} 1 & 0 \\ 0 & -1 \end{bmatrix} \quad \begin{bmatrix} 0 & 1 \\ 1 & 0 \end{bmatrix} \quad \begin{bmatrix} -1 & 0 \\ 0 & 1 \end{bmatrix} \quad \begin{bmatrix} 0 & -1 \\ -1 & 0 \end{bmatrix}$$

□

Definition 3.6.4 *A generating group $U(S)$ of S is a subgroup of the symmetry group $\Gamma(S)$ that is minimally sufficient to generate S from an arbitrary initial vector. Such a map from $U(S)$ to S induces a group structure on S, making it isomorphic to the generating group.*

Example 3.9 (continued)

It is seen that 4-PSK can be generated not only by the rotation group R_4 as above (this group is isomorphic to the group Z_4 of integers mod 4), but also by the group of reflections about either axis, corresponding to the four diagonal matrices of D_4. This observation shows that a signal set may have more than one generating group. □

Example 3.10

The 8-PSK constellation is GU. One of its generating groups is R_8, the set of rotations by multiples of $\pi/4$. The symmetry group of S is $\Gamma(S) = V R_8$, the set of all compositions of elements of R_8 with elements of a two-element reflection group V consisting of the identity and a reflection about the line between any point and the origin. This group is isomorphic to the dihedral group D_8. D_8 is the symmetry group of the only other two-dimensional uniform constellation with eight points, *asymmetric 8-PSK* (see Figure 3.9). □

We quote from Forney [3.4]:

A geometrically uniform signal set S has the important property of looking the same from any of its points. This property implies that any arbitrary point of S may be taken as the "center of the universe," and that all the geometric properties relative to that point do not depend on which point is chosen.

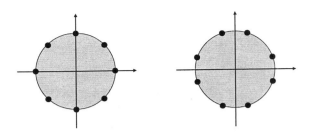

Figure 3.9: *8-PSK and asymmetric 8-PSK.*

3.6.1   Error probability

By combining the Bhattacharyya bound (3.20) with the union bound (3.15), we obtain for a GU constellation the *union-Bhattacharyya bound*

$$P(e) \le \sum_{\mathbf{x}\in\mathcal{S}\setminus\widehat{\mathbf{x}}} \exp\left\{-\|\mathbf{x}-\widehat{\mathbf{x}}\|^2/4N_0\right\} \tag{3.63}$$

The last bound can be given a convenient form by defining the *distance enumerator function* of the signal set $\mathcal{S}$ as the following function of the indeterminate X:

$$D(X) \triangleq \sum_{\mathbf{x}\in\mathcal{S}\setminus\widehat{\mathbf{x}}} A_i X^{\|\mathbf{x}-\widehat{\mathbf{x}}\|^2} \tag{3.64}$$

where A_i is the number of signals $\mathbf{x}$ at Euclidean distance $\|\mathbf{x}-\widehat{\mathbf{x}}\|$ from the reference signal $\widehat{\mathbf{x}}$. This function does not depend on $\widehat{\mathbf{x}}$ because of the GU assumption. With this definition, (3.63) takes the compact form

$$P(e) \le D(\exp\{-1/4N_0\}) \tag{3.65}$$

A tighter bound can be obtained by observing the exact value of the pairwise error probability (3.18). Using the inequality (see Problem section)

$$Q(\sqrt{x+y}) \le Q(\sqrt{x})\,e^{-y/2}, \qquad x \ge 0, y \ge 0 \tag{3.66}$$

and observing that for $\widehat{\mathbf{x}} \ne \mathbf{x}$ we have $\|\mathbf{x}-\widehat{\mathbf{x}}\| \ge d_{\mathrm{E,min}}$, we can bound the PEP as follows:

$$Q\left(\sqrt{\frac{\|\mathbf{x}-\widehat{\mathbf{x}}\|^2}{2N_0}}\right) = Q\left(\sqrt{\frac{d_{\mathrm{E,min}}^2 + (\|\mathbf{x}-\widehat{\mathbf{x}}\|^2 - d_{\mathrm{E,min}}^2)}{2N_0}}\right)$$

$$\le Q\left(\frac{d_{\mathrm{E,min}}}{2N_0}\right) e^{d_{\mathrm{E,min}}^2/4N_0} e^{-\|\mathbf{x}-\widehat{\mathbf{x}}\|^2/4N_0}$$

Thus,

$$
\begin{aligned}
P(e) &\leq \sum_{\mathbf{x} \in \mathcal{S} \backslash \hat{\mathbf{x}}} Q\left(\frac{\|\mathbf{x} - \hat{\mathbf{x}}\|}{\sqrt{2N_0}}\right) \\
&\leq Q\left(\frac{d_{\mathrm{E,min}}}{\sqrt{2N_0}}\right) e^{d_{\mathrm{E,min}}^2/4N_0} \sum_{\mathbf{x} \in \mathcal{S} \backslash \hat{\mathbf{x}}} e^{-\|\mathbf{x}-\hat{\mathbf{x}}\|^2/4N_0} \\
&= Q\left(\frac{d_{\mathrm{E,min}}}{\sqrt{2N_0}}\right) e^{d_{\mathrm{E,min}}^2/4N_0} D(\exp\{-1/4N_0\})
\end{aligned}
$$

3.7 Algebraic structure in $\mathcal{S}$: Binary codes

Here we show how, by introducing a suitable algebraic structure in a subset $\mathcal{S}$ of $\mathcal{X}^n$, it is possible to obtain constellations that can be easily described and analyzed.

Let $\mathbb{F}_2$ denote the set $\{0, 1\}$ with mod-2 operations, and $\mathbb{F}_2^n$ the set of binary n-tuples with mod-2 operations extended componentwise.

Definition 3.7.1 *An $[n, M, d]$ binary code $\mathcal{C}$ is a set of M binary n-tuples, its* words, *such that d is the minimum Hamming distance between any two of them.*

Based on this definition, a code $\mathcal{S}$ in the signal space can be generated by applying to each component of each word of the binary code $\mathcal{C}$ the map $\{0, 1\} \to \{\pm\sqrt{\mathcal{E}}\}$ (see Figure 3.10). Under this map, the set $\mathbb{F}_2^n$ is transformed into the set $\{\pm\sqrt{\mathcal{E}}\}^n$ of the 2^n vertices of the n-dimensional hypercube of side $\sqrt{\mathcal{E}}$ centered at the origin. All signals in $\mathcal{S}$ have the same energy $n\mathcal{E}$. The binary code $\mathcal{C}$ maps to the subset $\mathcal{S}$ of M vertices of this hypercube. Figure 3.10 shows a three-dimensional cube and two codes, one with $M = 8$ and $d = 1$, and the other with $M = 4$ and $d = 2$.

Notice also that (3.11) holds here, so that $\mathcal{S}$ has

$$
d_{\mathrm{E,min}}^2 = 4\mathcal{E}d
$$

Moreover, since $M \leq 2^n$, using (3.29) we can see that the bandwidth efficiency of $\mathcal{S}$ satisfies

$$
\frac{R_b}{W} = 2\frac{\log M}{n} \leq 2
$$

which shows that binary codes are suitable for the power-limited regime. The asymptotic power efficiency (3.30) takes the value

$$
\gamma = \frac{4\mathcal{E}d}{4\mathcal{E}_b} = \frac{\mathcal{E}}{\mathcal{E}_b}d = \frac{\log M}{n}d \tag{3.67}
$$

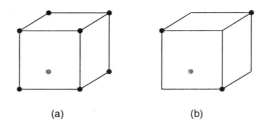

(a)   (b)

Figure 3.10: *Geometric representation of two codes with* $n = 3$. *(a)* $[3, 8, 1]$ *code, whose words are all the binary triples. (b)* $[3, 4, 2]$ *code, whose words are the binary triples with an even number of "1" components.*

where the last equality is derived from $\mathcal{E}_b = n\mathcal{E}/\log M$. The asymptotic coding gain of the code is given by the ratio of γ above to the asymptotic power efficiency of the baseline constellation 2-PAM. The latter is 1, so γ is also the asymptotic coding gain.

To understand the limitations of binary coding, Table 3.1, taken from [3.1], shows upper bounds to M for various values of the block length n and of the minimum Hamming distance d.

Definition 3.7.2 *If* $\mathcal{C}$ *has the form*

$$\mathcal{C} = \mathbb{F}_2^k \mathbf{G},$$

where $\mathbf{G}$ *is a* $k \times n$ *binary matrix with* $n \leq k$ *and rank* k, *called the* generator matrix *of* $\mathcal{C}$, *then* $\mathcal{C}$ *is called an* (n, k, d) linear *binary code. The code words of a linear code have the form* $\mathbf{uG}$, *where* $\mathbf{u}$ *is any binary* k-*tuple of binary source digits.*

It follows from the definition that the words of code $\mathcal{C}$ are all the linear combinations (with coefficients 0 or 1) of the k rows of $\mathbf{G}$; the code forms a linear subspace of $\mathbb{F}_2^n$ with 2^k elements. Figure 3.11 represents the coding chain for linear binary codes. The binary source symbols are grouped in blocks of k, then sent into the *encoder*. This transforms each of them in a word $\mathbf{c} \in \mathcal{C}$. The mapper transforms $\mathbf{c}$ into an element $\mathbf{x}$ in $\mathcal{S}$.

The code rate is $\rho = k/n$ bit/dimension, which yields a bandwidth efficiency $2k/n$, with a loss k/n with respect to the efficiency of $\mathcal{X} = \{\pm 1\}$, i.e., 2-PAM (that is, the use of this code entails a bandwidth expansion by a factor of n/k). The asymptotic power efficiency (3.67) (and hence the asymptotic coding gain with

	$d = 4$	6	8	10	12	14
$n = 6$	4	2				
7	8	2				
8	16	2	2			
9	20	4	2			
10	40	6	2	2		
11	72	12	2	2		
12	144	24	4	2	2	
13	256	32	4	2	2	
14	512	64	8	2	2	2
15	1024	128	16	4	2	2
16	2048	256	32	4	2	2
17	3276	340	37	6	2	2
18	6552	680	72	10	4	2
19	13104	1288	144	20	4	2
20	26208	2372	279	40	6	2
21	43689	4096	512	48	8	4
22	87378	6941	1024	88	12	4
23	173491	13774	2048	150	24	4
24	344308	24106	4096	280	48	6
25	599185	48148	6425	549	56	8
26	1198370	86132	10336	1029	98	14
27	2396740	162400	17804	1764	169	28
28	4793480	291269	32205	3200	288	56

Table 3.1: *Upper bounds to the number M of code words for binary codes with length n and minimum Hamming distance d.*

$u_i \in \{0,1\}$ $\mathbf{u} \in \{0,1\}^k$ $\mathbf{c} = \mathbf{u}G \in \mathcal{C} \subseteq \{0,1\}^n$ $\mathbf{x} \in \mathcal{S} \subseteq \{\pm\sqrt{\mathcal{E}}\}^n$

Figure 3.11: *Transmission of linearly encoded symbols.*

respect to binary PAM) becomes

$$\gamma = \frac{k}{n}d \qquad\qquad (3.68)$$

Example 3.11

Let the generator matrix of a $(4, 3, d)$ binary code be

$$\mathbf{G} = \begin{bmatrix} 1 & 0 & 0 & 1 \\ 0 & 1 & 0 & 1 \\ 0 & 0 & 1 & 1 \end{bmatrix}$$

The eight words of this code are obtained by multiplying all the binary triples by $\mathbf{G}$:

$$(0\ 0\ 0)\mathbf{G} = (0\ 0\ 0\ 0) \qquad (0\ 0\ 1)\mathbf{G} = (0\ 0\ 1\ 1) \qquad (0\ 1\ 0)\mathbf{G} = (0\ 1\ 0\ 1)$$

$$(0\ 1\ 1)\mathbf{G} = (0\ 1\ 1\ 0) \qquad (1\ 0\ 0)\mathbf{G} = (1\ 0\ 0\ 1) \qquad (1\ 0\ 1)\mathbf{G} = (1\ 0\ 1\ 0)$$

$$(1\ 1\ 0)\mathbf{G} = (1\ 1\ 0\ 0) \qquad (1\ 1\ 1)\mathbf{G} = (1\ 1\ 1\ 1)$$

and are characterized by having an even number of "1" components. Binary codes with this property are called *single-parity-check* codes. One can easily verify that $d = 2$ in this case, so the asymptotic coding gain is $6/4 \Rightarrow 1.76$ dB.   □

Capacity limits

Using the capacity formula, we can derive the minimum value of $\mathcal{E}_b/N_0$ that allows transmission with vanishingly small error probability under the constraint that the code rate is ρ. Since for one-dimensional signals we have SNR$= 2\rho\mathcal{E}_b/N_0$, from

$$\rho \leq \frac{1}{2} \log\left(1 + 2\rho\frac{\mathcal{E}_b}{N_0}\right)$$

we obtain the capacity limit

$$\frac{\mathcal{E}_b}{N_0} \geq \frac{2^{2\rho} - 1}{2\rho} \tag{3.69}$$

As an example, for infinitely reliable transmission with rate-$1/2$ codes we need $\mathcal{E}_b/N_0 > 0$ dB. For unconstrained rate (i.e., for $\rho \to 0$), we obtain the known result $\mathcal{E}_b/N_0 > \ln 2 = -1.6$ dB. Some additional values are tabulated in Table 4.1.

The capacity limit is often used to evaluate the quality of a practical code. The value of $\mathcal{E}_b/N_0$ necessary to achieve a small error probability (e.g., 10^{-5}) is compared to (3.69) for the same rate. If these two values are close enough, we say that the code performs "close to capacity."

Systematic codes

Definition 3.7.3 *A linear code is called* systematic *if its generator matrix has the form* $\mathbf{G} = [\mathbf{I}_k \vdots \mathbf{P}]$, *where* $\mathbf{P}$ *is a* $k \times (n-k)$ *matrix. The words of these codes have the form*

$$\mathbf{uG} = [\mathbf{u} \vdots \mathbf{uP}] \tag{3.70}$$

that is, their first k *positions are a copy of the source* k-*tuple.*

The assumption that a code generated by $\mathbf{G}$ is systematic does not entail any loss of generality. In fact, by combining linearly the rows of $\mathbf{G}$ (and possibly permuting its columns, which corresponds to changing the order of the components of all code words), we can give $\mathbf{G}$ the form (3.70). A general algorithm to do this is Gauss-Jordan elimination (Section B.3, Appendix B).

Parity-check matrix

Definition 3.7.4 *An alternative definition of a linear code is through the concept of an* $(n-k) \times n$ parity-check *matrix* $\mathbf{H}$. *A code* $\mathcal{C}$ *is linear if*

$$\mathbf{H}\mathcal{C}' = \mathbf{0},$$

where the prime ' *denotes transposition of all code words.*

This definition follows from the observation that, if $\mathcal{C}^{\perp}$ denotes the $(n-k)$-dimensional vector space orthogonal to $\mathcal{C}$ (which is another code called the *dual* of $\mathcal{C}$), and $\mathbf{h}_1, \ldots, \mathbf{h}_{n-k}$ are n-vectors spanning it, then for any i and for any binary n-tuple $\mathbf{y}$ we have $\mathbf{h}_i \mathbf{y}' = 0$ if and only if $\mathbf{x} \in \mathcal{C}$. The matrix $\mathbf{H}$ whose rows are the vectors above is the parity-check matrix of $\mathcal{C}$, and the vector $\mathbf{Hy}'$ is called the *syndrome* of $\mathbf{r}$: this is the null vector if and only if $\mathbf{y}$ is a code word. $\mathbf{H}$ describes the $n-k$ linear constraints ("parity checks") that a binary n-tuple must satisfy to be a code word. To avoid redundant constraints, the $n-k$ rows of $\mathbf{H}$ are assumed to be linearly independent.

Example 3.12

Consider a linear binary $(n, n-1, d)$ code with

$$\mathbf{H} = [1\ 1\ \ldots\ 1]$$

The syndrome of the binary n-vector $\mathbf{y} = (y_1, y_2, \ldots, y_n)$ is given by

$$\mathbf{Hy}' = y_1 + y_2 + \ldots + y_n$$

which is zero if and only if there are an even number of 1s among the components of $\mathbf{y}$. $\mathcal{C}$ has $d = 2$ and is a single-parity-check code. $\square$

Example 3.13

The $(n, 1, n)$ binary *repetition code*, consisting of two words, namely, the all-zero and the all-1 binary vectors, has the parity-check matrix

$$\mathbf{H} = \left[\begin{array}{cc} & 1 \\ \mathbf{I}_{n-1} & \vdots \\ & 1 \end{array} \right]$$

where $\mathbf{I}_k$ denotes the $k \times k$ identity matrix. For $n = 4$, we obtain

$$\mathbf{H} = \left[\begin{array}{cccc} 1 & 0 & 0 & 1 \\ 0 & 1 & 0 & 1 \\ 0 & 0 & 1 & 1 \end{array} \right]$$

and the syndrome $\mathbf{H}\mathbf{y}'$ is null if and only if $y_1 + y_4 = y_2 + y_4 = y_3 + y_4 = 0$, i.e., $y_1 = y_2 = y_3 = y_4$. $\square$

Observation 3.7.1   From the definitions of $\mathbf{G}$ and $\mathbf{H}$ it follows that

$$\mathbf{H}\mathbf{G}' = \mathbf{0} \qquad \text{and} \qquad \mathbf{G}\mathbf{H}' = \mathbf{0},$$

where $\mathbf{0}$ denotes an all-zero matrix with suitable dimensions.

Error detection

The concept of syndrome of an n-tuple is convenient to define a procedure called *error detection*. If hard decisions $\hat{x}_i$ (see Section 1.2.1) are separately made on the components of a code word observed at the output of a noisy channel, the resulting vector $\tilde{\mathbf{x}} \triangleq (\hat{x}_1, \ldots, \hat{x}_n)$ may not be a code word, thus indicating that "the channel has made errors." To verify if a binary n-tuple is a code word, it suffices to compute its syndrome.

Hamming distance and Hamming weight

From the definition of a linear code, it follows that d, the minimum Hamming distance of a code $\mathcal{C}$, equals the minimum *Hamming weight* of the words of $\mathcal{C}$, i.e., the minimum number of "1"s contained in each nonzero code word. In fact, if $w(\mathbf{c})$ denotes the Hamming weight of $\mathbf{c}$, we have, for the Hamming distance of the code words $\mathbf{c}'$, $\mathbf{c}''$:

$$d_{\mathrm{H}}(\mathbf{c}', \mathbf{c}'') = w(\mathbf{c}' + \mathbf{c}'') = w(\mathbf{c})$$

where $\mathbf{c} \triangleq \mathbf{c}' + \mathbf{c}''$ is (because of the linearity of the code) another word of $\mathcal{C}$ (see the Problem section). Thus,

$$\min_{\mathbf{c}' \neq \mathbf{c}''} d_{\mathrm{H}}(\mathbf{c}', \mathbf{c}'') = \min_{\mathbf{c} \neq \mathbf{0}} w(\mathbf{c}) \qquad (3.71)$$

Now, rewrite the parity-check matrix $\mathbf{H}$ of $\mathcal{C}$ in the form

$$\mathbf{H} = [\mathbf{h}_1 \ \cdots \ \mathbf{h}_n]$$

where $\mathbf{h}_i$, $i = 1, \ldots, n$ is a binary $(n - k)$-vector. The condition for $\mathbf{c}$ to be a code word, i.e., $\mathbf{Hc}' = \mathbf{0}$, can be expressed as

$$\sum_{i=1}^{n} \mathbf{h}_i c_i = \mathbf{0} \qquad (3.72)$$

This equation expresses the $n - k$ linear parity checks that the symbols of $\mathbf{c}$ must satisfy. Thus, since $\mathbf{c}$ has $w(\mathbf{c})$ ones, (3.72) shows that $w(\mathbf{c})$ columns of $\mathbf{H}$ sum to the null vector. This fact, in combination with (3.71), implies that, for a linear code, d is the minimum number of columns of $\mathbf{H}$ to be added together to obtain $\mathbf{0}$.

Example 3.14

Consider the parity-check matrix of the *Hamming* code $\mathcal{C}$ such that

$$\mathbf{H} = \begin{bmatrix} 0 & 0 & 0 & 1 & 1 & 1 & 1 \\ 0 & 1 & 1 & 0 & 0 & 1 & 1 \\ 1 & 0 & 1 & 0 & 1 & 0 & 1 \end{bmatrix}$$

Here the columns of $\mathbf{H}$ consist of all binary nonzero triples. We verify that no two columns sum to $\mathbf{0}$, while, for example, $\mathbf{h}_1 + \mathbf{h}_2 + \mathbf{h}_3 = \mathbf{0}$. Thus, $d = 3$. This code has parameters $(7, 4, 3)$. $\square$

Coset decomposition of a code $\mathcal{C}$

As briefly mentioned before, a linear binary code forms a vector space over the Galois field $\mathbb{F}_2^n$; it is also a commutative additive group. We may define an (n, k', d') *subcode* $\mathcal{C}'$ of the (n, k, d) code $\mathcal{C}$ (with $k' \leq k$, $d' \geq d$) as a subset of $\mathcal{C}$ that is itself a linear code. The *cosets* of the subcode $\mathcal{C}'$ in $\mathcal{C}$ are the codes $\mathcal{C}' + \mathbf{c}$ with $\mathbf{c} \in \mathcal{C}$, i.e., the cosets of the subgroup $\mathcal{C}'$. Every coset of $\mathcal{C}'$ contains the same number of code words. Notice that only the coset $\mathcal{C}' + \mathbf{0}$ is a *linear* code.

Given $\mathcal{C}$ and its subcode $\mathcal{C}'$, two code words are in the same coset if and only if their sum is in $\mathcal{C}'$ and the union of the cosets forms a partition of $\mathcal{C}$. The situation is summarized in the following Theorem:

Theorem 3.7.1 *Any two cosets are either disjoint or identical.*

Proof.   Let $\mathbf{x}$ belong to the cosets $\mathcal{C}' + \mathbf{a}$ and $\mathcal{C}' + \mathbf{b}$, $\mathbf{b} \neq \mathbf{a}$. Then $\mathbf{c} + \mathbf{a} = \widehat{\mathbf{c}} + \mathbf{b}$, with $\mathbf{c}, \widehat{\mathbf{c}} \in \mathcal{C}'$, and consequently $\mathbf{a} = \mathbf{c} + \widehat{\mathbf{c}} + \mathbf{b} \in \mathcal{C}' + \mathbf{b}$, which implies $\mathcal{C}' + \mathbf{a} \subseteq \mathcal{C}' + \mathbf{b}$. In a similar way we obtain $\mathcal{C}' + \mathbf{b} \subseteq \mathcal{C}' + \mathbf{a}$, which yields $\mathcal{C}' + \mathbf{a} = \mathcal{C}' + \mathbf{b}$, thus proving the theorem.   □

Example 3.15

The "universe" $(n, n, 1)$ code $\mathcal{C}$ admits as its coset the single-parity-check $(n, n - 1, 2)$ code $\mathcal{C}'$. The two cosets of $\mathcal{C}'$ in $\mathcal{C}$ are the two $(n, n-1, 2)$ codes whose words have even parity and odd parity, respectively. We may decompose $\mathcal{C}$ in the form

$$\mathcal{C} = \mathcal{C}' \cup \{\mathcal{C}' + \widehat{\mathbf{c}}\}$$

where $\widehat{\mathbf{c}}$ is any fixed n-tuple with weight 1.   □

3.7.1   Error probability and weight enumerator

Consider the calculation of error probability for the constellation $\mathcal{S}$ obtained from the linear binary code $\mathcal{C}$. We have, using the inequality $Q(x) \leq \exp(-x^2/2)$:

$$P(e \mid \mathbf{x}) \leq \sum_{\widehat{\mathbf{x}} \neq \mathbf{x}} P(\mathbf{x} \to \widehat{\mathbf{x}}) \leq \sum_{\widehat{\mathbf{x}} \neq \mathbf{x}} e^{-\|\mathbf{x} - \widehat{\mathbf{x}}\|^2 / 4N_0} \tag{3.73}$$

Now, since we are considering binary antipodal signaling, from (3.11) we have that Euclidean distance and Hamming distance are related through $\|\mathbf{x} - \widehat{\mathbf{x}}\|^2 =$

$4\mathcal{E}d_{\mathrm{H}}(\mathbf{c}, \widehat{\mathbf{c}})$, where $\mathbf{c}$ and $\widehat{\mathbf{c}}$ are the binary code words mapped into $\mathbf{x}$, $\widehat{\mathbf{x}}$, respectively. Thus, we can write, recalling the linearity assumption for $\mathcal{C}$,

$$P(e \mid \mathbf{c}) \leq \sum_{\widehat{\mathbf{c}} \neq \mathbf{c}} e^{-d_{\mathrm{H}}(\mathbf{c}, \widehat{\mathbf{c}})\mathcal{E}/N_0} = \sum_{\mathbf{c}^* \neq 0} e^{-w(\mathbf{c}^*)\mathcal{E}/N_0} \qquad (3.74)$$

The value of last summation does not depend on $\mathbf{c}$, and hence $P(e \mid \mathbf{c}) = P(e)$ for any choice of $\mathbf{c}$. In conclusion,

$$P(e) \leq \sum_{\mathbf{c} \neq 0} \exp(-w(\mathbf{c})\mathcal{E}/N_0) \qquad (3.75)$$

Recall now that n binary symbols, each with energy $\mathcal{E}$, carry k information bits, so $\mathcal{E}_b = (n/k)\mathcal{E}$, and hence

$$P(e) \leq \sum_{\mathbf{c} \neq 0} \exp(-w(\mathbf{c})(k/n)\mathcal{E}_b/N_0) \qquad (3.76)$$

Consider next the set of weights of the words of the linear code $\mathcal{C}$. The *weight enumerator* of $\mathcal{C}$ is the polynomial in the indeterminate X defined as

$$W(X) \triangleq \sum_{\mathbf{c} \in \mathcal{C}} X^{w(\mathbf{c})} = \sum_i A_i X^i \qquad (3.77)$$

where the last summation index runs through the set of values taken on by the code word weights, and A_i is the number of words whose weight is i.

From (3.75)–(3.76), we have the simple expression

$$P(e) \leq W\left(e^{-\mathcal{E}/N_0}\right) - 1 = W\left(e^{-k/n \cdot \mathcal{E}_b/N_0}\right) - 1 \qquad (3.78)$$

Example 3.16

The "universe" code $(n, n, 1)$, whose words are all the binary n-tuples, has weight enumerator (see the Problem section)

$$W(X) = (1 + X)^n$$

and consequently

$$P(e) \leq \left(1 + e^{-\mathcal{E}/N_0}\right)^n - 1$$

The single-parity-check $(n, n - 1, 2)$ code, whose words are all the n-tuples with an even number of 1s, has weight enumerator (see the Problem section)

$$W(X) = \frac{(1 + X)^n + (1 - X)^n}{2}$$

□

3.8 ★ Symbol MAP decoding

So far, we have examined decoding of code $\mathcal{C}$ based on the maximum-likelihood rule, which consists of maximizing over $\mathbf{x} \in \mathcal{C}$ the function $p(\mathbf{y} \mid \mathbf{x})$. This rule minimizes the word error probability under the assumption that all code words are equally likely. If the latter assumption is removed, then to minimize the word error probability we should use instead the maximum a posteriori (MAP) rule, which consists of maximizing the function $p(\mathbf{x} \mid \mathbf{y})$. Now, assume that we want to minimize the *information symbol* error probability, that is, the probability that a source symbol u_i is received erroneously: in this case, we obtain *symbol maximum a posteriori* (MAP) decoding by maximizing the *a posteriori probabilities* (APP)

$$p(u_i \mid \mathbf{y}), \qquad i = 1, \ldots, k$$

Specifically, we decode u_i into 0 or 1 by comparing the probabilities that the ith bit of the source sequence that generates $\mathbf{x}$ is equal to 0 or 1, given the received vector $\mathbf{y}$ and the fact that code $\mathcal{C}$ is used. The APPs $p(u_i \mid \mathbf{y})$ can be expressed as

$$p(u_i \mid \mathbf{y}) = \sum_{\mathbf{x} \in \mathcal{C}_i(u_i)} p(\mathbf{x} \mid \mathbf{y}) \tag{3.79}$$

and $\mathcal{C}_i(u_i)$ denotes the subset of code words generated by the source words whose ith component is u_i. Observe that, since we are interested in maximizing $p(u_i \mid \mathbf{y})$ over u_i, we can omit constants that are the same for $u_i = 0$ and $u_i = 1$. For equally likely code words $\mathbf{x}$ and a memoryless channel, we may write

$$p(u_i \mid \mathbf{y}) \propto \sum_{\mathbf{x} \in \mathcal{C}_i(u_i)} p(\mathbf{y} \mid \mathbf{x}) = \sum_{\mathbf{x} \in \mathcal{C}_i(u_i)} \prod_{j=1}^{n} p(y_j \mid x_j) \tag{3.80}$$

Example 3.17

Consider an AWGN channel with noise variance σ^2, so that

$$p(y_j \mid x_j) \propto \exp\{-(y_j - x_j)^2/2\sigma^2\} \triangleq f(x_j, y_j)$$

Assume the single-parity-check $(3, 2, 2)$ binary code of Table 3.2 is used, and let the received symbols y_1, y_2, y_3 and the noise variance be such that

$$\begin{aligned}
f(+1, y_1) &= 0.98 & f(-1, y_1) &= 0.20 \\
f(+1, y_2) &= 0.99 & f(-1, y_2) &= 0.18 \\
f(+1, y_3) &= 0.55 & f(-1, y_3) &= 0.67
\end{aligned}$$

source symbols	coded symbols
00	$+1 \ +1 \ +1$
01	$+1 \ -1 \ -1$
10	$-1 \ +1 \ -1$
11	$-1 \ -1 \ -1$

Table 3.2: *Words of the single-parity-check* $(3, 2, 2)$ *binary code.*

We can compute $p(\mathbf{y} \mid \mathbf{x})$ for each code word:

$$+1 \ +1 \ +1 \longrightarrow 0.98 \times 0.99 \times 0.55 = 0.5336$$
$$+1 \ -1 \ -1 \longrightarrow 0.98 \times 0.18 \times 0.67 = 0.1182$$
$$-1 \ -1 \ +1 \longrightarrow 0.20 \times 0.18 \times 0.55 = 0.0198$$
$$-1 \ +1 \ -1 \longrightarrow 0.20 \times 0.99 \times 0.67 = 0.1327$$

so that

$$p(u_1 = 0 \mid \mathbf{y}) \propto 0.5336 + 0.1182 = 0.6518$$
$$p(u_1 = 1 \mid \mathbf{y}) \propto 0.0198 + 0.1327 = 0.1525$$
$$p(u_2 = 0 \mid \mathbf{y}) \propto 0.5336 + 0.1327 = 0.6663$$
$$p(u_2 = 1 \mid \mathbf{y}) \propto 0.1182 + 0.0198 = 0.1380$$

and symbol MAP decoding yields $\hat{u}_1 = 0$, $\hat{u}_2 = 0$. □

The brute-force approach consisting of direct computation of (3.79) is generally inefficient. In Chapters 5 and 8 we shall examine more efficient algorithms.

3.9 Bibliographical notes

For details on modulation schemes, see, e,g., [3.2]. Our discussion on the coding gain and the definitions of normalized SNR and normalized data rate follow [3.5, 3.13]. The concept of Shannon Bandwidth was introduced by Massey [3.9]. Example 3.6 is borrowed from [3.12]. The section on geometrically uniform signals is taken from [3.4].

A thorough treatment of algebraic codes can be found in the classical work [3.8], a *monumentum ære perennius* to algebraic coding theory. Recent results are collected in [3.6, 3.7], while [3.3] is oriented towards communications applications.

3.10   Problems

1. For transmission of the M-signal constellation $\mathcal{S} = \{\mathbf{x}\}$ over the AWGN channel with noise power spectral density $N_0/2$, derive the chain of upper bounds

$$
\begin{aligned}
P(e) &\leq \frac{1}{M} \sum_{\mathbf{x}} \sum_{\hat{\mathbf{x}} \neq \mathbf{x}} Q\left(\frac{d_{\mathrm{E}}(\mathbf{x}, \hat{\mathbf{x}})}{\sqrt{2N_0}}\right) \\
&\leq (M-1)Q\left(\frac{d_{\mathrm{E,min}}}{\sqrt{2N_0}}\right)
\end{aligned}
$$

the chain of lower bounds

$$
\begin{aligned}
P(e) &\geq \frac{1}{M} \sum_{\mathbf{x}} Q\left(\frac{\min_{\hat{\mathbf{x}} \neq \mathbf{x}} d_{\mathrm{E}}(\mathbf{x}, \hat{\mathbf{x}})}{\sqrt{2N_0}}\right) \\
&\geq \frac{\alpha_{\mathrm{min}}}{M} Q\left(\frac{d_{\mathrm{E,min}}}{\sqrt{2N_0}}\right) \\
&\geq \frac{2}{M} Q\left(\frac{d_{\mathrm{E,min}}}{\sqrt{2N_0}}\right)
\end{aligned}
$$

(where α_{min} is the number of signals having at least one neighbor at distance $d_{\mathrm{E,min}}$), and the approximation

$$
P(e) \stackrel{\sim}{\leq} \alpha Q\left(\frac{d_{\mathrm{E,min}}}{\sqrt{2N_0}}\right)
$$

(where α is the average number of "nearest neighbors," i.e., of signals at a distance $d_{\mathrm{E,min}}$ from any element of $\mathcal{S}$).

2. Consider the one-dimensional quaternary constellation $\mathcal{S} = \{\pm a, \pm b\}$, $0 \leq a \leq b$, subject to the constraint of unit average energy. Compute the exact error probability of this constellation over the AWGN channel, and find the values of a, b that minimize it as a function of N_0. For which value of N_0 does the geometry of the minimum-$P(e)$ constellation maximize the minimum Euclidean distance? What happens as $N_0 \to 0$?

3. Assuming an AWGN channel with noise power spectral density $N_0/2$, compute the bit error probability $P_b(e)$ of 4-PSK, with Gray coding as in Figure 3.1, as a function of $\mathcal{E}_b/N_0$.

4. The general *Bhattacharyya bound* on the pairwise error probability of a general channel described by the conditional probability density function $p(\mathbf{y} \mid \mathbf{x})$ can be derived as follows. The maximum-likelihood demodulation rule consists of picking the signal $\widehat{\mathbf{x}}$ that maximizes $p(\mathbf{y} \mid \mathbf{x})$. Now, the pairwise error probability is given by

$$P(\mathbf{x} \to \widehat{\mathbf{x}}) = \mathbb{P}\left\{p(\mathbf{y} \mid \widehat{\mathbf{x}}) \geq p(\mathbf{y} \mid \mathbf{x}) \mid \mathbf{x}\right\}$$

and can also be expressed in the form

$$P(\mathbf{x} \to \widehat{\mathbf{x}}) = \int_{\mathbf{y}} f(\mathbf{y}) p(\mathbf{y} \mid \mathbf{x}) \, d\mathbf{y}$$

where

$$f(\mathbf{y}) \triangleq \begin{cases} 1 & p(\mathbf{y} \mid \widehat{\mathbf{x}}) \geq p(\mathbf{y} \mid \mathbf{x}) \\ 0 & \text{otherwise} \end{cases}$$

Show, by finding a suitable upper bound to the function $f(\mathbf{y})$, that the Bhattacharyya bound holds:

$$P(\mathbf{x} \to \widehat{\mathbf{x}}) \leq \int_{\mathbf{y}} \sqrt{p(\mathbf{y} \mid \widehat{\mathbf{x}}) p(\mathbf{y} \mid \mathbf{x})} \, d\mathbf{y}$$

Notice also how the right-hand side of this inequality is connected to the Bhattacharyya distance.

5. **(a)** Derive the *Chernoff bound*

$$\mathbb{P}[X \leq 0] \leq \min_{\lambda > 0} \mathbb{E}\left[e^{-\lambda X}\right]$$

where X is a real random variable. This is especially useful when X is the sum of independent random variables. (**Hint:** Write $\mathbb{P}[X \leq 0] = \mathbb{E}[f(X)]$ for a suitable choice of the function $f(\cdot)$, and bound this function from above.)

(b) Use the Chernoff bound to obtain an upper bound to the pairwise error probability for the AWGN channel. Use the following result, valid for $X \sim \mathcal{N}(\mu, \sigma^2)$:

$$\mathbb{E}[\exp(\lambda X)] = \exp(-\lambda \mu + \lambda^2 \sigma^2 / 2)$$

6. Draw the equivalent of Figure 3.6 for QAM constellations. Use the following approximation to error probability:

$$P(e) \approx 4Q\left(\sqrt{3\frac{\text{SNR}}{|\mathcal{X}| - 1}}\right)$$

where $\text{SNR} = \mathcal{E}/N_0$, $\mathcal{E}$ the average energy of the constellation.

7. Redraw Figure 3.4 by plotting C vs. $\mathcal{E}_b/N_0$.

8. Generalize (3.69) by assuming that a nonzero error probability p is tolerated. (**Hint:** A finite error probability can be obtained by this simple scheme: transmit with zero error probability a fraction ρ^* of binary symbols, and randomly guess the balance.) It can be shown, by using Shannon's *Rate–Distortion Theory* [3.10], that the above simple scheme can be improved by transmitting with zero error probability a fraction $1 + p\log p + (1-p)\log(1-p)$ of binary symbols. This leads to the results shown in Figure 1.5.

9. Verify that for a linear code:

   (a) The all-zero n-tuple is a code word.

   (b) The sum of two code words is a code word.

10. A $(5, 3, d)$ linear binary code is defined through the correspondence given in the following table:

u_1	u_2	u_3	x_1	x_2	x_3	x_4	x_5
1	1	0	1	0	1	0	1
1	0	1	0	1	0	1	0
0	1	0	0	1	1	0	0

   Find generator and parity-check matrices for this code.

11. Consider the linear binary *Reed-Muller* code generated by the 4×8 matrix

$$\mathbf{G} = \begin{bmatrix} 1 & 1 & 1 & 1 & 1 & 1 & 1 & 1 \\ 0 & 0 & 0 & 0 & 1 & 1 & 1 & 1 \\ 0 & 0 & 1 & 1 & 0 & 0 & 1 & 1 \\ 0 & 1 & 0 & 1 & 0 & 1 & 0 & 1 \end{bmatrix}$$

   Prove that this code has $d = 4$, and compute its asymptotic coding gain.

12. Recall that a linear binary code is called *systematic* if $\mathbf{G}$ has the form $\mathbf{G} = [\mathbf{I}_k \vdots \mathbf{P}]$, where $\mathbf{P}$ is a $k \times (n-k)$ matrix. The words of these codes have the form

$$\mathbf{uG} = [\mathbf{u} \vdots \mathbf{uP}]$$

    that is, the first k positions of any code word coincide with the source vector $\mathbf{u}$.

    (a) Prove that the parity check matrix of a systematic code has the form
    $$\mathbf{H} = [\mathbf{P}' \vdots \mathbf{I}_{n-k}].$$

    (c) How should one proceed if the parity-check matrix in (b) is not full-rank?

13. Prove that, for an (n, d, k) systematic linear code $\mathcal{C}$, the following *Singleton inequality* holds:
$$d \leq n - k + 1$$

    (if equality holds, $\mathcal{C}$ is called a *maximum distance* [MD] code.) **(Hint:** Consider words having a single "1" in their first k positions.)

14. Prove inequality (3.66).

15. Exhibit three simple examples of binary maximum-distance codes with length n. It can be proved that these are the only binary MD codes: there are, however, *nonbinary* MD codes, the most celebrated among them being the *Reed–Solomon* codes [3.8].

16. Prove that the weight enumerator of the $(n, n, 1)$ universe code is $W(X) = (1 + X)^n$, and that the enumerator of the $(n, n-1, 2)$ single-parity-check code is
$$W(X) = \frac{(1+X)^n + (1-X)^n}{2}$$

17. Prove that the linear binary codes are geometrically uniform.

18. Consider transmission over the AWGN channel with an $(n, n-1, 2)$ binary single-parity-check code, and the following decoding algorithm. First, make separate hard decisions $\hat{x}_i$ on the received components of $\mathbf{y}$, based on their polarities, and form the n-tuple $\tilde{\mathbf{x}} \triangleq (\hat{x}_1, \ldots, \hat{x}_n)$. If $\tilde{\mathbf{x}}$ is a code word, then the decoder chooses $\hat{\mathbf{x}} = \tilde{\mathbf{x}}$. Otherwise, the decoder inverts the polarity of the component of $\mathbf{y}$ having the smallest absolute value, and proceeds as above. Prove that this "Wagner rule" yields ML decoding.

References

[3.1]   E. Agrell, A. Vardy, and K. Zeger, "A table of upper bounds for binary codes," *IEEE Trans. Inform. Theory*, Vol. 47, No. 7, pp. 3004–3006, November 2001.

[3.2]   S. Benedetto and E. Biglieri, *Digital Transmission Principles with Wireless Applications*. New York: Kluwer/Plenum, 1999.

[3.3]   R. E. Blahut, *Algebraic Codes for Data Transmission*. Cambridge, UK: Cambridge University Press, 2003.

[3.4]   G. D. Forney, Jr., "Geometrically uniform codes," *IEEE Trans. Inform. Theory*, Vol. 37, No. 5, pp. 1241–1260, September 1991.

[3.5]   G. D. Forney, Jr., and G. Ungerboeck, "Modulation and coding for linear Gaussian channels," *IEEE Trans. Inform. Theory*, Vol. 44, No. 6, pp. 2384–2415, October 1998.

[3.6]   W. C. Huffman and V. Pless, *Fundamentals of Error-Correcting Codes*. Cambridge, UK: Cambridge University Press, 2003.

[3.7]   S. Lin and D. J. Costello, Jr, *Error Control Coding* (2nd edition). Upper Saddle River, NJ: Pearson Prentice Hall, 2004.

[3.8]   F. J. MacWilliams and N. J. A. Sloane, *The Theory of Error-Correcting Codes*. Amsterdam: North-Holland, 1977.

[3.9]   J. L. Massey, "Towards an information theory of spread-spectrum systems," in: S. G. Glisic and P. A. Leppänen (eds.), *Code Division Multiple Access Communications*, pp. 29–46. Boston, MA: Kluwer, 1995.

[3.10] R. J. McEliece, *The Theory of Information and Coding* (2nd edition). Cambridge, UK: Cambridge University Press, 2002.

[3.11] D. Slepian, "On bandwidth," *IEEE Proc.*, Vol. 64, No. 3, pp. 292–300, March 1976.

[3.12] A. J. Viterbi, *Principles of Coherent Communication*. New York: McGraw-Hill, 1966.

[3.13] L. Zheng and D. N. C. Tse, "Diversity and multiplexing: A fundamental tradeoff in multiple antenna channels," *IEEE Trans. Inform. Theory*, Vol. 49, No. 5, pp. 1073–1096, May 2003.

4

all four of them, in their quartan agues

Fading channels

Wireless communications' main feature is the randomness of signal attenuation (fading). In this chapter we study how this randomness affects the error probability and the capacity of transmission over channels affected by fading. We also examine how the channel randomness can be used to improve performance: in fact, if the same information is conveyed through several transmission paths, this diversity allows an increase of the reliability of its reception.

4.1   Introduction

The "narrowband" channel model we consider here assumes that the bandwidth of the signal is much narrower than the coherence bandwidth of the channel (see Section 2.4). In this case, then all frequency components in the transmitted signal are affected by the same random attenuation and phase shift, and the channel is frequency flat. This entails that the fading affects the transmitted signal *multiplicatively*: that is, if $x(t)$ is transmitted and $z(t)$ denotes the additive noise, the received signal has the form

$$y(t) = R(t)e^{j\Theta(t)}x(t) + z(t) \tag{4.1}$$

Examine now the rate of variation with time of the random process $R(t)e^{j\Theta(t)}$ modeling the fading. A possible situation is that this process is constant during the transmission of an elementary signal, and varies from signal to signal. We refer to this channel model as to the frequency-flat, slow fading channel.

If we can further assume that the fading is so slow that we can estimate the phase Θ with sufficient accuracy, and hence compensate for it (*coherent demodulation*: see, e.g., [4.7]), then model (4.1) can be further simplified to

$$y(t) = Rx(t) + z(t) \tag{4.2}$$

and hence, in the framework of Chapter 3, the input–output relationship describing the channel behavior is

$$y = Rx + z, \qquad x \in \mathcal{X} \tag{4.3}$$

For coded signals, the above equation becomes

$$\mathbf{r} = \mathbf{R}\mathbf{x} + \mathbf{z} \tag{4.4}$$

where $\mathbf{r}, \mathbf{x}, \mathbf{z}$ are column n-vectors and $\mathbf{R}$ is a diagonal matrix. In the following, most of the calculations will be made under the assumption that R, and the elements of the main diagonal of $\mathbf{R}$, have a Rayleigh pdf with $\mathbb{E}[R^2] = 1$, that is,

$$p_R(r) = 2re^{-r^2}, \qquad r \geq 0 \tag{4.5}$$

and we refer to this model as the *Rayleigh fading* channel.

4.1.1   Ergodicity of the fading channel

The choice of the correlations among the components of the main diagonal of $\mathbf{R}$ characterize different channel models, to the extent that their channel capacities not only are different but also should be defined in different ways. The definition of

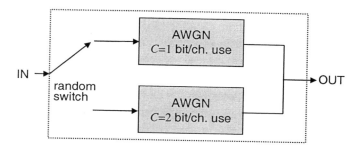

Figure 4.1: *A fading channel.*

channel capacity for the fading channel is connected to its *ergodicity*. Assume first that the components of the main diagonal of matrix $\mathbf{R}$ in (4.4) are independent RVs. Then, when a long code word is transmitted, its symbols are likely to experience all states of the channel, so the code word in a sense "averages" the channel effect, and the channel is ergodic. Consequently, since the capacity of an AWGN channel with constant attenuation R is $\log(1 + R^2 \, \mathrm{SNR})$ bit/dimension pair, the ergodic fading channel capacity is given by

$$C = \mathbb{E}\left[\log(1 + R^2 \, \mathrm{SNR})\right] \qquad \text{bit/dimension pair} \qquad (4.6)$$

Consider instead the opposite situation of a channel whose fading is so slow that it remains constant for the whole duration of a code word. The model for this channel has a matrix $\mathbf{R}$ in (4.4) whose main-diagonal components are all equal, and independent from word to word. This channel is nonergodic, as no code word will be able to experience all the states of the channel, and hence (4.6) is not valid anymore.

Let us elaborate on the assumption of ergodicity for a fading channel used with coding. The example that follows will illustrate the difficulties of defining a capacity for nonergodic channels. Consider, for motivation sake, the following simple model of a fading channel. A source is connected, through a random switch taking on both positions with equal probabilities, to one of a pair of channels; these are AWGN with constant attenuation. Channel 1's attenuation is such that its capacity is 1 bit/channel use, while Channel 2 has capacity 2 bit/channel use (Figure 4.1).

Suppose initially that the switch changes its position every symbol period. In this case, average capacity makes sense because a long code word will experience both channels with equal probabilities, and time average and ensemble average will be equal due to ergodicity.

Assume next that the switch remains in the same position for the whole duration of a code word. Then the channel is nonergodic, i.e., we cannot exchange time averages with ensemble averages as far as its capacity is concerned. In these conditions, ensemble average does not yield channel capacity: in fact, if we transmit at a rate slightly lower than the average capacity $C = (C_1 + C_2)/2 = 1.5$, one half of the code words (those experiencing channel 1) would be transmitted beyond capacity and hence have a large error probability. To achieve high reliability through long code words, we need to transmit at a rate lower than 1 bit/channel use, which by consequence may be interpreted as the true capacity of this channel. Another interpretation consists of assuming that *the capacity is a random variable*, which in this example takes on values 1 and 2 with equal probabilities. This latter interpretation turns out to be more fruitful in practice. In fact, consider the more realistic situation of a fading channel modeled through a switch choosing from a continuum of AWGN channels whose attenuation R, $0 \leq R < \infty$, has a Rayleigh probability density function. With the first interpretation, the capacity of this channel would be zero, because there is *no nonzero rate* at which long code words can be transmitted with a vanishingly small error probability. With the second interpretation, the mutual information of this channel (which we call *instantaneous* mutual information) is the random variable $C(R) \triangleq \log(1 + R^2 \mathrm{SNR})$ bit/dimension pair. Suppose we are transmitting at rate ρ bits per channel use. When the transmission rate exceeds the capacity, we say that an *information outage* occurs, an event that has probability

$$P_{\mathrm{out}} = \mathbb{P}\left[C(R) \leq \rho\right] = 1 - \exp[-(\mathrm{SNR})^{-1}(2^\rho - 1)] \qquad (4.7)$$

and corresponds to the long code word transmitted being received unreliably. We see that, in this case, only the zero rate $\rho = 0$ is compatible with infinitely reliable transmission ($P_{\mathrm{out}} = 0$), so capacity is zero. This situation corresponds, as before, to the worst channel state ($R = 0$ here). In general, outage probability expresses a tradeoff between rate and error probability.

In these conditions, for a nonergodic channel we may define an *ε-outage capacity* as the maximum rate ρ that can be transmitted with an outage probability $P_{\mathrm{out}} = \varepsilon$. Notice also that the outage probability provides an estimate of word error probability when the transmitted words are long enough. In fact, powerful error-control codes provide nearly error-free frames at transmission rates below instantaneous mutual information, and mostly erroneous frames at rates above it.

4.1.2   Channel-state information

As we know, in fading channels more than one source of randomness is included, viz., the fading process and the additive noise. It may happen that the receiver or the transmitter or both have at least a partial knowledge of the realization of the fading process, which we call the *channel-state information* (CSI). The coding and decoding strategies depend crucially on the availability of CSI.

CSI at the receiver can be obtained through the insertion, in the transmitted signal, of suitable pilot tones or of pilot symbols, i.e., known symbols transmitted periodically. These allow the demodulator to take advantage of its knowledge of the channel fading gain and hence to adjust its parameters to optimize operation. On the other hand, CSI at the transmitter allows it to adjust its transmitted power, or its information rate, so as to adapt the transmission to channel conditions. Transmitter CSI can be relayed from the receiver through a feedback path, or, when transmissions in both directions are multiplexed in time, the signal from the opposite link can be used to measure the channel state. Transmit-power control turns out to be a most effective technique to mitigate fading. A practical problem with this approach comes from the difficulty of obtaining a reliable estimate of the CSI. In fact, unless uplink and downlink transmissions occur at the same frequency and in time intervals spaced by less than the coherence time of the channel, the CSI has to be relayed from the receiver back to the transmitter, which decreases the throughput and increases the complexity of the system.

4.2   Independent fading channel

In the fading model (4.3), the only difference with respect to an AWGN channel resides in the fact that R, instead of being a constant attenuation, is an RV, whose value affects the amplitude, and hence the energy, of the received signal. Here we assume that the fading values R are independent and identically distributed, and also that the values taken by R are known at the receiver: we describe this situation by saying that we have *perfect CSI*.

Detection of an uncoded elementary constellation with perfect CSI can be performed in exactly the same way as for the AWGN channel: in fact, the constellation structure is perfectly known, as is the attenuation incurred by the signal. The optimum demodulation rule in this case consists again of choosing the signal in S that minimizes the Euclidean distance

$$|y - Rx| \qquad\qquad (4.8)$$

A consequence of this fact is that the error probability with perfect CSI can be evaluated as follows. We first compute the error probability $P(e \mid R)$ obtained by assuming R constant in (4.3). Next we take the expectation of $P(e \mid R)$, with respect to the random variable R. The calculation of $P(e \mid R)$ is performed as if the channel were AWGN, but with a constellation scaled by a factor R, i.e., with an energy associated with each signal changed from $|x|^2$ into $R^2|x|^2$.

The conditional error probability $P(e \mid x, R)$ can be bounded above by the union bound

$$P(e \mid x, R) \le \sum_{\hat{x} \neq x} P(x \to \hat{x} \mid R) \tag{4.9}$$

where $P(x \to \hat{x} \mid R)$ denotes the conditional pairwise error probability, i.e., the probability that the distance of the received signal from $\hat{x}$ is smaller than that from the transmitted signal x when R is the channel state. We have explicitly

$$P(x \to \hat{x} \mid R) = Q\left(\frac{R|x - \hat{x}|}{\sqrt{2N_0}}\right) \tag{4.10}$$

and hence

$$P(x \to \hat{x}) = \mathbb{E}_R P(x \to \hat{x} \mid R)$$

where the expectation is taken with respect to the fading RV R. Under the assumption of Rayleigh fading, the expectation above can be given a closed form, and we have

$$P(x \to \hat{x}) = \frac{1}{2}\left(1 - \sqrt{\frac{|x - \hat{x}|^2/4N_0}{1 + |x - \hat{x}|^2/4N_0}}\right) \tag{4.11}$$

Using the approximation, valid for $z \to \infty$,

$$1 - \sqrt{\frac{z}{1+z}} \sim \frac{1}{2z} \tag{4.12}$$

we obtain, as $N_0 \to 0$,

$$P(x \to \hat{x}) \sim \frac{1}{|x - \hat{x}|^2/N_0} \tag{4.13}$$

Example 4.1

With binary antipodal signals with common energy $\mathcal{E}$, we have $|x - \hat{x}|^2 = 4\mathcal{E}$, and hence

$$P(e) = \mathbb{E}_R\, Q\left(\sqrt{2R^2\frac{\mathcal{E}}{N_0}}\right) = \frac{1}{2}\left[1 - \sqrt{\frac{\mathcal{E}/N_0}{1 + \mathcal{E}/N_0}}\right] \sim \frac{1}{4\mathcal{E}/N_0} \qquad (4.14)$$

This equation shows how $P(e)$ turns out to be inversely proportional to SNR. Comparison of the error probability of this constellation for the AWGN channel and the Rayleigh fading channel (see Figure 4.2) shows that there is a considerable gap in performance between the two channels, which increases with SNR. As we shall see in the following, coding can be used to reduce this gap by a considerable amount. □

Example 4.2

If, over the AWGN channel, $P(e)$ can be approximated by $\alpha Q(\sqrt{2\gamma\mathcal{E}_b/N_0})$, then over a fading channel we can use the approximation

$$P(e) \approx \alpha\mathbb{E}_R Q\left(\sqrt{2\gamma R^2\frac{\mathcal{E}_b}{N_0}}\right)$$

As a special case, for a Rayleigh fading channel we have, in closed form,

$$P(e) \approx \frac{\alpha}{2}\left(1 - \sqrt{\frac{\gamma\mathcal{E}_b/N_0}{1 + \gamma\mathcal{E}_b/N_0}}\right)$$

and asymptotically

$$P(e) \sim \frac{\alpha}{4\gamma\mathcal{E}_b/N_0}$$

This shows that, on the Rayleigh fading channel, all modulation schemes are equally bad, as their error probabilities decrease equally slowly as the SNR increases. □

4.2.1 Consideration of coding

Consider now coded transmission. We have, using the inequality $Q(x) \leq e^{-x^2/2}$, or, equivalently, the Chernoff bound (see the Problem section of Chapter 3):

$$\begin{aligned} P(\mathbf{x} \to \widehat{\mathbf{x}} \mid \mathbf{R}) &= Q\left(\frac{\|\mathbf{R}(\mathbf{x} - \widehat{\mathbf{x}})\|}{\sqrt{2N_0}}\right) \\ &\leq \exp(-\|\mathbf{R}(\mathbf{x} - \widehat{\mathbf{x}})\|^2/4N_0) \end{aligned} \qquad (4.15)$$

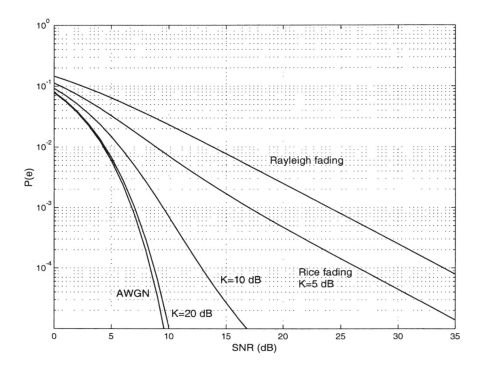

Figure 4.2: *Comparison of error probabilities of antipodal binary modulation over the AWGN and the Rayleigh fading channel. Here SNR= $\mathbb{E}(R^2)\mathcal{E}/N_0 = \mathcal{E}/N_0$. The effect of independent Rice fading with parameter K is also shown.*

Now, the assumption of independent Rayleigh fading yields

$$
\begin{aligned}
P(\mathbf{x} \to \widehat{\mathbf{x}}) \;\leq\; & \mathbb{E}_{R_1,\dots,R_n}\left[\exp\left(-\sum_{i=1}^{n} R_i^2 |x_i - \hat{x}_i|^2/4N_0\right)\right] \\
= \; & \prod_{i=1}^{n} \mathbb{E}_{R_i}\left[\exp\left(-R_i^2 |x_i - \hat{x}_i|^2/4N_0\right)\right] \\
= \; & \prod_{i=1}^{n} \frac{1}{1 + |x_i - \hat{x}_i|^2/4N_0}
\end{aligned}
\tag{4.16}
$$

Further, observe that for some index i we may have $\hat{x}_i = x_i$, although $\widehat{\mathbf{x}} \neq \mathbf{x}$. Specifically, $\widehat{\mathbf{x}}$ will differ from $\mathbf{x}$ in exactly $d_{\mathrm{H}}(\mathbf{x}, \widehat{\mathbf{x}})$ components, whose indices are collected in a set $\mathcal{J}$. Thus, as $N \to 0$ we can write

$$P(\mathbf{x} \to \widehat{\mathbf{x}}) \leq \prod_{i \in \mathcal{J}} \frac{1}{|x_i - \hat{x}_i|^2 / 4N_0}$$

$$= \left[\prod_{i \in \mathcal{J}} |x_i - \hat{x}_i|^2 \right]^{-1} \left(\frac{1}{4N_0} \right)^{-d_H(\mathbf{x}, \widehat{\mathbf{x}})} \tag{4.17}$$

By using the union bound as we did in the previous chapter, we see that the error probability is dominated by the pairwise errors with the smallest $d_H(\mathbf{x}, \widehat{\mathbf{x}})$, denoted by $d_{H,min}$. For small noise, this is the exponent of SNR^{-1}.

Notice also the effect of the *product distance*

$$d_\Pi(\mathbf{x}, \widehat{\mathbf{x}}) \triangleq \prod_{i \in \mathcal{J}} |x_i - \hat{x}_i|^2$$

This does not depend on SNR, and its effect is to shift horizontally the curve of PEP vs. SNR. The smallest among the product distances can be called *coding gain*.

Result (4.17) shows that a sensible criterion for the selection of a code for the independent Rayleigh fading channel with high SNR (low N_0) is the maximization of the minimum Hamming distance between any two code words. This selection criterion differs considerably from the one obtained for the AWGN channel. The minimum Hamming distance of the code is sometimes referred to as *code diversity*. Originally, the term *diversity* denoted the independent replicas of the transmitted signal made available to the receiver through multiple antennas, transmission through separate channels or separate polarizations, etc., to be described later in this chapter. Since the *diversity order* appears as the exponent of SNR^{-1} in the expression of error probability, and Hamming distance is also an exponent to error probability when coding is used, the term *code diversity* was coined. Among codes with the same diversity, a sensible choice is to choose the one with the largest coding gain.

Example 4.3

Consider transmission of the 4-PSK constellation shown in Figure 4.3(a). This figure corresponds to the code with $n = 2$ and the four words

$$(\cos(2k+1)\pi/4, \sin(2k+1)\pi/4)_{k=0}^3$$

(assume unit energy). Constellation (b) is obtained by rotating (a) by $\pi/8$, and has the same performance as (a) over the AWGN channel [4.7, Section 4.2.2]. However, while (a) has a minimum Hamming distance $d_{H,min} = 1$, it can be easily

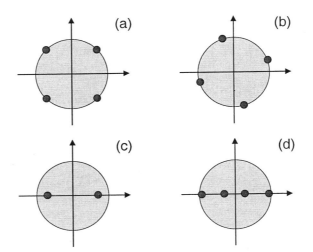

Figure 4.3: *(a) Standard 4PSK constellation. (b) 4PSK constellation after a rotation by $\pi/8$. (c)–(d) Effect of a deep fade affecting the signal component corresponding to the vertical axis on constellations (a) and (b), respectively.*

checked that (b) has $d_{\mathrm{H,min}} = 2$ and hence offers a better performance for high SNR over the independent Rayleigh fading channel. To illustrate this point, consider a two-dimensional transmission whereby the two components of each signal are independently faded. Let a deep fade affect only the second signal component: hence, constellation (a) collapses to a pair of points, thus losing one bit of information, while constellation (b) retains a separation among points even in a single dimension. It is instructive to evaluate the product distances of the rotated constellation, and to determine the rotation angle that minimizes the coding gain.   □

4.2.2   Capacity of the independent Rayleigh fading channel

Consider first the assumption of availability of channel-state information at the receiver only. With this model we observe once again that, for every channel use, conditionally on the value of R, the channel is Gaussian with attenuation R. Assume that the sequence of fading values R forms an ergodic process, which is verified if the fading values are iid. The Shannon capacity is now the average capacity, which can be calculated by using the following equation:

$$C = \mathbb{E} \log(1 + R^2 \, \mathrm{SNR}) \qquad \text{bit/dimension pair} \qquad (4.18)$$

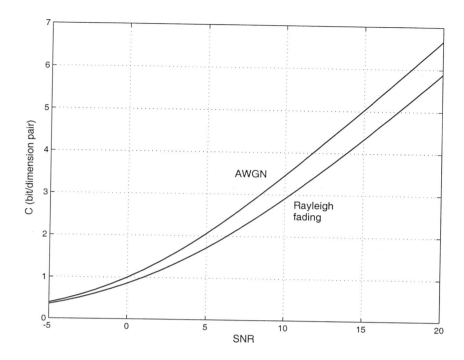

Figure 4.4: *Capacity of the AWGN channel and of the independent Rayleigh fading channel.*

where the expectation is taken with respect to the RV R. Using Jensen inequality $\mathbb{E}\left[f(X)\right] \leq f(\mathbb{E}\left[X\right])$, valid for any concave $\cap$ function f and any RV X, we can see that, with $\mathbb{E}\,R^2 = 1$, (4.18) is always less than the capacity of the AWGN channel with the same average power.

If R is Rayleigh distributed, calculation of the expectation in (4.18) yields

$$C = -\frac{1}{2\ln 2}\exp\left(\frac{1}{\text{SNR}}\right)\text{Ei}\left(-\frac{1}{\text{SNR}}\right) \qquad \text{bit/dimension pair} \qquad (4.19)$$

where

$$\text{Ei}\left(x\right) \triangleq \int_{-\infty}^{x}\frac{e^t}{t}\,dt$$

The average capacity (4.19) is plotted in Figure 4.4. It is seen that the gap between the AWGN channel capacity and the capacity of the independent Rayleigh fading channel is much smaller than the one exhibited by the error-probability curves (Figure 4.2): this suggests that coding can be very beneficial to compensate for fading. The gap widens as SNR increases, and reaches its asymptotic maximum of 2.5 dB (see Problem 1).

We can also duplicate, for the independent Rayleigh fading channel, the calculations leading to (3.69) and deriving the minimum SNR necessary to achieve infinitely reliable transmission with a rate-ρ code. Some values of $\mathcal{E}_b/N_0$ solving the equation

$$\rho = -\frac{1}{2\ln 2} \exp\left(\frac{1}{2\rho\mathcal{E}_b/N_0}\right) \text{Ei}\left(-\frac{1}{2\rho\mathcal{E}_b/N_0}\right)$$

are shown in Table 4.1, along with the corresponding values for the AWGN channel.

ρ	0	1/8	1/4	1/3	1/2	2/3	3/4
$\mathcal{E}_b/N_0$ (AWGN)	-1.6	-1.21	-0.82	-0.55	0.	0.57	0.86
$\mathcal{E}_b/N_0$ (Rayleigh)	-1.6	-0.89	-0.23	0.18	1.0	1.77	2.16

Table 4.1: *Minimum $\mathcal{E}_b/N_0$ values allowing infinitely reliable transmission with rate-ρ codes over the AWGN and the independent Rayleigh fading channel.*

As for the codes that achieve capacity, as in the AWGN case a simple standard (Gaussian) code will suffice; however, it should be observed that since ergodicity is invoked here, the code words must be long enough to experience all the channel states. This implies that, according to our discussion in Chapter 2, their duration should be much greater than the coherence time of the channel.

As for the constellation-constrained capacity (evaluated under the usual assumption that the signals of the elementary constellation $\mathcal{X}$ are used with equal probabilities), we can compute it by defining the following probability density function, derived from (3.53):

$$p_R(y \mid x) = c\, e^{-\|y - Rx\|^2/N_0} \tag{4.20}$$

and using this expression for capacity (see 3.51):

$$C = \log_2 |\mathcal{X}| - \mathbb{E}_{x,y,R}\left[\log_2 \frac{\displaystyle\sum_{x' \in \mathcal{X}} p_R(y \mid x')}{p_R(y \mid x)}\right] \tag{4.21}$$

This capacity can be evaluated numerically (for small $|\mathcal{X}|$) or via Monte Carlo simulation. Figure 4.5 compares the capacity of the ergodic Rayleigh fading chan-

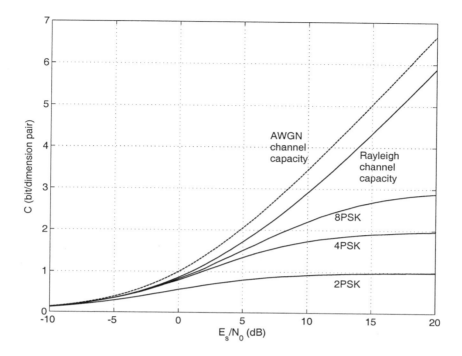

Figure 4.5: *Constellation-constrained capacity of the independent Rayleigh fading channel with coherent detection of binary, quaternary, and octonary PSK and perfect channel-state information. The unconstrained Rayleigh-fading channel and AWGN channel capacities are also shown for comparison.*

nel with that of the AWGN channel for some two-dimensional signal constellations. (Compare Figure 4.5 with Figure 3.4, where the constrained capacity for the AWGN is shown for the same constellations).

No channel-state information

In the absence of channel-state information (regarding both envelope and phase of the fading), we observe the received signal

$$y = Re^{j\Theta}x + n$$

where $R\exp(j\Theta) \sim \mathcal{N}_c(0, 1)$. Thus, the conditional pdf of y given x, R, and Θ, is Gaussian:

$$p(y \mid x, R, \Theta) = \frac{1}{\pi N_0} \exp\left[-|y - Re^{j\Theta}x|^2/N_0\right] \qquad (4.22)$$

We average the pdf above over the random variable $R\exp(j\Theta)$ by observing that

$$\zeta \triangleq y - Re^{j\Theta}x \sim \mathcal{N}_c(y, |x|^2)$$

so the conditional pdf describing the channel behavior takes the form

$$
\begin{aligned}
p(y \mid x) &= \mathbb{E}[p(y \mid x, R, \Theta)] \\
&= \mathbb{E}\left[\frac{1}{\pi N_0}e^{-|\zeta|^2/N_0}\right] \\
&= \frac{1}{\pi}\frac{1}{N_0 + |x|^2}\exp\left[-\frac{|y|^2}{N_0 + |x|^2}\right]
\end{aligned}
$$

where the following result, valid for a real RV $X \sim \mathcal{N}(\mu, \sigma^2)$, has been used:

$$\mathbb{E}\left[\alpha^{-1/2}e^{-X^2/\alpha}\right] = \frac{1}{\sqrt{\alpha + 2\sigma^2}}\,e^{-\mu^2/(\alpha+2\sigma^2)}$$

Over this channel we cannot use a modulation scheme whose signals have a constant magnitude and information-carrying phases. For these, the received signal becomes independent of the transmitted signal even in the absence of noise. We observe, for example, that binary antipodal signaling, which has $\mathcal{X} = \{\pm 1\}$, fails, as $p(y \mid +1) = p(y \mid -1)$.

By adding a constraint on the average transmitted power, the capacity of this channel can be computed. The surprising result here is that the capacity-achieving input distribution is *discrete*. No general closed-form is known for this distribution; however, asymptotic results are available. Specifically, for low SNR the capacity-achieving distribution has only *two* mass points, with one of the masses located at zero, and hence the optimum modulation scheme is on–off.

The resulting capacity of a Rayleigh fading channel is shown in Figure 4.6, and compared with the capacity of the AWGN channel and of the Rayleigh fading channel with channel-state information at the receiver. The lack of CSI at the receiver is seen to be especially penalizing at high SNR values.

Channel-state information at the transmitter and receiver

Here we assume again that the sequence of fading values R forms an ergodic process. Conditional on the value of R, the capacity of the channel is $\log(1 + R^2\,\mathrm{SNR})$ bit/dimension pair. Now the value of R is known at the transmitter, which may take appropriate actions to counteract the fading effects. Suppose that we allow the transmitted power, and hence the SNR, to vary with R, so we write $S(R)$ in lieu

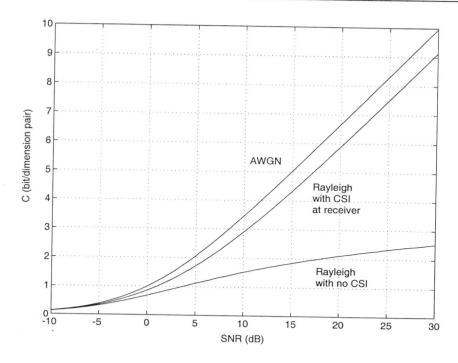

Figure 4.6: *Capacity of the AWGN channel and of the independent Rayleigh fading channel with and without channel-state information.*

of SNR. (One may prefer thinking of $S(R)$ as a transmitted power, which is consistent with the following if a unit noise power is assumed.) Subject to an average power constraint $\mathbb{E}\left[S(R)\right] \leq \bar{S}$, the channel capacity can be defined as [4.12]:

$$C(S) \triangleq \max_{S(R):\mathbb{E}\left[S(R)\right]=\bar{S}} \mathbb{E}\log\left(1 + R^2 S(R)\right) \qquad (4.23)$$

The power-adaptation policy $S(R)$ that yields the maximum in (4.23) is obtained by using standard Lagrange-multiplier techniques to solve the constrained-maximization problem implied by (4.23). The results is the "water-filling" formula

$$S(R) = \begin{cases} \dfrac{1}{R_0^2} - \dfrac{1}{R^2}, & R \geq R_0 \\ 0, & R < R_0 \end{cases} \qquad (4.24)$$

for some cutoff value R_0. From (4.24) we see in particular that, if R is below this cutoff, then the best strategy is to transmit no data.

R_0 is determined by the average-power constraint and the fading statistics. Substitution of (4.24) into the average power constraint yields the equation that R_0 must satisfy:

$$\int_{R_0}^{\infty} \left(\frac{1}{R_0^2} - \frac{1}{r^2} \right) p_R(r)\, dr = \bar{S} \tag{4.25}$$

Substitution of (4.24) into (4.23) yields the closed-form expression

$$C(S) = \int_{R_0}^{\infty} \log \left(\frac{r^2}{R_0^2} \right) p_R(r)\, dr \tag{4.26}$$

This capacity is achieved by a Gaussian code book where every symbol is generated as $\sim \mathcal{N}(0, 1)$, then scaled in amplitude according to the power adaptation policy (4.24). Notice also that, while the capacity with CSI at the receiver only never exceeds that of the AWGN channel (we observed this in Subsection 4.2.2), no such inequality holds for the capacity with CSI at transmitter and receiver [4.11]. Figure 4.7 shows the effect on capacity of CSI at transmitter. The capacity of AWGN is also shown. We can observe that, when the SNR is very low, the capacity with CSI at the transmitter and receiver exceeds that of the AWGN channel. We also observe that, unless the SNR is very low, CSI at the transmitter increases capacity very little.

4.3 Block-fading channel

In a typical wireless system, Doppler spreads may range from 1 to 100 Hz, corresponding to coherence times from 0.01 to 1 s, while transmission rates range from $2 \cdot 10^4$ to $2 \cdot 10^6$ elementary signals per second. Thus, blocks with a length L ranging from $2 \cdot 10^4 \times 0.01 = 200$ symbols to $2 \cdot 10^6 \times 1 = 2 \cdot 10^6$ symbols are affected by approximately the same fading gain.[1] If coding is used, then in order to make the fading gains affecting the symbols of x independent, *interleaving* must be introduced.

Interleaving consists of making the channel approximately memoryless by permuting the order of the transmitted elementary signals. These are dispersed over different coherence intervals, and hence are affected by independent fades. More specifically, consider an $a \times b$ matrix. After being generated, the elementary signals are written rowwise into the matrix, then read columnwise and transmitted.

---

[1] In practice, to claim with a fairly high degree of confidence that the fading gain is almost constant throughout the block, the maximum block duration should be limited to a fraction (e.g., 1/4) of the coherence time.

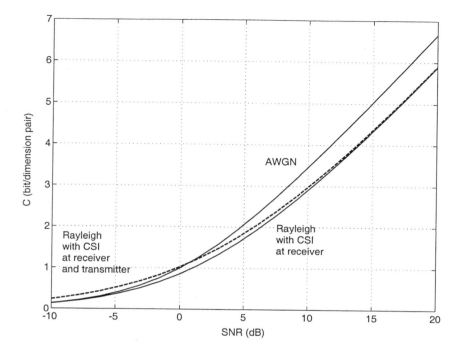

Figure 4.7: *Capacity of the independent Rayleigh fading channel with channel state information at the receiver only and with channel-state information at the receiver and transmitter. The capacity of the AWGN channel is also shown for comparison.*

Figure 4.8: *A 4×8 interleaving matrix.*

For example, if the interleaving matrix is 4×8 as shown in Figure 4.8, the signals $x_1, x_2, \ldots, x_{32}$ enter the channel in the order $x_1, x_9, x_{17}, x_{25}, x_2, x_{10}, x_{18}$, etc. After reception, the original order of the signals is reconstituted by deinterleaving, i.e., by having them written columnwise and read rowwise. Thus, if

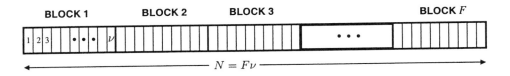

Figure 4.9: *The block-fading channel model: the code word length is split into F blocks of length ν each; the fading is constant over each block.*

the same fading value affects four signals that are adjacent on the channel, say for instance $x_2, x_{10}, x_{18}, x_{26}$, we see that it affects signals that, in their original order, are separated by at least eight positions. In order for the fading to affect each coded symbol independently, the size of the matrix must be made large enough. Notice that interleaving and deinterleaving *involve a delay* due to the time required to read/write the matrices: this delay is proportional to the size of the matrix.

Now, with a code with length n, if we want each component of x to be affected by an independent fading gain the time interval spanned by a single interleaved code word x must be made at least nL. Thus, the (interleaving) delay of the system is large and, above all, characterized by L, a parameter that is not under control of the code designer.

The trade-off involved in interleaving (which improves performance, as we shall see, but increases delay) is dependent on applications. In delay-tolerant systems (data transmission, broadcasting, etc.) deep interleaving is possible, and hence the independent-fading assumption is reasonable. In delay-constrained systems, like those transmitting real-time speech, separation of coded symbols by more than the coherence time of the channel is not possible, and therefore a length-n code word is affected by a number of independent fading gains that is less than n. In this case, each word is split into a number of blocks that is a fraction of n, and over each block the channel fading is correlated so highly that we may model it as constant. Note that codes designed ad hoc for correlated fading are rather unpractical, since optimality criteria would depend on the fading Doppler bandwidth, which in turn depends on the mobile speed.

When delay constraints are present, the block-fading model turns out to be the sensible choice in many instances. This model assumes that the fading gain process is piecewise constant and can be described through a sequence of independent random variables, each of which is the fading gain in a block of ν elementary signals. A code word of length n is spread over F blocks of length ν symbols each, so that $n = F\nu$ (see Figure 4.9). If $\nu = n$, and hence $F = 1$, we have a channel in which the entire code word is affected by the same fading gain. If $\nu = 1$, and hence $F = n$ (ideal interleaving), each symbol is affected by an independent fading gain, which shows that the independent fading channel model examined above is a special case of this model.

The delay constraint to which the communication system is subject determines the maximum number F of independently faded blocks over which a code word of length $n = F\nu$ can be spread. The choice $F \to \infty$ makes the channel ergodic and allows channel capacity, in Shannon's sense, to be defined.

This block-fading model can be used as an approximation whenever the fading-gain process can be approximated by a piecewise-constant process. It is an exact model whenever the coherence time T_c is large enough (stationary or almost-stationary users) and the code word is spread over a finite number F of blocks transmitted over independent channels. These, in turn, can be generated by transmitting blocks over frequency bands separated by at least B_c (slow frequency hopping where ν symbols are transmitted per hop), or over time intervals separated by at least T_c. The first choice occurs in GSM, with $F = 8$ (full-rate GSM) or $F = 4$ (half-rate GSM), while the second occurs in IS-136.

4.3.1 Mathematical formulation of the block-fading model

To describe the block-fading model mathematically, we define the code word $\mathbf{X}$ as the $F \times \nu$ matrix whose mth row contains the mth block $\mathbf{x}_m$, $m = 1, \ldots, F$, with length ν. The mth block is sent over a constant-fading channel with gain R_m. The channel output matrix corresponding to $\mathbf{X}$ is given by

$$\mathbf{Y} = \mathbf{RX} + \mathbf{Z} \tag{4.27}$$

where

$$\mathbf{X} = \begin{bmatrix} \mathbf{x}_1 \\ \vdots \\ \mathbf{x}_F \end{bmatrix}$$

and $\mathbf{R} \triangleq \text{diag}(R_1, \ldots, R_F)$, while $\mathbf{Z}$ is an $F \times \nu$ matrix of independent Gaussian noise RVs.

As observed before, the fully interleaved channel and the block-fading channel are formally very similar. The former is obtained from the latter by choosing $\nu = 1$, and hence $F = n$, while the latter corresponds to the former with $\mathbf{X}$ interpreted as a code word of length F whose components are chosen from $\mathcal{X}^\nu$, the ν-fold Cartesian product of $\mathcal{X}$—that is, their dimensionality is ν times the dimensionality of the elementary signals x. As usual, $\mathcal{X}$ is a two-dimensional signal set, such as QAM or PSK, so that each row of the matrix $\mathbf{X}$ can be seen to be a signal of a 2ν-dimensional signal set.

4.3.2   Error probability for the coded block-fading channel

Based on the model developed in the previous section, we focus here on the transmission of a coded modulation scheme over a block-fading channel with perfect CSI at the receiver. As discussed before, a code word of length n is split into F blocks, each of length ν. Each block represents a signal in $\mathcal{X}^{\nu}$: that is, we do not consider explicitly the "fine grain" of the code, i.e., the fact that a block is actually a sequence of elementary signals. Here, a "channel use" indicates transmission of an element x of $\mathcal{X}^{\nu}$, affected by an independent fading value and by AWGN.

As we have done previously, upper bounds and approximations to error probability can be constructed by using the pairwise error probability (PEP) $P(\mathbf{X} \to \widehat{\mathbf{X}})$. This is the probability of mistaking the transmitted code word $\mathbf{X}$ for a different coded block $\widehat{\mathbf{X}}$ when these two are the only possible outcomes of the decoder. Then, simple coding optimization criteria can be based on the analysis of PEP.

PEP analyses carried out for the independent fading model can be repeated here, *mutatis mutandis*, for the new signal set $\mathcal{X}^{\nu}$. Denote by $\|\mathbf{B}\|$ the Frobenius norm of $\mathbf{B}$ (see Section B.5 of Appendix B). Based on (4.27), the PEP is given by

$$P(\mathbf{X} \to \widehat{\mathbf{X}}) = \mathbb{P}(\|\mathbf{Y} - \mathbf{R}\mathbf{X}\|^2 > \|\mathbf{Y} - \mathbf{R}\widehat{\mathbf{X}}\|^2 \mid \mathbf{X}) \qquad (4.28)$$

so that, using once again the bound $Q(x) \leq \exp(-x^2/2)$, we obtain

$$P(\mathbf{X} \to \hat{\mathbf{X}}) = \mathbb{E}_{\mathbf{R}}\left[Q\left(\frac{\|\mathbf{R}(\mathbf{X} - \widehat{\mathbf{X}})\|}{\sqrt{2N_0}}\right)\right] \leq \mathbb{E}_{\mathbf{R}}\left[e^{-\|\mathbf{R}(\mathbf{X}-\widehat{\mathbf{X}})\|^2/4N_0}\right] \quad (4.29)$$

where the expectation is with respect to the fading sequence $R_1, \ldots, R_F$.

For Rayleigh fading, independent from block to block, we have

$$\mathbb{E}_{\mathbf{R}}\left[\exp\{-\|\mathbf{R}(\mathbf{X} - \widehat{\mathbf{X}})\|^2/4N_0\}\right] \qquad (4.30)$$

$$= \mathbb{E}_{R_1,\ldots,R_F}\left[\exp\left\{-\sum_{i=1}^{F} R_i^2\|\mathbf{x}_i - \widehat{\mathbf{x}}_i\|^2/4N_0\right\}\right] \qquad (4.31)$$

so that

$$\begin{aligned}
P(\mathbf{X} \to \widehat{\mathbf{X}}) &\leq \prod_{i=1}^{F} \frac{1}{1 + \|\mathbf{x}_i - \widehat{\mathbf{x}}_i\|^2/4N_0} \\
&= \prod_{i \in \mathcal{I}} \frac{1}{1 + \|\mathbf{x}_i - \widehat{\mathbf{x}}_i\|^2/4N_0} \\
&\leq \prod_{i \in \mathcal{I}} \frac{1}{\|\mathbf{x}_i - \widehat{\mathbf{x}}_i\|^2/4N_0} \qquad (4.32)
\end{aligned}$$

where $\mathcal{J}$ is the set of indices i such that $\|\mathbf{x}_i - \widehat{\mathbf{x}}_i\| \neq 0$, that is, such that the blocks $\mathbf{x}_i$ and $\widehat{\mathbf{x}}_i$ differ. Denote by $D_F(\mathbf{X}, \widehat{\mathbf{X}})$ the number of rows in which $\mathbf{X}$ and $\widehat{\mathbf{X}}$ differ (we call this the *Hamming block distance* between $\mathbf{X}$ and $\widehat{\mathbf{X}}$). We can write

$$P(\mathbf{X} \to \widehat{\mathbf{X}}) \leq \left[\prod_{i \in \mathcal{J}} \|\mathbf{x}_i - \widehat{\mathbf{x}}_i\|^2 \right]^{-1} \left(\frac{1}{4N_0} \right)^{-D_F(\mathbf{X}, \widehat{\mathbf{X}})} \qquad (4.33)$$

(Notice the similarity of the above equation with (4.17).)

Hamming block distance and its significance.   Result (4.33) shows the important fact that the error probability is (asymptotically for high SNR) inversely proportional to the *product* of the squared Euclidean distances between the signals transmitted in a block, and, to a more relevant extent, to a negative power of the signal-to-noise ratio whose exponent is the Hamming block-distance between $\mathbf{X}$ and $\widehat{\mathbf{X}}$.

By using the union bound as we did in the previous chapter, we see that the error probability is dominated by the pairwise errors with the smallest $D_F(\mathbf{X}, \widehat{\mathbf{X}})$, say $D_{F,\min}$. This in turn is a nondecreasing function of F, as can be easily verified: thus, the presence of a delay constraint impairs the exponent of error probability.

Example 4.4

As an example, consider a block code with length 16, $\mathcal{X} = \{\pm 1\}$, and the two code words

$$+1 \ +1 \ +1 \ +1 \ +1 \ +1 \ +1 \ +1 \ +1 \ +1 \ +1 \ +1 \ +1 \ +1 \ +1 \ +1$$

$$-1 \ -1 \ +1 \ -1 \ -1 \ +1 \ -1 \ +1 \ +1 \ +1 \ +1 \ +1 \ +1 \ -1 \ -1 \ +1$$

It can be seen that the Hamming block distances between these two code words corresponding to different values of F are $D_{16} = 7$ (corresponding to independent fading, i.e., full interleaving), $D_8 = 6$, $D_4 = 3$, $D_2 = 2$, and $D_1 = 1$ (corresponding to no interleaving). $\square$

Example 4.5

In Chapter 3 we have exhibited a signal constellation (the *orthogonal signals*) that asymptotically, for large alphabet size, has an error probability that tends to zero provided that $\mathcal{E}_b/N_0 > \ln 2$. This threshold value corresponds to the limiting value given by the capacity formula for infinite-bandwidth AWGN channel. We shall now see that the same constellation, when used on the block-fading channel with no interleaving, cannot achieve zero error probability for any finite $\mathcal{E}_b/N_0$.

Since for the AWGN channel the error probability of M-ary orthogonal signals can be written in the form

$$\lim_{M \to \infty} P_M(e) = [\mathcal{E}_b/N_0 < \ln 2]$$

(where $[\mathcal{A}]$ takes value 0 if the proposition $\mathcal{A}$ is true, and 0 otherwise), with the fading gain R affecting all the transmitted components of signal $\mathbf{x}$, we have

$$\lim_{M \to \infty} P_M(e \mid R) = [R^2 \mathcal{E}_b/N_0 < \ln 2]$$

With R a Rayleigh random variable, we have

$$\begin{aligned}
\lim_{M \to \infty} P_M(e) &= \mathbb{E}\left[R^2 \mathcal{E}_b/N_0 < \ln 2\right] \\
&= \mathbb{E}\left[R < \sqrt{\ln 2/(\mathcal{E}_b/N_0)}\right] \\
&= 1 - \exp\{-\ln 2/(\mathcal{E}_b/N_0)\} \qquad (4.34)
\end{aligned}$$

Thus, no infinitely reliable transmission is possible with finite SNR, contrary to what happens with ergodic fading. $\square$

Singleton bound on Hamming block-distance.   It can be shown that the maximum Hamming block-distance achievable on an F-block fading channel can be obtained as follows (*Generalized Singleton bound*: see the Problem section). Assume again that the code words are composed of F blocks from $\mathcal{X}^\nu$, and let ρ denote the code rate, expressed in bits per elementary signal. The following inequality holds (see the Problem section):

$$D_{F,\min} \le 1 + \left\lfloor F\left(1 - \frac{\rho}{\log |\mathcal{X}|}\right) \right\rfloor \qquad (4.35)$$

We notice that the right-hand side of (4.35) increases with F (showing that interleaving is beneficial) and with $|\mathcal{X}|$ (showing that binary elementary constellations are worse). It decreases as the rate ρ increases.

Codes meeting the equality in (4.35) are called *maximum-distance* (MD) codes.

4.3.3   Capacity considerations

As mentioned before, when the ergodicity assumption cannot be invoked, the channel capacity may be viewed as a random entity that depends on the instantaneous parameters of the fading process. In this situation we associate an outage probability with every given rate ρ; since the capacity associated with the ith block, $1 \le i \le F$, is the random variable $C(R_i) = \log(1 + R_i^2\, \text{SNR})$, the average instantaneous mutual information is $(1/F) \sum_{i=1}^{F} C(R_i)$, and hence the outage probability is given by [4.16]

$$P_{\text{out}}(\rho) = \mathbb{P}\left(\frac{1}{F} \sum_{i=1}^{F} C(R_i) \le \rho \right)$$

where ρ is the average transmission rate.

Consider first $F = 1$: a stringent delay constraint prevents transmitting information in more than a single block, and the same fading value affects the entire code word. With Rayleigh fading, the probability that $C(R) \le \rho$ is given by

$$P_{\text{out}}(\rho) = 1 - \exp[-(\text{SNR})^{-1}(2^\rho - 1)] \tag{4.36}$$

as derived earlier in this chapter. Observe that, at high SNR,

$$P_{\text{out}}(\rho) \approx \frac{2^\rho - 1}{\text{SNR}}$$

which decreases as SNR^{-1}. We have previously observed (Example 4.2) that the error probability of any uncoded modulation scheme decreases as SNR^{-1}. Thus, we cannot expect coding to improve significantly the error performance of this channel with $F = 1$.

With $F = 2$, which corresponds to a decoding-delay constraint that is slightly relaxed, a code word is transmitted in two separate blocks. Using independent Gaussian symbols on the two blocks, the random capacity is $(1/2)[C(R_1) + C(R_2)]$, and consequently the outage probability, with Rayleigh fading and under the assumption of equal SNR on both blocks, can be computed as

$$
\begin{aligned}
P_{\text{out}}(\rho) &= \tag{4.37}\\
&= \mathbb{P}[(1/2)\log(1 + R_1^2\,\text{SNR}) + (1/2)\log(1 + R_2^2\,\text{SNR}) \le \rho]\\
&= \mathbb{P}[(1 + R_1^2\,\text{SNR})(1 + R_2^2\,\text{SNR}) \le 2^{2\rho}]\\
&= \int_0^{(2^{2\rho}-1)/\text{SNR}} e^{-x}\left[1 - \exp\left\{-\left(\frac{2^{2\rho}}{1 + x\,\text{SNR}} - 1\right)(\text{SNR})^{-1}\right\}\right]\, dx
\end{aligned}
$$

The outage probabilities for $F = 1$ and $F = 2$ are plotted in Figure 4.10.

The calculations done so far assumed that channel-state information was available at the receiver only. If no CSI is made available before transmission, the outage-probability results still hold, as the following simple argument shows. Our

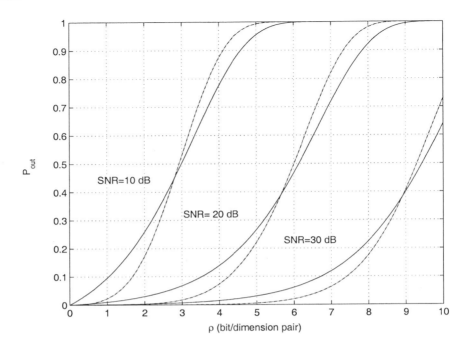

Figure 4.10: *Outage probability for the block-fading channel with Rayleigh fading,* $F = 1$ *(continuous line) and* $F = 2$ *(dash-dot line).*

assumption with block fading is that the channel state remains constant as the block length n of the code increases. Hence, we can estimate it, using a training sequence whose length is proportional to $\sqrt{n}$, which entails no decrease in the code rate as $n \to \infty$.

Consider now the more complex case of channel-state information being available at both transmitter and receiver, so that a strategy can be used to compensate for the effects of fading. This is described by the function $S(R)$, which yields for the ith block the random capacity $\log(1 + R_i^2 S(R_i))$. With $F = 1$ the optimal strategy is to invert the channel, that is, to choose $S(R) \propto 1/R^2$. If the constraint to be satisfied has the form $\mathbb{E}[S(R)] = \bar{S}$, then channel inversion yields

$$S(R) = \frac{R^{-2}}{\mathbb{E}[R^{-2}]} \bar{S}$$

which corresponds to the capacity

$$C = \log\left(1 + \frac{\bar{S}}{\mathbb{E}[R^{-2}]}\right) \tag{4.38}$$

Thus, if $\mathbb{E}[R^{-2}] < \infty$, capacity is nonzero, and hence the channel can support a finite rate with a zero outage probability. When this occurs, C is called *zero-outage capacity*.

Example 4.6

With Rayleigh fading we have $\mathbb{E}[R^{-2}] = \infty$, that is, channel inversion requires transmission of an infinite average power. Thus, zero outage cannot be reached with finite power even in the presence of CSI at the transmitter. (Later on in this chapter we shall see that the introduction of receiver diversity yields a positive zero-outage capacity.) □

When channel inversion is not possible with finite power because $\mathbb{E}[R^{-2}] = \infty$, an alternative strategy is to choose $S(R)$ inversely proportional to R^2 only when $R > R^*$, where R^* is a suitable threshold, and $S(R) = 0$ otherwise. In [4.11] it is shown that the latter power-control policy is actually optimum, in the sense that it minimizes the outage probability under a long-term power constraint, i.e., a constraint on the power expenditure as averaged over many code words. We may interpret this power allocation technique by saying that, if fading is very bad, then the power needed to compensate for it by channel inversion would affect the average power too much. Under these conditions, it is better to turn off transmission and accept an outage.

The above solution can be generalized to higher values of F. In this case, the random capacity corresponding to the fading values $R_1, \ldots, R_F$ is given by

$$C(R_1, \ldots, R_F, S_1, \ldots, S_F) = \frac{1}{F} \sum_{i=1}^{F} \log\left(1 + R_i^2 S_i(R_i)\right) \qquad (4.39)$$

and the outage probability is again minimized under a long-term power constraint by turning off transmission over μ blocks (where $\mu \in \{0, 1, \ldots, F\}$) whenever the point in the F-dimensional space with coordinates $R_1, \ldots, R_F$ falls in certain regions whose structure depends on the channel statistics and on the power constraint [4.11].

Figures 4.11 and 4.12 show the outage probabilities obtained by choosing the optimum power-allocation strategy or constant power (which corresponds to no CSI at transmitter) in a transmission with rate $\rho = 0.4$ bit/dimension pair. With $F = 1$, it is seen that infinite SNR is required to have $P_{\text{out}} = 0$, which implies that the zero-outage capacity is zero here. On the other hand, the power savings obtained by using the optimum power-allocation strategy are dramatic (e.g., 22 dB

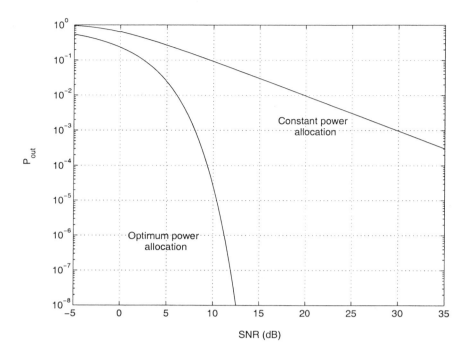

Figure 4.11: *Outage probability for the block-fading channel with Rayleigh fading and $F = 1$. The rate is $\rho = 0.4$ bit/dimension pair. Channel-state information is available at transmitter and receiver.*

for $P_{\text{out}} = 10^{-3}$), especially when we observe that CSI provides little advantage in terms of ergodic capacity. With $F = 2$, the power savings decrease, and we have $P_{\text{out}} = 0$ for finite SNR, indicating that the zero-outage capacity is nonzero.

4.3.4   Practical coding schemes for the block-fading channel

From (4.35) we observe that, for high SNR, binary signal sets ($|\mathcal{X}| = 2$) are not the most effective on block-fading channels.[2] Thus, codes constructed over high-level alphabets should be considered [4.14, 4.15]. The simplest coding scheme for achieving diversity F over the block-fading channel is repetition coding. For $\nu = 1$, this has $|\mathcal{X}|$ words, $\rho = \log_2 |\mathcal{X}|/F$, and $D_{\min} = F$. Short MD codes for block-fading channels with $M = 2, 6$, and 8 can be formed by either shortening

---

[2]Compounding this, when the SNR is low, higher-order constellations may result in additional losses due their sensitivity to synchronization inaccuracies.

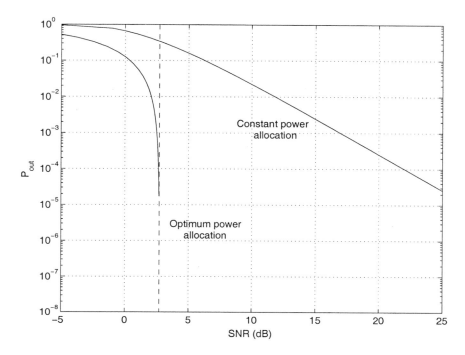

Figure 4.12: *Outage probability for the block-fading channel with Rayleigh fading and $F = 2$. The rate is $\rho = 0.4$ bit/dimension pair. Channel-state information is available at transmitter and receiver.*

or lengthening Reed–Solomon codes [4.14]. An extended MD Hamming code for $F = 6$ can also be exhibited. A computer search of trellis codes suitable for this channel has also been performed [4.14].

4.4 Introducing diversity

We have seen that the effect of fading on the performance of uncoded transmission requires delivering a power higher, and in some cases much higher, than that for an AWGN channel to achieve the same error probability. For example, passing from AWGN to Rayleigh fading transforms an exponential dependency of error probability on SNR into an inverse linear one. To combat fading, and hence to reduce transmit-power needs, a very effective technique consists of introducing *diversity* in the channel. Based on the observation that, on a fading channel, the SNR at the receiver is a random variable, the idea is to transmit the same signal

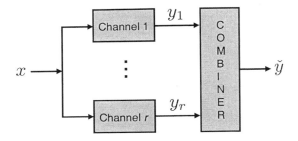

Figure 4.13: *Illustrating diversity and combining.*

through r separate fading channels. These are chosen so as to provide the receiver with r independent (or close-to-independent) replicas of the same signal, giving rise to independent SNRs. If r is large enough, then, at any time instant, there is a high probability that at least one of the signals received from the r "diversity branches" is not affected by a deep fade and hence that its SNR is above a critical threshold. By suitably combining the received signals, the fading effect will be mitigated.

Many techniques have been advocated for generating the independent channels on which the diversity principle is based, and several methods are known for combining the signals $y_1, \ldots, y_r$ obtained at their outputs into a single signal $\check{y}$ (Figure 4.13). The most important among them can be categorized as follows.

Space diversity. This consists of receiving the signal through r separate antennas, whose spacing is wide enough with respect to their coherence distance so as to obtain sufficient decorrelation. This technique can be easily implemented and does not require extra spectrum occupancy. (In Chapter 10 we shall examine in detail a situation in which multiple transmitting and receiving antennas are simultaneously employed.)

Polarization diversity. If a radio channel exhibits independent fading for signals transmitted on orthogonal polarizations, then diversity can be obtained by using a pair of cross-polarized antennas in the receiver. Notice that only two diversity branches are available here, while any value of r can in principle be obtained with space diversity. On the other hand, cross-polarized antennas do not need the large physical separation necessary for space diversity. In scattering environments tending to depolarize a signal, there is no need for separate transmission.

Frequency diversity. This is obtained by sending the same signal over different carrier frequencies whose separation must be larger than the coherence bandwidth of the channel. Clearly, frequency diversity is not a bandwidth-efficient solution.

Time diversity. If the same signal is transmitted in different time slots separated by an interval longer than the coherence time of the channel, time diversity can be obtained. Since, in mobile radio systems, slow-moving receivers have a large coherence time, time diversity in these conditions could only be introduced at the price of large delays.

4.4.1 Diversity combining techniques

Three main combining techniques, viz., selection, maximal ratio, and equal gain, will be described here. Each of them can be used in conjunction with any of the diversity schemes just listed. Some analyses will follow; however, it should be clear from the onset that the relative advantage of a diversity scheme will be lower as the channel moves away from Rayleigh fading towards Rice fading. In fact, increasing the Rice factor K causes the various diversity branches to exhibit a smaller difference in their instantaneous SNRs. This typically occurs when a fixed path becomes available in addition to scatter paths (see Section 2.3.1). Notice that the increased quality of the Rice channel may more than make up for the decreased diversity.

We assume here that transmission is uncoded (the case of coded transmission will be dealt with, in a more general framework, in Chapter 10). When the elementary signal two-dimensional x is transmitted, the received signal at the output of the r diversity branches can be modeled as an r-vector: the model for this single-input, multiple-output (SIMO) channel is

$$\mathbf{y} = \mathbf{h}x + \mathbf{z} \tag{4.40}$$

where $\mathbf{y}$ is a vector whose r components are the observed channel outputs, $\mathbf{h}$ is a random r-vector modeling the fading affecting the r diversity branches (its entries $h_i = R_i e^{j\theta_i}$ are independent complex RVs under our assumptions), and $\mathbf{z} \sim \mathcal{N}_c(0, N_0 \mathbf{I}_r)$ describes the white noise, assumed independent from branch to branch.

The optimum (maximum-likelihood) detection of x given the observation $\mathbf{y}$ and perfect knowledge of the value taken on by $\mathbf{h}$ (the channel-state information) consists of looking for the signal x that minimizes the norm $\|\mathbf{y} - \mathbf{h}x\|$. A simpler way of proceeding consists of transforming, through a *combiner*, the vector $\mathbf{y}$ into a scalar $\check{y}$ that is used for demodulation as if it were obtained at the output of a single-input, single-output channel. Here we examine some of these combining techniques, focusing on the SNR obtained. Specifically, consider the SNR in the ith diversity branch. This SNR is $R_i^2 \mathcal{E}/N_0$. A combination technique generates an output whose SNR is $\check{R}^2 \mathcal{E}/N_0$, with $\check{R}$ a function of $R_1, \ldots, R_r$. The error probability can consequently be reduced to the calculation of the expectation $\mathbb{E}_{\check{R}} \, f(\check{R}^2 \mathcal{E}/N_0)$, where $f(\mathcal{E}/N_0)$ is the error probability over the AWGN channel, consistently with our discussion in Section 4.2.

Maximal-ratio combining

A family of combination techniques consists of forming a linear combination of the signals at the output of the r diversity branches. The problem here is to select the coefficients of the linear combination according to a suitable optimization criterion. Formally, before detection the received signal $\mathbf{y}$ is linearly transformed into the scalar $\check{y}$ by using an r-vector $\mathbf{a}$ to obtain $\check{y} \triangleq \mathbf{a}^\dagger \mathbf{y}$.

With *maximal-ratio combining*, the vector $\mathbf{a}$ is chosen so as to maximize the SNR at the combiner's output. Since $\mathbf{a}^\dagger \mathbf{y} = \mathbf{a}^\dagger \mathbf{h} x + \mathbf{a}^\dagger \mathbf{z}$, the ratio of signal energy to noise power spectral density at the output of the combiner is

$$\frac{\check{\mathcal{E}}}{\check{N}_0} = \frac{|\mathbf{a}^\dagger \mathbf{h}|^2 \mathcal{E}}{\|\mathbf{a}^\dagger\|^2 N_0} \lesssim \|\mathbf{h}\|^2 \frac{\mathcal{E}}{N_0} \qquad (4.41)$$

where the Schwarz inequality $|\mathbf{a}^\dagger \mathbf{h}|^2 \leq \|\mathbf{a}\|^2 \|\mathbf{h}\|^2$ has been used. Now, since (4.41) holds with equality if and only if $\mathbf{a} = \kappa \mathbf{h}$ for some complex scalar κ, then the SNR (4.41) is maximized by choosing $\check{y} = \mathbf{h}^\dagger \mathbf{y}$. This yields

$$\frac{\check{\mathcal{E}}}{\check{N}_0} = \sum_{i=1}^{r} R_i^2 \frac{\mathcal{E}}{N_0}$$

Maximum-likelihood detection can be performed by minimizing $\|\mathbf{y} - \mathbf{h}x\|$ over x. Now, observe that

$$|\check{y} - \mathbf{h}^\dagger \mathbf{h} x|^2 = |\check{y}|^2 - 2\|\mathbf{h}\|^2 \Re[\check{y}^* x] + \|\mathbf{h}\|^4 |x|^2$$

and that the term $|\check{y}|^2$ is irrelevant when it comes to searching for the signal x that minimizes the metric. Moreover, multiplication of a metric by a positive quantity yields an equivalent metric. Thus, we may use $-2\Re[\check{y}^* x] + \|\mathbf{h}\|^2 |x|^2$ instead of the original metric. This is equivalent to the ML metric: in fact,

$$\|\mathbf{y} - \mathbf{h}x\|^2 = \|\mathbf{y}\|^2 - 2\Re[\mathbf{y}^\dagger \mathbf{h}x] + \|\mathbf{h}\|^2 |x|^2$$

is equivalent to $-2\Re[\check{y}^*x] + \|\mathbf{h}\|^2|x|^2$.

Note that, since $\check{R}^2 = \sum_{i=1}^r |h_i|^2$ with $\mathbb{E}[|h_i|^2] = 1$, we have $\mathbb{E}[\check{R}^2] = r$. This is due to the increase of the power captured by the r separate receivers in a SIMO transmission system, and shows that r-branch diversity with maximal-ratio combining increases the average SNR by a factor r. In addition to this effect, the shape of the pdf of the fading gain changes as r increases. To isolate this last effect from the previous one, we may consider the normalized SNR, which is

$$\frac{\check{\mathcal{E}}'}{\tilde{N}_0} \triangleq \frac{1}{r}\frac{\check{\mathcal{E}}}{\tilde{N}_0} = \frac{\mathcal{E}}{N_0}\frac{1}{r}\sum_{i=1}^r R_i^2 \tag{4.42}$$

Using the law of large numbers, we see that as $r \to \infty$ we have $\check{\mathcal{E}}'/\tilde{N}_0 \to \mathcal{E}/N_0$, which is the SNR one would obtain on an AWGN channel without fading: we see how diversity serves the dual role of capturing more power and at the same time reducing channel fluctuations.

To evaluate the error performance of maximal-ratio combining, observe that, since $\check{R} = \|\mathbf{h}\|$, the pairwise error probability becomes

$$P(x \to \hat{x}) = \mathbb{E}\,P(x \to \hat{x} \mid \check{R}) = \mathbb{E}\,Q\left(\check{R}\frac{|x - \hat{x}|}{\sqrt{2N_0}}\right) \tag{4.43}$$

with the expectation being taken with respect to $\check{R}$. In general, a PEP can be evaluated by first writing down the corresponding PEP for the AWGN channel, next multiplying the SNR by the factor $\check{R}^2$, and finally taking the expectation.

If the components of vector $\mathbf{h}$ are independent complex Gaussian RVs with mean zero and common variance $\mathbb{E}[|h_i|^2] = 1$, $i = 1, \ldots, r$, then $\check{R}^2$ is a chi-square distributed random variable with $2r$ degrees of freedom. Its probability density function is

$$p_{\check{R}^2}(x) = \frac{1}{(r-1)!}\,x^{r-1}e^{-x}, \qquad x \geq 0 \tag{4.44}$$

The following expectation can be computed in closed form as follows (see Section D.1, Appendix D):

$$\mathbb{E}\,Q\left(\beta\check{R}\right) = \left(\frac{1-\mu}{2}\right)^r \sum_{k=0}^{r-1}\binom{r+k-1}{k}\left(\frac{1+\mu}{2}\right)^k \tag{4.45}$$

where

$$\mu \triangleq \sqrt{\frac{\beta^2}{2+\beta^2}} \tag{4.46}$$

Using this calculation, the PEP (4.43) can be given a closed form by identifying

$$\beta \triangleq \frac{|x - \hat{x}|}{\sqrt{2N_0}} \qquad (4.47)$$

For large-enough values of β, we have $(1+\mu)/2 \sim 1$ and, using (4.12), $(1-\mu)/2 \sim 1/2\beta^2$. Moreover,

$$\sum_{k=0}^{r-1} \binom{r+k-1}{k} = \binom{2r-1}{r}$$

which yields

$$\mathbb{E}\, Q\,(\beta \check{R}) \sim \binom{2r-1}{r} \left(\frac{1}{2\beta^2}\right)^r \qquad (4.48)$$

Since β^2 is proportional to SNR, diversity of order r makes the error probability decrease as SNR^{-r}.

Example 4.7

Consider binary antipodal modulation with coherent detection transmitted over an independent Rayleigh fading channel and detected with r-branch diversity and maximal-ratio combining. Since $|x - \hat{x}|^2 = 4\mathcal{E}_b$, its error probability is given by

$$P(e) = \mathbb{E}\, Q\left(\sqrt{\check{R}^2 \frac{2\mathcal{E}_b}{N_0}}\right) \qquad (4.49)$$

and this expectation can be computed by using (4.45), with

$$\mu = \sqrt{\frac{\mathcal{E}_b/N_0}{1 + \mathcal{E}_b/N_0}}$$

Figure 4.14 shows this error probability for various values of r. The approximation (4.48) is also shown.   □

Example 4.8

Consider a block-fading channel with $F = 1$, diversity order r, Rayleigh-distributed fading, and channel-state information available at both transmitter and receiver. The equivalent fading channel has a gain R whose pdf is given by (4.44), where $\alpha = 1/r$

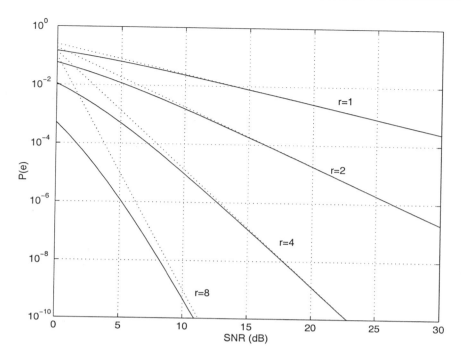

Figure 4.14: *Error probability of binary antipodal transmission with coherent detection and r-branch diversity with maximal-ratio combining. Exact values (continuous lines) and approximate values (dotted lines) are shown.*

in order to satisfy the normalization condition $\mathbb{E}[R^2] = 1$. Direct calculation shows that

$$\mathbb{E}[R^{-2}] = \frac{r}{r-1} \qquad (4.50)$$

so that channel inversion can be performed with finite average power, provided that $r > 1$. The resulting zero-outage capacity is given by

$$C = \log\left(1 + \frac{r-1}{r}\bar{S}\right) \qquad \text{bit/dimension pair} \qquad (4.51)$$

Capacity values for $r = 2, 4, 8$ are shown in Figure 4.15. We observe here the interesting fact that as $r \to \infty$ the capacity tends to that of the AWGN channel: over this channel, zero-outage capacity coincides with ergodic capacity. This finding can be interpreted by saying that, in addition to capturing more power, receiver diversity stabilizes channel fluctuations by reducing the amount of effective fading. $\square$

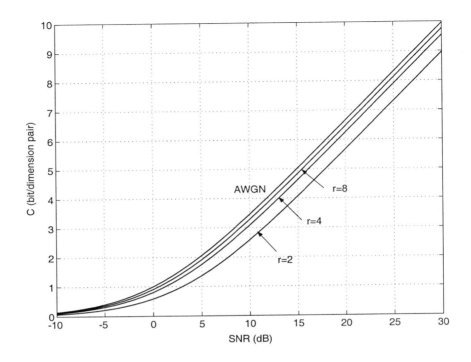

Figure 4.15: *Zero-outage capacity of a block-fading channel with* $F = 1$, *Rayleigh-distributed fading and channel-state information available at both transmitter and receiver. The AWGN channel capacity is also shown for comparison. The signal-to-noise ratio is* $\mathcal{E}/N_0$.

Equal-gain combining

Maximal-ratio combining requires knowledge of channel-state information, which in this case corresponds to the r values of the gains of each diversity branch. Should this full information be unavailable, one may use a combining technique in which $\tilde{y} = \mathbf{a}^\dagger \mathbf{y}$, with $\mathbf{a}$ the vector whose components are $a_i = e^{j\theta_i}$, the phases of the components of $\mathbf{h}$. Notice that channel-state information might be estimated for other purposes: for example, unequal-energy constellations (typically, QAM) need channel gains for automatic gain control. If this is the case, maximal-ratio combining is the natural choice.

From (4.41), we see that

$$\frac{\check{\mathcal{E}}}{\check{N}_0} = \frac{\mathcal{E}}{N_0} \left[\frac{1}{r} \sum_{i=1}^{r} R_i \right]^2 \qquad (4.52)$$

which shows that analysis depends on the statistics of a sum of Rayleigh (or Nakagami, ...) random variables. Obtaining a closed-form pdf of this sum is a notoriously difficult problem [4.5] (see the Bibliography section for results in this area).

Selection combining

This consists of selecting at each time, among the r diversity branches, the one with the largest value of signal-to-noise ratio. Its simplicity makes it often used in practice in conjunction with antenna diversity. In fact, the receiver only needs a single complete receive chain, which can be switched among the individual antennas. Also, like with equal-gain combining, CSI is not required.

We assume that each diversity branch is affected by the same Gaussian noise power, so selecting the branch with the largest instantaneous SNR is tantamount to selecting the branch with the largest instantaneous power, and hence the largest fading gain. Hence, this combining technique is equivalent to choosing the diversity branch whose fading gain is

$$\check{R} \triangleq \max\{R_1, \ldots, R_r\} \tag{4.53}$$

This combining technique yields the following ratio of signal energy to noise power spectral density:

$$\frac{\check{\mathcal{E}}}{\check{N}_0} = \frac{\mathcal{E}}{N_0} \max_i R_i^2$$

Under our assumption of independent diversity path gains, the cumulative distribution function of $\check{R}$ is given by

$$F_{\check{R}}(\check{r}) \triangleq \mathbb{P}(\check{R} \leq \check{r}) = \mathbb{P}(R_1 \leq \check{r}, \ldots, R_r \leq \check{r}) = [F_R(\check{r})]^r \tag{4.54}$$

(in fact, $\check{R}$ is less than $\check{r}$ if and only if $R_1, \ldots, R_r$ are all less than $\check{r}$). The derivative of (4.54) yields the pdf of $\check{R}$.

If the channel gains R_i are independent Rayleigh-distributed RVs, then

$$F_{\check{R}}(\check{r}) = [1 - \exp(-\check{r}^2)]^r$$

and we have, integrating by parts:

$$
\begin{aligned}
\mathbb{E}\, Q(\beta \check{R}) &= \int_0^\infty Q(\beta x)\, d[(1 - \exp(-x^2))^r] \\
&= \frac{\beta}{\sqrt{2\pi}} \int_0^\infty (1 - \exp(-x^2))^r \exp(-\beta^2 x^2/2)\, dx \\
&= \frac{\beta}{\sqrt{2\pi}} \sum_{k=0}^r \binom{r}{k} (-1)^k \int_0^\infty \exp[-(k + \beta^2/2)x^2]\, dx \\
&= \frac{1}{2} \sum_{k=0}^r \binom{r}{k} (-1)^k \frac{\beta}{\sqrt{\beta^2 + 2k}}
\end{aligned}
\tag{4.55}
$$

As $\beta \to \infty$, we have the asymptotic expression (see Problem 10):

$$
\mathbb{E}\, Q(\beta \check{R}) \sim \frac{1}{2}(2r - 1)!! \left(\frac{1}{\beta^2}\right)^r
\tag{4.56}
$$

where we see that, as with maximal-ratio combining, the error probability decreases as SNR^{-r}.

Example 4.9

> With binary antipodal modulation as in Example 4.7, but with selection combining, the error probability is obtained from (4.55) with $\beta = \sqrt{2\mathcal{E}_b/N_0}$. The results for various values of r are shown in Figure 4.16. □

4.5   Bibliographical notes

An extensive discussion, summarizing the state of the art in information-theoretic analyses of the fading channel, can be found in [4.9]. For a fading channel with no channel-state information at the transmitter, the fact that the capacity under an average power constraint is achieved by a discrete input distribution was proved in [4.1] (see also [4.19]). Under the same conditions, if the constraint is on the peak amplitude, then the capacity-achieving distribution is discrete with a finite number of points [4.18]. The calculation of outage probability for the block-fading channel was done in [4.16] (see also [4.13]). Zero-outage capacity (or delay-limited capacity) is discussed in [4.9, 4.11] and references therein. An extensive analysis of diversity techniques can be found in [4.21, Chap. 5]; for recent work, see [4.20].

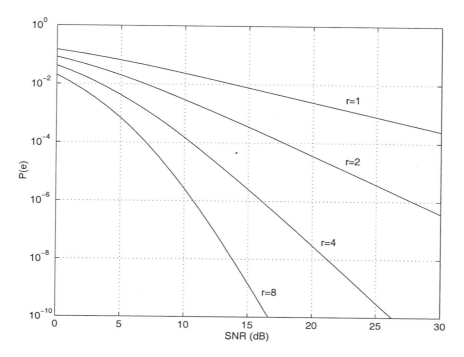

Figure 4.16: *Error probability of binary antipodal transmission with coherent detection and r-branch diversity with selection combining. The signal-to-noise ratio is $\mathcal{E}/N_0$.*

For a historical perspective, see the now-classic paper by Brennan, originally published in 1959, and recently reprinted in [4.10]. Recent analyses of equal-gain combining can be found in [4.2–4.6, 4.17, 4.22, 4.23].

4.6 Problems

1. Prove that, as SNR→ ∞, the SNRs needed to obtain the same ergodic capacity for the AWGN channel and the independent Rayleigh fading channel differ by 2.5 dB.

2. Prove that the ergodic capacity of the independent Rayleigh fading channel with r-branch diversity and maximal-ratio combining tends to the capacity of the AWGN channel as $r \rightarrow \infty$ if the normalized SNR of (4.42) is used in the calculations.

3. Repeat the calculation of Example 4.7 with Nakagami-distributed fading. Show that for high SNRs the error probability for $M \to \infty$ is inversely proportional to $(\mathcal{E}_b/N_0)^{-m}$, where m is the Nakagami-distribution parameter.

4. Consider the code with eight words and $n = 8$ whose words are all the permutations of $(\sqrt{\mathcal{E}}, 0, 0, 0, 0, 0, 0, 0)$. Discuss its error performance for transmission on a high-SNR block-fading channel with $F = 1, 2, 4$ and on a high-SNR independent Rayleigh fading channel.

5. Prove the "generalized Singleton bound" (4.35).

6. Consider transmission of QPSK over a Rayleigh fading channel with two-branch diversity. Compute the asymptotic (for high SNR) error probability with (a) maximal-ratio combining and (b) selection combining. **Hint:** You may use the integral

$$\int_0^\infty e^{-a^2 x^2} \, dx = \frac{\sqrt{\pi}}{2a}$$

and the asymptotic expansion, valid for $x \to 0$:

$$\frac{1}{\sqrt{1+x}} \sim 1 - \frac{1}{2}x + \frac{3}{8}x^2$$

7. Consider an ergodic Rayleigh fading channel with r independent fading diversity branches. Compute its average capacity with maximal-ratio combining and with selection combining.

8. Consider an ergodic Rayleigh fading channel with r independent fading diversity branches and maximal-ratio combining. Compute its average capacity with channel state information at the receiver only and with channel state information at the receiver and transmitter. Choose $r = 1, 2,$ and 3 (the results for $r = 1$ are shown in Figure 4.7).

9. Compute the outage probability of a block-fading channel with r-branch diversity and maximal-ratio combining. Assume Rayleigh fading and $F = 1$. Show in particular that $P_{\text{out}} \to 0$ as $r \to \infty$. If the normalized SNR (4.42) is used, show that, as $r \to \infty$, P_{out} is equal to either 0 or 1 according to the value of the rate ρ.

10. Derive (4.56) from (4.55). Use the expansions

$$\frac{1}{\sqrt{1+x}} = \sum_{j=0}^\infty (-1)^j x^j \frac{(2j-1)!!}{(2j)!!}$$

and

$$x^n = \sum_{m=0}^{n} S(n,m) x^{(m)}$$

where $S(n,m)$ are the Stirling number of the second kind, with $S(m,m) = 1$, and

$$x^{(m)} \triangleq \begin{cases} x(x-1)\cdots(x-m+1), & m \geq 1 \\ 1, & m = 0 \end{cases}$$

11. Define the ϵ-outage capacity C_ϵ of a nonergodic fading channel as the largest rate ρ such that $P_{\text{out}}(\rho)$ is less than a fixed amount ϵ. Examine the behavior of C_ϵ. In particular, show that

   (a) At high SNR, C_ϵ differs from the AWGN capacity by an additive term dependent on ϵ but independent of SNR.

   (b) At low SNR, C_ϵ differs from the AWGN capacity by a multiplicative term dependent on ϵ but independent of SNR.

   Compute the two terms above for the Rayleigh fading channel.

12. *MMSE combining* consists of transforming the received signal $\mathbf{y}$ into the scalar $\mathbf{b}^\dagger \mathbf{y}$, where the r-vector $\mathbf{b}$ minimizes the mean-square error $\varepsilon^2 \triangleq \mathbb{E}[|\mathbf{b}^\dagger \mathbf{y} - x|^2]$. Find the vector $\mathbf{b}$, and Compare this combining technique with MRC.

References

[4.1] I. C. Abou-Faical, M. D. Trott, and S. Shamai (Shitz), "The capacity of discrete-time memoryless Rayleigh-fading channels," *IEEE Trans. Inform. Theory*, Vol. 47. No. 4, pp. 1290–1301, May 2001.

[4.2] A. A. Abu-Dayya and N. C. Beaulieu, "Microdiversity on Rician fading channels," *IEEE Trans. Commun.*, Vol. 42, No. 6, pp. 2258–2267, June 1994.

[4.3] M.-S. Alouini and M. K. Simon, "Performance analysis of coherent equal gain combining over Nakagami-m fading channels," *IEEE Trans. Vehic. Technol.*, Vol. 50, No. 6, pp. 1149–1463, November 2001.

[4.4] A. Annamalai, C. Tellambura, and V. K. Bhargava, "Exact evaluation of maximal-ratio and equal-gain diversity receivers for M-ary QAM on Nakagami fading channels," *IEEE Trans. Commun.*, Vol. 47, No. 9, pp. 1335–1344, September 1999.

[4.5]   N. C. Beaulieu, "An infinite series for the computation of the complementary probability distribution function of a sum of independent random variables and its application to the sum of Rayleigh random variables," *IEEE Trans. Commun.*, Vol. 38, No. 9, pp. 1463–1474, September 1990.

[4.6]   N. C. Beaulieu and A. A. Abu-Dayya, "Analysis of equal gain diversity on Nakagami fading channels," *IEEE Trans. Commun.*, Vol. 39, No. 2, pp. 225–234, February 1991.

[4.7]   S. Benedetto and E. Biglieri, *Digital Transmission Principles with Wireless Applications*. New York: Kluwer/Plenum, 1999.

[4.8]   E. Biglieri, G. Caire, and G. Taricco, "Coding vs. spreading over block fading channels," in: Francis Swarts (Ed.), *Spread Spectrum: Developments for the New Millennium*. Boston, MA: Kluwer Academic, 1998.

[4.9]   E. Biglieri, J. Proakis, and S. Shamai (Shitz), "Fading channels: Information-theoretic aspects," *IEEE Trans. Inform. Theory*, Vol. 44, No. 6, pp. 2169–2692, October 1998.

[4.10] D. G. Brennan, "Linear diversity combining techniques," *IEEE Proc.*, Vol. 91, No. 2, pp. 331–356, February 2003.

[4.11] G. Caire, G. Taricco, and E. Biglieri, "Optimal power control for the fading channel," *IEEE Trans. Inform. Theory*, Vol. 45, No. 5, pp. 1468–1489, July 1999.

[4.12] A. J. Goldsmith and P. P. Varaiya, "Capacity of fading channels with channel side information," *IEEE Trans. Inform. Theory*, Vol. 43, No. 6, pp. 1986–1992, November 1997.

[4.13] G. Kaplan and S. Shamai (Shitz), "Error probabilities for the block-fading Gaussian channel," *A.E.Ü.*, Vol. 49, No. 4, pp. 192–205, 1995.

[4.14] R. Knopp, *Coding and Multiple Access over Fading Channels*, Ph.D. thesis, École Polytechnique Fédérale de Lausanne, Lausanne, Switzerland, 1997.

[4.15] R. Knopp and P. A. Humblet, "On coding for the block fading channel," *IEEE Trans. Inform. Theory*, Vol. 46, No. 1, pp. 189–205, January 2000.

[4.16] L. Ozarow, S. Shamai, and A. D. Wyner, "Information theoretic considerations for cellular mobile radio," *IEEE Trans. Vehic. Technol.*, Vol. 43, No. 2, pp. 359–378, May 1994.

[4.17] X. Qi, M.-S. Alouini, and Y.-C. Ko, "Closed-form analysis of dual-diversity equal-gain combining over Rayleigh fading channels," *IEEE Trans. Wireless Commun.*, Vol. 2, No. 6, pp. 1120–1125, November 2003.

[4.18] S. Shamai (Shitz) and I. Bar-David, "The capacity of average and peak-power-limited quadrature Gaussian channels," *IEEE Trans. Inform. Theory*, Vol. 41, pp. 1060–1071, July 1995.

[4.19] G. Taricco and M. Elia, "Capacity of fading channel with no side information," *Electron. Lett.*, Vol. 33, No. 16, pp. 1368–1370, July 31, 1997.

[4.20] M. Z. Win and J. H. Winters, "Analysis of hybrid selection/maximal-ratio combining in Rayleigh fading," *IEEE Trans. Commun.*, Vol. 47, No. 12, pp. 1773–1776, December 1999.

[4.21] M. D. Yacoub, *Foundations of Mobile Radio Engineering*. Boca Raton, FL: CRC Press, 1993.

[4.22] Q. T. Zhang, "Probability of error for equal-gain combiners over Rayleigh channels: Some closed-form solutions," *IEEE Trans. Commun.*, Vol. 45, No. 3, pp. 270–273, March 1997.

[4.23] Q. T. Zhang, "A simple approach to probability of error for equal-gain combiners over Rayleigh channels," *IEEE Trans. Vehic. Technol.*, Vol. 48, No. 4, pp. 1151–1154, July 1999.

but the whole of him nevertheless in Owenmore's five quarters

Trellis representation of codes

In Chapter 3 we described a technique to introduce an algebraic structure into a signal constellation S obtained as a subset of X^n, $X = \{\pm 1\}$. This technique generates linear binary codes. Each of these codes can be represented in two ways: either as the set of all linear combinations of the rows of a generator matrix or as the set of all binary n-tuples that satisfy some parity-check equations. The present chapter describes an exceedingly convenient representation of linear block codes as the set of all n-tuples corresponding to paths traversing a trellis. This representation can be used for optimal decoding based on the Viterbi algorithm. The complexity of this trellis representation is also examined, and minimal trellises are introduced. The complexity of trellis representations can be further reduced by introducing tail-biting trellises.

5.1   Introduction

A convenient description of algebraic codes as well as of codes on a signal space is through a *trellis*: this provides at the same time a compact method of cataloging code words and, via the Viterbi algorithm (to be described soon), an efficient decoding algorithm.

A trellis is a directed graph where each distinct code word corresponds to a distinct path across it, starting from an initial vertex and ending into a final vertex. A trellis is described by a set Σ of vertices (called *states*), a set B of labeled edges (called *branches*), and a set L of labels, with the property that every state has a well-defined trellis time. Formally, we have the following definition:

Definition 5.1.1 *A trellis* $\mathcal{T} = (\Sigma, B, L)$ *of depth* n *is a branch-labeled directed graph with the following property: the state set* Σ *can be decomposed as a union of disjoint subsets*

$$\Sigma = \Sigma_0 \cup \Sigma_1 \cup \cdots \cup \Sigma_n \tag{5.1}$$

such that every branch in $\mathcal{T}$ *that begins at a state in* Σ_i *ends at a state in* Σ_{i+1}, *and every state in* $\mathcal{T}$ *lies on at least one path from a state in* Σ_0 *to a state in* Σ_n.

The trellis $\mathcal{T}$ represents a block code of length n over $\mathcal{X}$ if $L = \mathcal{X}$ (that is, its branches are labeled by the elements of $\mathcal{X}$), and the set of all the sequences of branch labels is the set of code words. It is convenient to define a "time" in the trellis, by assuming that the code symbols are transmitted and received sequentially in time. The graphical representation of a trellis has the horizontal axis associated with time, and the vertical axis with states. Branches connect states at two adjacent time instants.

Although we shall restrict our attention to linear codes here, a trellis can also be used to describe nonlinear codes. Two exceedingly simple (but by no means insignificant) cases of trellis descriptions of binary codes are described in the examples that follow.

Example 5.1 ($(n, 1, n)$ repetition code)

Here there are only two code words, each corresponding to an n-tuple of equal elements. The corresponding trellis is shown in Figure 5.1.   □

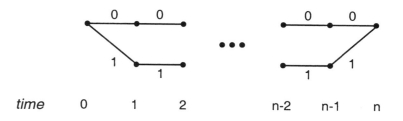

Figure 5.1: *Trellis representation of the binary $(n, 1, n)$ repetition code.*

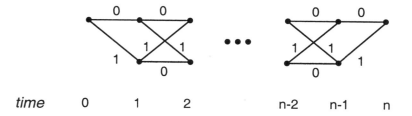

Figure 5.2: *Trellis representation of the $(n, n - 1, 2)$ binary single-parity-check code.*

Example 5.2 ($(n, n - 1, 2)$ binary single-parity-check code)

The trellis cataloging the 2^{n-1} words, characterized by having an even number of 1s, is shown in Figure 5.2.   □

5.2 Trellis representation of a given binary code

Here we focus our attention on the construction of the trellis describing a given linear binary code. Assume that the (n, k, d) code is *systematic*, i.e., its generator matrix has the form

$$\mathbf{G} = [\mathbf{I}_k \vdots \mathbf{P}]$$

where $\mathbf{I}_k$ is the $k \times k$ identity matrix, and $\mathbf{P}$ is a generic $k \times (n-k)$ binary matrix.[1] The first k positions of the words of a systematic code include all possible binary k-tuples. Draw the 2^k paths associated with these k-tuples, starting from a common initial node, as shown in Figure 5.3.

---

[1] Recall from Chapter 3 that this assumption entails no loss of generality.

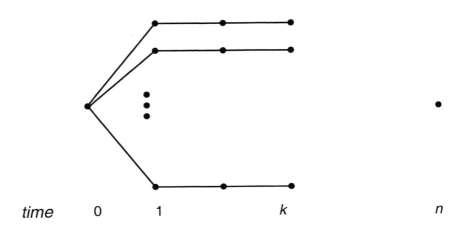

time   0   1   k   n

Figure 5.3: *Initial part of the trellis representing an (n, k, d) linear binary code in systematic form.*

Next, "close" the trellis by associating with each code word at time k its corresponding $n - k$ redundant symbols (which are uniquely determined by the submatrix **P**). It can be easily seen that the trellis of Figure 5.1 corresponds to this construction.

A related construction is based on the parity-check matrix **H** of code $\mathcal{C}$: recall that, given a binary n-tuple **x**, this is a word of $\mathcal{C}$ if and only if $\mathbf{Hx'} = 0$. The $(n - k)$-tuple $\mathbf{Hx'}$ is the syndrome of **x** and cannot take on more than 2^{n-k} values. Now, choose one state for each syndrome value, and label it by this value. The path corresponding to the code word $(x_1, x_2, \ldots, x_n)$ is obtained as follows. Start from the zero state at time 0. Let the syndrome of $(x_1, 0, 0, \cdots, 0)$ be σ_1. Then draw a branch from state 0 to state σ_1 and label it x_1. Next, consider $(x_1, x_2, 0, \cdots, 0)$. Let its syndrome be σ_2. Draw a new branch from σ_1 to σ_2 and label it x_2, and so on. The total number of trellis states at any time cannot exceed the number of syndrome values, viz., $|\Sigma_i| \leq 2^{n-k}$, $i = 0, \ldots, n$. Figure 5.2 shows an example of this construction: here $\mathbf{H} = [1\ 1\ \ldots\ 1]$, and hence the syndrome of an n-tuple is the modulo-2 sum of its elements. Its value is either 0 or 1.

The two constructions just described lead to the following result: for linear binary codes, the maximum number of trellis states at any given time is bounded above by $\min\{2^k, 2^{n-k}\}$.

5.3 ★ Decoding on a trellis: Viterbi algorithm

Consider transmission of the code word $\mathbf{x} = (x_1, \ldots, x_n)$. We receive $\mathbf{y}$, a corrupted version of $\mathbf{x}$, at the output of a channel whose conditional probability density function (pdf) factors into the product involving elementary signals, i.e.,

$$p(\mathbf{y} \mid \mathbf{x}) = \prod_{i=1}^{n} p(y_i \mid x_i) \tag{5.2}$$

A channel satisfying (5.2) is called *stationary memoryless*, because it transforms each elementary symbol in a way that does not depend on the other symbols and on the transmission instant. The task of the decoder is to make a decision on the transmitted $\mathbf{x}$ upon observation of $\mathbf{y}$. The maximum-likelihood decoding rule consists of choosing the signal $\mathbf{x} \in \mathcal{S}$ that maximizes $p(\mathbf{y} \mid \mathbf{x})$. Now, in general, we cannot maximize $p(\mathbf{y} \mid \mathbf{x})$ by maximizing separately $p(y_i \mid x_i)$ because the components of $\mathbf{x}$ are interrelated by the code structure (for example, in a single-parity-check code, the components of a code word are such that the number of 1s is even). Only if the signals are uncoded are we allowed to make "symbol-by-symbol" decisions without any loss of optimality.

On the other hand, the solution of the maximization problem above may be computationally very intensive if the code has little structure, or its structure is not taken into account: in fact, with M code words we would have to compute M values of $p(\mathbf{y} \mid \mathbf{x})$ and find their maximum (brute-force approach). Now, M may be so large as to make this approach impractical.

We now show that a way of exploiting the structure of the code for decoding is by taking advantage of its trellis representation. We can do this under the assumption that the "metric," i.e., the quantity whose maximization is equivalent to the maximization of $p(\mathbf{y} \mid \mathbf{x})$, is *additive* over the x_i, that is,

$$m(\mathbf{y} \mid \mathbf{x}) = \sum_{i=1}^{n} m(y_i \mid x_i) \tag{5.3}$$

For a stationary memoryless channel, additivity is satisfied by choosing the metric $m(\mathbf{y} \mid \mathbf{x}) = \ln p(\mathbf{y} \mid \mathbf{x})$. In fact,

$$
\begin{aligned}
\ln p(\mathbf{y} \mid \mathbf{x}) &= \ln \prod_{i=1}^{n} p(y_i \mid x_i) = \sum_{i=1}^{n} \ln p(y_i \mid x_i) \\
&= \sum_{i=1}^{n} m(y_i \mid x_i) = m(\mathbf{y} \mid \mathbf{x})
\end{aligned}
$$

Note that the metric can assume different forms, according to the specific problem at hand. For example, with AWGN channels (for the fading channels with CSI known at the receiver, examined in the previous chapter, the extension is straightforward), we have

$$p(\mathbf{y} \mid \mathbf{x}) = c\,e^{-\|\mathbf{y}-\mathbf{x}\|^2}$$

(c a normalization constant), and hence, by disregarding the constant $\ln c$ and assuming for simplicity that the elementary constellation $\mathcal{X}$ contains real or complex signals

$$m(\mathbf{y} \mid \mathbf{x}) = -\|\mathbf{y} - \mathbf{x}\|^2 = -\sum_{i=1}^{n} |y_i - x_i|^2$$

Thus, we may choose

$$m(y_i \mid x_i) = -|y_i - x_i|^2$$

Moreover, observe that

$$|y_i - x_i|^2 = |y_i|^2 + |x_i|^2 - 2\Re(y_i x_i^*)$$

where $\Re$ denotes real part. Here the term $|y_i|^2$ is irrelevant to the maximization of the metric and hence can be removed from consideration. Similarly, if the signals in the elementary constellation have one and the same energy, the term $|x_i|^2$ is also irrelevant, and the metric is reduced to $\Re y_i x_i^*$. If in addition $x_i = \pm\sqrt{\mathcal{E}}$, no product is necessary for the computation of the metric.

The Viterbi algorithm (VA) decodes a code described by a trellis when the metric induced by the channel is additive.

A *branch metric* is associated with each branch of the trellis, in the form of a label. Since the metrics are additive, the metric associated with a pair of adjoining branches is the sum of the two metrics. Consequently, the total metric associated with a path traversing the whole trellis from left to right is the sum of the labels of the branches forming the path. The problem here is to find the path traversing the trellis with the maximum total metric.

We start our description of the VA with the illustration of its key step, commonly called ACS (for Add, Compare, Select). Consider Figure 5.4. It shows the trellis states at time k (denoted σ_k) and at time $k + 1$ (denoted σ_{k+1}). The branches joining pairs of paths are labeled by the corresponding branch metrics, while the states σ_k are labeled by the *accumulated state metrics*, to be defined soon. The ACS step consists of the following: For each state σ_{k+1}, examine the branches leading to it and stemming from states σ_k (there are two such branches in Figure 5.4). For these branches, ADD the metric accumulated at the state from which it stems to the metric of the branch itself. Then COMPARE the results of these sums,

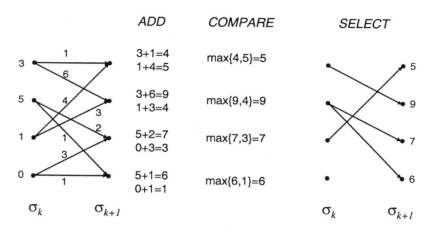

Figure 5.4: *The ACS step of the Viterbi algorithm.*

and SELECT the branch associated with the maximum value (and consequently discard, for each state, all other branches entering it; and, if two or more of the quantities being compared are equal, choose one at random). The maximum value is associated with the state, and forms its accumulated metric. This value is retained only for the next ACS step, then discarded.

The VA consists of repeating the ACS step from the starting state to the ending state of the trellis. After each ACS step, the VA retains, for each state, one value of accumulated metric and one path, called the *survivor* corresponding to the state. Thus, at any time k we are left, for each σ_k, with a single survivor path traversing the trellis from the initial state to σ_k, and with one value of accumulated metric. This survivor path is the maximum-metric path to the corresponding state. After n ACS steps, at the termination of the trellis we obtain a single n-branch path and a single accumulated metric. These are the maximum-metric path and the maximum-metric value, respectively.[2]

Figure 5.5 illustrates the determination of a maximum-metric path through a four-state trellis via the VA.

To prove the optimality of the VA, it suffices to observe the following. Assume that the optimum path passes through a certain state σ at time k. Then *its first k branches must be the same as for the survivor corresponding to σ.* In fact, if they were not, the optimum path would begin with a path passing through σ and having

---

[2]In practice, since the maximum-metric path for each state can be found by retracing the branch decisions, one may choose to store only the branch decisions, rather than the entire path.

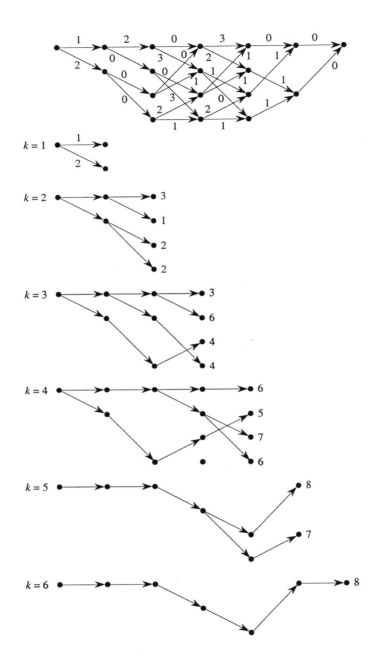

Figure 5.5: *Determination of the maximum-metric path through a trellis with* $n = 6$ *and four states via the Viterbi algorithm.*

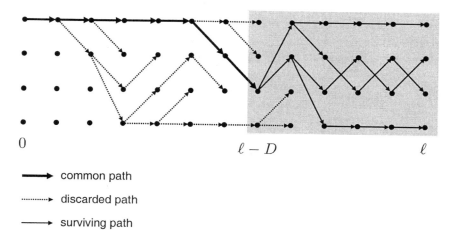

common path

discarded path

surviving path

Figure 5.6: *Surviving paths merging into a common one at time $\ell - D$.*

a metric lower than the survivor of σ, which is a contradiction. In other words, no path discarded in favor of a survivor can provide a contribution to the total metric larger than the survivors.

The computational complexity of the VA grows with n only linearly. More specifically, at time i the VA requires $|\Sigma_i|$ storage locations, one for each state, with each location storing an accumulated metric and a surviving path. In terms of the number of computations, assuming for simplicity two branches per state, at time i the VA must make B_i additions, where B_i is the number of branches in the trellis section at time i, and $|\Sigma_i|$ comparisons.

It must be observed that in some cases the number of surviving paths that the decoder must store and search may be too large for practical applications. In this case the Viterbi algorithm may be abandoned in favor of suboptimal algorithms, which search only a fraction of trellis paths (*sequential* algorithms, M-algorithm) [5.2].

5.3.1   Sliding-window Viterbi algorithm

When the transmitted block of symbols is very long, it might be unrealistic to assume that the whole data sequence should be received before making a decision on it: in fact, this would entail a large memory and a long delay. Now, by tracing back all surviving paths at a given time ℓ, it often occurs (especially with large SNR) that they all stem from a single path, originating at time 0 and splitting at a time $\ell - D$ (Figure 5.6).   This observation leads to the concept of the *sliding-*

window Viterbi algorithm. This algorithm consists of forcing a decision at time ℓ on the symbol received at time $\ell - D - 1$; this decision is based on the comparison of the metrics accumulated across the sliding window of Figure 5.6. Thus, after an initial latency D, at every time a decision is made on a single symbol. The ensuing loss of optimality is reduced as D increases, because for large D there is a high probability that all paths surviving at ℓ have a common part extending before $\ell - D$.

5.4   The BCJR algorithm

A symbol-by-symbol maximum a posteriori probability (MAP) algorithm for codes described on a trellis is known as the *BCJR algorithm*, from the names of their proposers [5.3]. Having observed $\mathbf{y}$, we want to compute the soft decisions $p(x_i \mid \mathbf{y})$, whose maximization for $i = 1, \ldots, n$, yields decisions on symbols x_i that minimize symbol error probabilities. As noted in Section 3.8, since we are interested in maximizing $p(x_i \mid \mathbf{y})$ over x_i, we can omit constants that are the same for all values of x_i. Observing that symbol x_i is emitted as a transition takes place between states σ_{i-1} and σ_i, with possibly several transitions corresponding to the same symbol, we have

$$
\begin{aligned}
p(x_i \mid \mathbf{y}) &= p(\mathbf{y}, x_i)/p(\mathbf{y}) \\
&\propto p(\mathbf{y}, x_i) \\
&= \sum_{(\sigma_{i-1}, x_i, \sigma_i) \in \mathcal{T}_i} p(\mathbf{y}, \sigma_{i-1}, x_i, \sigma_i)
\end{aligned}
\tag{5.4}
$$

where the summation is extended to all pairs of states σ_{i-1}, σ_i joined by a branch labeled by symbol x_i in the trellis section $\mathcal{T}_i$ between times $i-1$ and i (Figure 5.7). Now, for any time i we write $\mathbf{y} = (\mathbf{y}_{<i}, y_i, \mathbf{y}_{>i})$, where $\mathbf{y}_{<i}$ and $\mathbf{y}_{>i}$ denote the components of vector $\mathbf{y}$ with indices $1, \ldots, i - 1$ and $i + 1, \ldots, n$, respectively. We call $\mathbf{y}_{<i}$ the *past* observations, $\mathbf{y}_{>i}$ the *future* observation, and y_i the *current* observation. We can write

$$
\begin{aligned}
&p(\mathbf{y}, \sigma_{i-1}, x_i, \sigma_i) \\
&= p(\mathbf{y}_{<i}, y_i, \mathbf{y}_{>i}, \sigma_{i-1}, x_i, \sigma_i) \\
&= p(\mathbf{y}_{<i}, y_i, \sigma_{i-1}, x_i, \sigma_i) p(\mathbf{y}_{>i} \mid \mathbf{y}_{<i}, y_i, \sigma_{i-1}, x_i, \sigma_i) \\
&= p(\mathbf{y}_{<i}, \sigma_{i-1}) p(y_i, x_i, \sigma_i \mid \mathbf{y}_{<i}, \sigma_{i-1}) p(\mathbf{y}_{>i} \mid \mathbf{y}_{<i}, y_i, \sigma_{i-1}, x_i, \sigma_i) \\
&= \underbrace{p(\mathbf{y}_{<i}, \sigma_{i-1})}_{\alpha} \underbrace{p(y_i, x_i, \sigma_i \mid \sigma_{i-1})}_{\gamma} \underbrace{p(\mathbf{y}_{>i} \mid \sigma_i)}_{\beta}
\end{aligned}
\tag{5.5}
$$

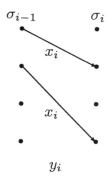

Figure 5.7: *The trellis section* $\mathcal{T}_i$.

The last equality comes from the trellis properties, which cause $\mathbf{y}_{>i}$ to depend on $(\mathbf{y}_{<i}, y_i, \sigma_{i-1}, x_i, \sigma_i)$ only through σ_i, and the pair (y_i, σ_i) to depend on $(\mathbf{y}_{<i}, \sigma_{i-1})$ only through σ_{i-1}. Observe now that, in (5.5), the term α depends on the past values of the observation, β on its future values, and γ on its current value. The BCJR algorithm works by recognizing that α and β can be given expressions that are recursive in time. To this purpose, define

$$\alpha_{i-1}(\sigma_{i-1}) \triangleq p(\mathbf{y}_{<i}, \sigma_{i-1}) \tag{5.6}$$

(this is the joint probability of observations $\mathbf{y}_{<i}$ and state σ_{i-1}) and

$$\beta_i(\sigma_i) \triangleq p(\mathbf{y}_{>i} \mid \sigma_i) \tag{5.7}$$

(the joint probability of observations $\mathbf{y}_{>i}$ given state σ_i). Moreover,

$$
\begin{aligned}
\gamma_{i-1,i}(\sigma_{i-1}, x_i, \sigma_i) &\triangleq p(y_i, x_i, \sigma_i \mid \sigma_{i-1}) \\
&= p(y_i \mid \sigma_{i-1}, x_i, \sigma_i) p(x_i, \sigma_i \mid \sigma_{i-1})
\end{aligned}
\tag{5.8}
$$

This is the *branch-transition* probability of transmitting x_i and observing y_i when a transition occurs between σ_{i-1} and σ_i. For stationary memoryless channels, it can be computed by observing that

$$p(y_i \mid \sigma_{i-1}, x_i, \sigma_i) = p(y_i \mid x_i)$$

and

$$p(x_i, \sigma_i \mid \sigma_{i-1}) = p(x_i) \left[(\sigma_{i-1}, x_i, \sigma_i) \in \mathcal{T}_i \right]$$

The function $\left[(\sigma_{i-1}, x_i, \sigma_i) \in \mathcal{T}_i \right]$ takes value 1 if the trellis section $\mathcal{T}_i$ is compatible with the transmission of symbol x_i when the transition $\sigma_{i-1} \to \sigma_i$ occurs, and value zero otherwise (we shall return on this notation in Chapter 8, when the BCJR algorithm will be rederived as a special case of a more general algorithm). In conclusion, we can write

$$\gamma_{i-1,i}(\sigma_{i-1}, x_i, \sigma_i) = p(y_i \mid x_i)p(x_i)\left[(\sigma_{i-1}, x_i, \sigma_i) \in \mathcal{T}_i\right] \tag{5.9}$$

and

$$p(\mathbf{y}, \sigma_{i-1}, x_i, \sigma_i) = \alpha_{i-1}(\sigma_{i-1})\gamma_{i-1,i}(\sigma_{i-1}, x_i, \sigma_i)\beta_i(\sigma_i) \tag{5.10}$$

so that, finally,

$$p(x_i \mid \mathbf{y}) \propto \sum_{\sigma_{i-1}}\sum_{\sigma_i} \alpha_{i-1}(\sigma_{i-1})\gamma_{i-1,i}(\sigma_{i-1}, x_i, \sigma_i)\beta_i(\sigma_i) \tag{5.11}$$

We now derive recursive formulas for the quantities defined in (5.6)–(5.8). We have the *forward recursion*

$$
\begin{aligned}
\alpha_i(\sigma_i) &= p(\mathbf{y}_{<i+1}, \sigma_i) \\
&= p(\mathbf{y}_{<i}, y_i, \sigma_i) \\
&= \sum_{\sigma_{i-1}}\sum_{x_i} p(\mathbf{y}_{<i}, y_i, \sigma_{i-1}, x_i, \sigma_i) \\
&= \sum_{\sigma_{i-1}}\sum_{x_i} p(\mathbf{y}_{<i}, \sigma_{i-1})p(y_i, x_i, \sigma_i \mid \sigma_{i-1}) \\
&= \sum_{\sigma_{i-1}}\sum_{x_i} \alpha_{i-1}(\sigma_{i-1})\,\gamma_{i-1,i}(\sigma_{i-1}, x_i, \sigma_i)
\end{aligned}
$$

with the initial condition $\alpha_0(\sigma_0) = 1$ (σ_0 the initial state of the trellis). Similarly, we have the *backward recursion*

$$
\begin{aligned}
\beta_{i-1}(\sigma_{i-1}) &= p(\mathbf{y}_{>i-1} \mid \sigma_{i-1}) \\
&= \sum_{\sigma_i}\sum_{x_i} p(y_i, \mathbf{y}_{>i}, x_i, \sigma_i \mid \sigma_{i-1}) \\
&= \sum_{\sigma_i}\sum_{x_i} p(y_i, x_i, \sigma_i \mid \sigma_{i-1})p(\mathbf{y}_{>i} \mid \sigma_i) \\
&= \sum_{\sigma_i}\sum_{x_i} \gamma_{i-1,i}(\sigma_{i-1}, x_i, \sigma_i)\,\beta_i(\sigma_i)
\end{aligned}
$$

with the initial value $\beta_n(\sigma_n) = 1$ (σ_n the ending state of the trellis). Combining the latter two recursions, we obtain the BCJR algorithm for the computation of a posteriori probabilities.

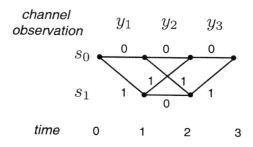

Figure 5.8: *Trellis for the* $(3, 2, 2)$ *single-parity-check code.*

Let us specialize the BCJR algorithm to binary systematic codes; in their trellis, the first k branches correspond to information symbols, and the last $n - k$ to parity-check symbols. Thus, two branches emanate from each node in the initial part of the trellis, and only one in the final part, where all transitions are determined by the information sequence $u_1, \ldots, u_k$. We may label the branches in the first k trellis sections by $u_1, \ldots, u_k$, and those in the remaining sections by $x_{k+1}, \ldots, x_n$. The branch transition probabilities $\gamma_{i-1,i}(\sigma_{i-1}, x_i, \sigma_i)$ can be computed as

$$\gamma_{i-1,i}(\sigma_{i-1}, x_i, \sigma_i) = \begin{cases} p(y_i \mid u_i)\, p(u_i)[(\sigma_{i-1}, u_i, \sigma_i) \in \mathcal{T}_i], & 1 \leq i \leq k \\ p(y_i \mid x_i)[(\sigma_{i-1}, x_i, \sigma_i) \in \mathcal{T}_i], & k + 1 \leq i \leq n \end{cases}$$

$$(5.12)$$

Example 5.3

As a simple illustrative example, we show how the BCJR algorithm can be used to compute the a posteriori probabilities $p(u_i \mid \mathbf{y})$ for the code and the channel observations of Example 3.17. The trellis representation of the code is shown in Figure 5.8. The observed data lead to the following values of the branch transitions:

$$\gamma_{0,1}(s_0, 0, s_0) = p(y_1 \mid 0) \cdot \frac{1}{2} = 0.49 \qquad \gamma_{0,1}(s_0, 1, s_1) = p(y_1 \mid 1) \cdot \frac{1}{2} = 0.10$$

$$\gamma_{1,2}(s_0, 0, s_0) = p(y_2 \mid 0) \cdot \frac{1}{2} = 0.495 \qquad \gamma_{1,2}(s_0, 1, s_1) = p(y_2 \mid 1) \cdot \frac{1}{2} = 0.09$$

$$\gamma_{1,2}(s_1, 1, s_0) = p(y_2 \mid 1) \cdot \frac{1}{2} = 0.09 \qquad \gamma_{1,2}(s_1, 0, s_1) = p(y_2 \mid 0) \cdot \frac{1}{2} = 0.495$$

$$\gamma_{2,3}(s_0, 0, s_0) = p(y_3 \mid 0) = 0.55 \qquad \gamma_{2,3}(s_1, 1, s_0) = p(y_3 \mid 1) = 0.67$$

(all other γs are zero, as they correspond to triples $(\sigma_{i-1}, x_i, \sigma_i)$ not consistent with the trellis structure). With the initialization $\alpha_0(s_0) = 1$, $\alpha_0(s_1) = 0$, the forward recursion yields the values

$$
\begin{aligned}
\alpha_1(s_0) &= \alpha_0(s_0)\gamma_{0,1}(s_0, 0, s_0) = 0.49 \\
\alpha_1(s_1) &= \alpha_0(s_0)\gamma_{0,1}(s_0, 1, s_1) = 0.10 \\
\alpha_2(s_0) &= \alpha_1(s_0)\gamma_{1,2}(s_0, 0, s_0) + \alpha_1(s_1)\gamma_{1,2}(s_1, 1, s_0) = 0.2516 \\
\alpha_2(s_1) &= \alpha_1(s_0)\gamma_{1,2}(s_0, 1, s_1) + \alpha_1(s_1)\gamma_{1,2}(s_1, 0, s_1) = 0.0936 \\
\alpha_3(s_0) &= \alpha_2(s_0)\gamma_{2,3}(s_0, 0, s_0) + \alpha_2(s_1)\gamma_{2,3}(s_1, 1, s_0) = 0.2011 \\
\alpha_3(s_1) &= \alpha_2(s_0)\gamma_{2,3}(s_0, 1, s_1) + \alpha_2(s_1)\gamma_{2,3}(s_1, 0, s_1) = 0
\end{aligned}
$$

With the initialization $\beta_3(s_0) = 1$, $\beta_3(s_1) = 0$, the backward recursion yields the values

$$
\begin{aligned}
\beta_2(s_0) &= \gamma_{2,3}(s_0, 0, s_0) = 0.55 \\
\beta_2(s_1) &= \gamma_{2,3}(s_1, 1, s_0) = 0.67 \\
\beta_1(s_0) &= \gamma_{1,2}(s_0, 0, s_0)\beta_2(s_0) + \gamma_{1,2}(s_0, 1, s_1)\beta_2(s_1) = 0.3326 \\
\beta_1(s_1) &= \gamma_{1,2}(s_1, 1, s_0)\beta_2(s_0) + \gamma_{1,2}(s_1, 0, s_1)\beta_2(s_1) = 0.3812 \\
\beta_0(s_0) &= \gamma_{0,1}(s_0, 0, s_0)\beta_1(s_0) + \gamma_{0,1}(s_0, 1, s_1)\beta_1(s_1) = 0.2011 \\
\beta_0(s_1) &= \gamma_{0,1}(s_1, 1, s_0)\beta_1(s_0) + \gamma_{0,1}(s_1, 0, s_1)\beta_1(s_1) = 0
\end{aligned}
$$

Application of (5.11) finally yields

$$
\begin{aligned}
p(u_1 = 0 \mid \mathbf{y}) &\propto \alpha_0(s_0)\gamma_{0,1}(s_0, 0, s_0)\beta_1(s_0) = 0.1629 \\
p(u_1 = 1 \mid \mathbf{y}) &\propto \alpha_0(s_0)\gamma_{0,1}(s_0, 1, s_1)\beta_1(s_1) = 0.0381 \\
p(u_2 = 0 \mid \mathbf{y}) &\propto \alpha_1(s_0)\gamma_{1,2}(s_0, 0, s_0)\beta_2(s_0) + \alpha_1(s_1)\gamma_{1,2}(s_1, 0, s_1)\beta_2(s_1) = 0.1666 \\
p(u_2 = 1 \mid \mathbf{y}) &\propto \alpha_1(s_0)\gamma_{1,2}(s_0, 1, s_1)\beta_2(s_1) + \alpha_1(s_1)\gamma_{1,2}(s_1, 1, s_0)\beta_2(s_0) = 0.0345
\end{aligned}
$$

Observe that the proportionality coefficients of the a posteriori probabilities here differ from those of Example 3.17. To reconcile the results obtained from the brute-force approach with those from the BCJR algorithm, one can verify that the ratios $p(u_i = 0 \mid \mathbf{y})/p(u_i = 1 \mid \mathbf{y})$ are equal. □

5.4.1　BCJR vs. Viterbi algorithm

Comparing the BCJR and Viterbi algorithms, we can see that both of them process the same channel observations. Their basic difference is in their outputs: while the VA decisions are hard, those of the BCJR algorithm are soft. This makes it crucial to use the BCJR algorithm if, after applying it, the a posteriori probabilities must be further processed before making hard decisions (see Chapter 9). Also, observe that the VA yields the most likely sequence (which must be a code word), while the BCJR algorithm yields the most likely symbol, along with its reliability, at each time (the BCJR sequence may not be a valid code word [5.4]).

5.5 Trellis complexity

In general, a given code can admit more than one trellis representation. If the trellis is to be used for decoding, it should be clear that among the various possible representations we should choose the one yielding the minimum complexity of the algorithm. We may define here a *state complexity profile* as the sequence of the numbers of trellis states $|\Sigma_i|$ at various times $i = 0, \ldots, n$. For example, the state complexity profile of both trellises of Figure 5.1 and of Figure 5.2 is $(1, 2, 2, \ldots, 2, 1)$. Another possible complexity measure (which measures more precisely the complexity of Viterbi Algorithm) is the *branch complexity profile*, defined as the sequence of the number of branches in the various trellis sections. This is $(2, 2, \ldots, 2)$ for the trellis of Figure 5.1 and $(2, 4, 4, \ldots, 4, 2)$ for the trellis of Figure 5.2. To characterize complexity by a single number we may use the largest entry in the state complexity profile, i.e., the maximum number of states

$$\max_{i=0,\ldots,n} |\Sigma_i|$$

We have the following definition:

Definition 5.5.1 *(**Minimal trellis**) A trellis $\mathcal{T}$ for a code $\mathcal{C}$ of length n is minimal if it satisfies the following property; for each i, the number of states in $\mathcal{T}$ at time i is less than or equal to the number of states at time i in any other trellis representing $\mathcal{C}$.*

Notice that this definition is rather strong. In fact, given a code $\mathcal{C}$, it is not obvious that there exists a minimal trellis for it: minimization of $|\Sigma_i|$ may be incompatible with the minimization of $|\Sigma_j|$, $j \neq i$. However, if $\mathcal{C}$ is linear, then a minimal trellis for $\mathcal{C}$ can be proved to exist [5.16]. The next section shows how to construct it.

5.6 Obtaining the minimal trellis for a linear code

Here we illustrate an algorithm, due to Forney [5.16], that yields the minimal trellis representing a given linear code.

We need some definitions and some additional theory first. Consider a code word $\mathbf{x} \in \mathcal{C}$, whose components are thought of as being generated sequentially in time. For any position $1 \le \ell \le n$ (corresponding to time, in the trellis) we refer to its first ℓ coordinates (already generated) as the *past*, and to the remaining $n - \ell$ coordinates (yet to be generated) as the *future*:

$$\big(\underbrace{x_1, \ x_2, \ \cdots , x_\ell,}_{\text{past}} \ \underbrace{x_{\ell+1}, \ \cdots , x_n}_{\text{future}}\big)$$

Let $\mathcal{C}$ be a binary linear (n, k, d) code. At time ℓ we can define the following four codes derived from $\mathcal{C}$:

① $\mathcal{C}_\ell^P$ (P for past) is the set of all code words $\mathbf{x} \in \mathcal{C}$ whose future is the null vector, i.e., $x_{\ell+1} = \cdots = x_n = 0$. $\mathcal{C}_\ell^P$ is a subcode of $\mathcal{C}$. We define its *dimension* k_ℓ^P as

$$k_\ell^P \triangleq \log_2 |\mathcal{C}_\ell^P| \qquad (5.13)$$

It is often convenient to look at this subcode as having length ℓ, which is obtained by deleting its zero components from $\ell + 1$ to n.

② $\mathcal{C}_\ell^F$ (F for future) is the set of all code words $\mathbf{x} \in \mathcal{C}$ whose past is the null vector, i.e., $x_1 = \cdots = x_\ell = 0$. $\mathcal{C}_\ell^F$ is a subcode of $\mathcal{C}$. We define its *dimension* k_ℓ^F as

$$k_\ell^F \triangleq \log_2 |\mathcal{C}_\ell^F| \qquad (5.14)$$

It is often convenient to look at this code as having length $n - \ell$, which is obtained by deleting its first ℓ zero components.

③ The "past projection code" $\mathcal{P}_\ell^P$ is the set of all codewords obtained by zeroing (or deleting) all components from $\ell + 1$ to n.

④ The "future projection code" $\mathcal{P}_\ell^F$ is the set of all codewords obtained by zeroing (or deleting) its first ℓ components.

Example 5.4

Consider the $(4, 3, 2)$ single-parity-check code, and $\ell = 2$. We have

$$\begin{aligned} \mathcal{C}_2^P &= (00, 11) \\ \mathcal{C}_2^F &= (00, 11) \\ \mathcal{P}_2^P &= (00, 01, 10, 11) \\ \mathcal{P}_2^F &= (00, 01, 10, 11) \end{aligned}$$

so that $\mathcal{C}_2^P$ and $\mathcal{C}_2^F$ are $(2,1,2)$ repetition codes, while $\mathcal{P}_2^P$, $\mathcal{P}_2^F$ are $(2,2,1)$ universe codes. $\square$

Now, the past coordinates influence the values that the future coordinates can take. This influence is characterized by the trellis state at time ℓ. With the definitions above, observe that the generator matrix of $\mathcal{C}$ can be decomposed as follows:

$$G = \begin{bmatrix} \mathbf{G}_\ell^P & \mathbf{0} \\ \mathbf{0} & \mathbf{G}_\ell^F \\ \mathbf{G}_\ell' & \mathbf{G}_\ell'' \end{bmatrix}$$

where $\mathbf{G}_\ell^P$ is the $k_\ell^P \times \ell$ generator matrix of $\mathcal{C}_\ell^P$, $\mathbf{G}_\ell^F$ is the $k_\ell^F \times (n - \ell)$ generator matrix of $\mathcal{C}_\ell^F$, and $\mathbf{G}_\ell'$, $\mathbf{G}_\ell''$ are additional matrices needed to obtain $\mathbf{G}$. Specifically, $\mathbf{G}_\ell^P$ and $\mathbf{G}_\ell'$ together generate $\mathcal{P}_\ell^P$, while $\mathbf{G}_\ell^F$ and $\mathbf{G}_\ell''$ together generate $\mathcal{P}_\ell^F$. The linear binary code generated by $[\mathbf{G}_\ell' \; \mathbf{G}_\ell'']$ is called the *state code* and is denoted by $\mathcal{S}_\ell$. Its dimension k_ℓ, also called the dimension of the *state space*, satisfies

$$k_\ell = k - k_\ell^P - k_\ell^F \tag{5.15}$$

so, from (5.13) and (5.14), $k_\ell = \log_2 |\mathcal{S}_\ell|$.

Example 5.4 (continued)

We have $\mathbf{G}_2^P = \mathbf{G}_2^F = [1\ 1]$. It can be verified that the matrix

$$G = \begin{bmatrix} 1 & 1 & 0 & 0 \\ 0 & 0 & 1 & 1 \\ 0 & 1 & 0 & 1 \end{bmatrix}$$

actually generates $\mathcal{C}$. The state code $\mathcal{S}_2$ is $(0000, 0101)$, and has dimension 1. $\square$

The above can be interpreted by saying that any code word $\mathbf{x} \in \mathcal{C}$ (which is a linear combination of rows of $\mathbf{G}$) can be expressed uniquely at any time ℓ as the sum of a *past code word* $\mathbf{x}^P \in \mathcal{C}_\ell^P$, a *future code word* $\mathbf{x}^F \in \mathcal{C}_\ell^F$, and a *state code word* $\mathbf{s} \in \mathcal{P}_\ell$. We write

$$\mathbf{x} = \mathbf{x}^P + \mathbf{s} + \mathbf{x}^F \tag{5.16}$$

for some $\mathbf{x}^P \in \mathcal{C}_\ell^P$ and $\mathbf{x}^F \in \mathcal{C}_\ell^F$, and we represent this decomposition in graphical form as shown in Figure 5.9. This corresponds to associating $\mathbf{x}$ with a trajectory in a trellis: for any given time ℓ, $\mathbf{x}^P$ is associated with the past trajectory from time

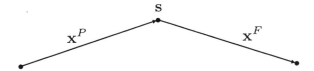

Figure 5.9: *Graphical representation of the decomposition of the code word* **x** *as the sum of the past code word* $\mathbf{x}^P$, *the future code word* $\mathbf{x}^F$, *and the state code word* **s**.

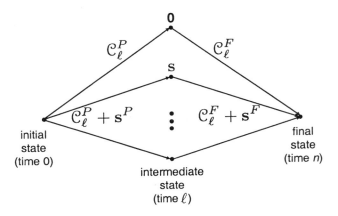

Figure 5.10: *Past and future of a code trellis at time* ℓ.

0 to time ℓ, $\mathbf{x}^F$ with the future trajectory from time ℓ to time n, and **s** with the state where the two trajectories meet at time ℓ.

Consider now the past projection of **x** in (5.16). This is given by the sum of the past projections of $\mathbf{x}^P$ and of **s**. Hence, the trellis branch joining the initial state to state **s** at time ℓ corresponds to the coset $\mathcal{C}_\ell^P + \mathbf{s}^P$, where $\mathbf{s}^P$ denotes the past projection of **s**. In a similar way, the future projection of **x** is given by the sum of the future projections of $\mathbf{x}^F$ and of **s**. Hence, the trellis branch joining state **s** at time ℓ with the final state at time n corresponds to the coset $\mathcal{C}_\ell^F + \mathbf{s}^F$, where $\mathbf{s}^F$ denotes the future projection of **s**. Thus, the code $\mathcal{C}$ may be described by a trellis diagram as shown in Figure 5.10. No other trellis for the same code can have less states at time ℓ. To prove the description shown in Figure 5.10 we must prove two things that form the so-called *Markov property* of the trellis:

   ① All code words **x** associated with the same state-code word **s** have the same set of possible future trajectories.

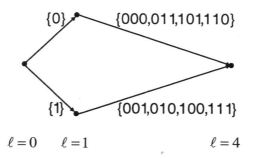

$$\ell = 0 \qquad \ell = 1 \qquad\qquad\qquad \ell = 4$$

Figure 5.11: *Code trellis at time 1.*

② If x, x' are associated with different state-code words s, s', then they have disjoint sets of possible future trajectories.

The first item is proved by observing that for all such x the possible future trajectories are given by the coset $\mathcal{C}_\ell^F + s^F$. In the conditions of the second item, the possible future trajectories are given by the cosets $\mathcal{C}_\ell^F + s^F$ and $\mathcal{C}_\ell^F + s'^F$, which are disjoint because the sum $s^F + s'^F$ is not a code word in $\mathcal{C}_\ell^F$. The Markov property can be suggestively summarized by saying that "the states of the trellis are equivalence classes of past histories modulo future possibilities."

If the decomposition illustrated in Figure 5.10 is repeated for every ℓ, a full trellis for code $\mathcal{C}$ is obtained. This will be illustrated by the examples that follow.

Example 5.5

Here we construct the minimal trellis for the $(4, 3, 2)$ single-parity-check code. For $\ell = 1$ we have $\mathcal{C}_1^P = \{0\}$, so $k_1^P = 0$, and $\mathcal{C}_1^F = \{000, 011, 101, 110\}$, so $k_1^F = 2$. Thus, $k_1 = 3 - 2 - 0 = 1$ and hence $|\Sigma_1| = 2$. The code trellis at time 1 is shown in Figure 5.11. Observe in particular that at time 1, i.e., after the first code symbol, the number of states is 2 because there are two possible pasts, each with a different future.

At time $\ell = 2$ we have $\mathcal{C}_2^P = \{00, 11\}$, so $k_2^P = 1$, and $\mathcal{C}_2^F = \{00, 11\}$, so $k_2^F = 1$. Thus, $k_2 = 3 - 1 - 1 = 1$ and hence $|\Sigma_2| = 2$. The code trellis at time 2 is shown in Figure 5.12.

At time 3 we have $\mathcal{C}_3^P = \{000, 011, 101, 110\}$, so $k_3^P = 2$, and $\mathcal{C}_3^F = \{0\}$, so $k_3^F = 0$. Thus, $k_3 = 3 - 2 - 0 = 1$ and hence $|\Sigma_3| = 2$. The code trellis at time 3 is shown in Figure 5.13.

We can now summarize our calculations by constructing the code trellis as shown in Figure 5.14. □

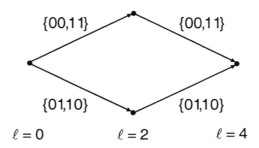

Figure 5.12: *Code trellis at time 2.*

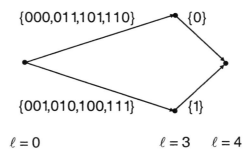

Figure 5.13: *Code trellis at time 3.*

5.7   Permutation and sectionalization

The structure and the complexity of the trellis of a code depend on the order of the code symbols. It turns out that the (seemingly innocuous) operation of permuting the symbols in each code word can drastically change the number of states in the minimal trellis representation of a given code $\mathcal{C}$. Unfortunately, finding a permutation that minimizes the complexity of a trellis representing a linear code is an intractable problem, since essentially the only way to solve it is to try all the permutations.

Sectionalization, which consists of grouping together two or more symbols labeling each branch, can also drastically change the structure and the complexity of the trellis representing a given code. Sectionalization shrinks the time axis at the expense of increasing $|\mathcal{X}|$: for example, a binary code with length $2n$ may be

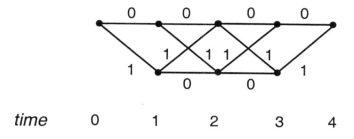

Figure 5.14: *Minimal trellis for the* $(4, 3, 2)$ *code.*

thought of as a quaternary code of length n if pairs of consecutive bits are grouped together.

Example 5.6

Here we construct a sectionalized trellis, based on a quaternary constellation, for the $(8, 4, 4)$ *Reed–Muller* code. The generator matrix of this code is

$$
\mathbf{G} = \begin{bmatrix}
1 & 1 & 1 & 1 & 0 & 0 & 0 & 0 \\
1 & 1 & 1 & 1 & 1 & 1 & 1 & 1 \\
1 & 0 & 1 & 0 & 1 & 0 & 1 & 0 \\
1 & 0 & 0 & 1 & 1 & 0 & 0 & 1
\end{bmatrix}
$$

The words of $\mathcal{C}$, i.e., the linear combinations of the rows of $\mathbf{G}$, are the sixteen 8-tuples that follow:

0	0	0	0	0	0	0	0
1	0	0	1	1	0	0	1
1	0	1	0	1	0	1	0
0	0	1	1	0	0	1	1
1	1	1	1	1	1	1	1
0	1	1	0	0	1	1	0
0	1	0	1	0	1	0	1
1	1	0	0	1	1	0	0
1	1	1	1	0	0	0	0
0	1	1	0	1	0	0	1
0	1	0	1	1	0	1	0
1	1	0	0	0	0	1	1
0	0	0	0	1	1	1	1
1	0	0	1	0	1	1	0
1	0	1	0	0	1	0	1
0	0	1	1	1	1	0	0

$\mathcal{C}_1^F$ (past: 00)	$\mathcal{C}_1^F + (010101)$ (past: 01)	$\mathcal{C}_1^F + (101010)$ (past: 10)	$\mathcal{C}_1^F + (111111)$ (past: 11)
00 0000	01 0101	10 1010	11 1111
00 1111	01 1010	10 0101	11 0000
11 1100	10 1001	01 0110	00 0011
11 0011	10 0110	01 1001	00 1100

Table 5.1: *The cosets of $\mathcal{C}_1^F$.*

Contrary to what we have done previously, we construct a trellis for this code by associating *two* binary symbols with each trellis branch. Consider first $\ell = 1$. There is only one word with zero components from time 2 to time 4, namely, the all-zero code word. Thus, we have

$$\mathcal{C}_1^P = \{(00)\}$$

There are four words beginning with two zeros, so

$$\mathcal{C}_1^F = \{(000000), (001111), (111100), (110011)\}$$

Thus $k_1^P = 0$ and $k_1^F = 2$, and from (5.15) we have

$$k_1 = 4 - 0 - 2 = 2$$

and hence four states at time 1. The cosets of $\mathcal{C}_1^P$ are 01, 10, and 11. The cosets of $\mathcal{C}_1^F$ are listed in Table 5.1.

The code trellis at this stage is shown in Figure 5.15. The trellis branches are labeled by the coset representatives: 00, 01, 10, and 11 for the past and 010101, 101010, and 111111 for the future.

The top node at time $\ell = 1$ corresponds to code words whose first two symbols (the representative of the subcode $\mathcal{C}_1^P$) are 00 and whose future (the representative of the subcode $\mathcal{C}_1^F$) is $\mathcal{C}_1^F$. The other states have as their past the cosets of the subcode $\mathcal{C}_1^P$ and as their future the cosets of the subcode $\mathcal{C}_1^F$. We have four states here because the pasts consist of all pairs of binary symbols.

Take now $\ell = 2$. There are two code words in $\mathcal{C}$ that begin with four zeros, i.e., 00000000 and 00001111. Thus we have

$$\mathcal{C}_2^P = \{(0000), (1111)\}.$$

There are also two words that end with four zeros, namely, 00000000 and 11110000. Thus

$$\mathcal{C}_2^F = \{(0000), (1111)\}$$

Since $k_2^P = 1$ and $k_2^F = 1$, from (5.15) we have

$$k_2 = 4 - 1 - 1 = 2.$$

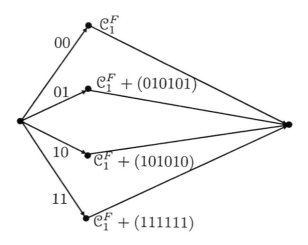

Figure 5.15: *Code trellis at time 1.*

Here the past consists of four binary symbols, but the code structure restricts them to four possible combinations. The cosets of $\mathcal{C}_2^P$ and $\mathcal{C}_2^F$ are listed in Tables 5.2 and 5.3. The trellis at time 4 is represented in Figure 5.16.

$\mathcal{C}_2^P$	$\mathcal{C}_2^P + (1100)$	$\mathcal{C}_2^P + (1010)$	$\mathcal{C}_2^P + (1001)$
0000	1100	1010	1001
1111	0011	0101	0110

Table 5.2: *The cosets of $\mathcal{C}_2^P$.*

$\mathcal{C}_2^F$	$\mathcal{C}_2^F + (1100)$	$\mathcal{C}_2^F + (1010)$	$\mathcal{C}_2^F + (1001)$
0000	1100	1010	1001
1111	0011	0101	0110

Table 5.3: *The cosets of $\mathcal{C}_2^F$.*

Take then $\ell = 3$. There are four code words in $\mathcal{C}$ ending with two zeros, and we have

$$\mathcal{C}_3^P = \{(000000), (110011), (111100), (001111)\}.$$

There is a single code word beginning with six zeros (the all-zero word), so

$$\mathcal{C}_3^F = \{(00)\}$$

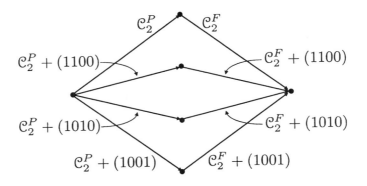

Figure 5.16: *Code trellis at time* 2.

Since $k_3^P = 2$ and $k_3^F = 0$, from (5.15) we have

$$k_3 = 4 - 2 - 0 = 2.$$

The cosets of $\mathcal{C}_3^F$ are 00, 01, 10, and 11. The cosets of $\mathcal{C}_3^P$ are listed in Table 5.4. The trellis at time 3 is represented in Figure 5.17.

$\mathcal{C}_3^P$	$\mathcal{C}_3^P + (010101)$	$\mathcal{C}_3^P + (101010)$	$\mathcal{C}_3^P + (111111)$
0000 00	0101 01	1010 10	1111 11
1100 11	1001 10	0110 01	0011 00
1111 00	1010 01	0101 10	0000 11
0011 11	0110 10	1001 01	1100 00

Table 5.4: *The cosets of* $\mathcal{C}_3^P$.

Thus, there are four states at $\ell = 2, 4$ and 6. Now, let σ_1 be a state at time 1, σ_2 a state at time 2, and σ_3 a state at time 3. If there is a code word in $\mathcal{C}$ that passes through σ_1, σ_2, and σ_3, then we join states σ_1, σ_1, and σ_3 in the complete code trellis. The result is shown in Figure 5.18.   □

5.8   Constructing a code on a trellis: The $|u|u + v|$ construction

We show here an example of a code that can be constructed directly on a trellis. Let $\mathcal{U}$ be an $(n, k_{\mathcal{U}}, d_{\mathcal{U}})$ linear code, and let $\mathcal{V}$ be a linear $(n, k_{\mathcal{V}}, d_{\mathcal{V}})$ sub-

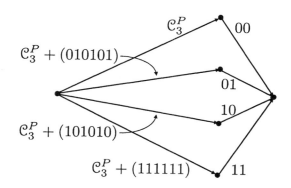

Figure 5.17: *Code trellis at time 3.*

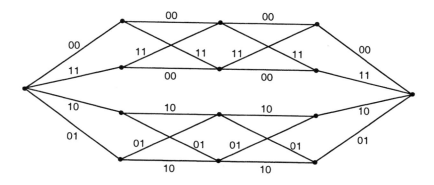

Figure 5.18: *Trellis of the (8,4,4) Reed–Muller code.*

code of $\mathcal{U}$ with minimum distance $d_V \geq d_U$. If $\mathbf{a} = (a_1, \cdots, a_m)$ and $\mathbf{b} = (b_1, \cdots, b_n)$, we denote here by $(\mathbf{a}, \mathbf{b})$ their concatenation, i.e., the $(m+n)$-vector $(a_1, \cdots, a_m, b_1, \cdots, b_n)$. The $|u|u + v|$ construction combines $\mathcal{U}$ and $\mathcal{V}$ to yield a code whose words are obtained by concatenating a word $\mathbf{u} \in \mathcal{U}$ with the sum $\mathbf{u} + \mathbf{v}$, where $\mathbf{v} \in \mathcal{V}$:

$$|u|u + v| = \{(\mathbf{u}, \mathbf{u} + \mathbf{v}) \mid \mathbf{u} \in \mathcal{U}, \mathbf{v} \in \mathcal{V}\}.$$

Theorem 5.8.1 $|u|u + v|$ *is a linear (n, k, d) code, with*

$$k = k_{\mathcal{U}} + k_{\mathcal{V}}$$

Its minimum Hamming distance is bounded below by

$$d \geq \min\{2d_{\mathcal{U}}, d_{\mathcal{V}}\} \tag{5.17}$$

Proof. For the first part of the theorem, see [5.12, p. 76]. For the second part, let $(\mathbf{u}, \mathbf{u} + \mathbf{v})$ be a non-zero code word in $|u|u+v|$. If $\mathbf{u} \in \mathcal{V}$, then both $\mathbf{u}$ and $\mathbf{u} + \mathbf{v}$ are words of $\mathcal{V}$, and either $\mathbf{u} \neq \mathbf{0}$ (in which case $w(\mathbf{u}) \geq d_\mathcal{V}$) or $\mathbf{u} + \mathbf{v} \neq \mathbf{0}$ (in which case $w(\mathbf{u} + \mathbf{v}) \geq d_\mathcal{V}$). Thus $w(\mathbf{u}, \mathbf{u} + \mathbf{v}) \geq d_\mathcal{V}$. If $\mathbf{u} \notin \mathcal{V}$ (and consequently $\mathbf{u} \neq \mathbf{0}$, because a linear code must contain the all-zero word), then $\mathbf{u}$ and $\mathbf{u} + \mathbf{v}$ are both nonzero words in $\mathcal{U}$, and $w(\mathbf{u}, \mathbf{u} + \mathbf{v}) \geq 2d_\mathcal{U}$.

Example 5.7

We are especially interested in codes $\mathcal{U}$, $\mathcal{V}$ for which $d_\mathcal{V} = 2d_\mathcal{U}$. For example, consider the $(4, 3, 2)$ single-parity-check code

$$\mathcal{U} = \begin{matrix} 0 & 0 & 0 & 0 \\ 1 & 0 & 0 & 1 \\ 1 & 0 & 1 & 0 \\ 0 & 0 & 1 & 1 \\ 1 & 1 & 1 & 1 \\ 0 & 1 & 1 & 0 \\ 0 & 1 & 0 & 1 \\ 1 & 1 & 0 & 0 \end{matrix}$$

and its $(4, 1, 4)$ "repetition" subcode $\mathcal{V} = \{(0000), (1111)\}$. The $|u|u+v|$ construction gives the $(8, 4, 4)$ code examined above, with words $(\mathbf{u}, \mathbf{u})$ and $(\mathbf{u}, \mathbf{u}+(1111))$. If we apply again this construction with $\mathcal{U}$ this $(8, 4, 4)$ code and $\mathcal{V}$ the $(8, 1, 8)$ repetition code, we obtain the $(16, 5, 8)$ Reed–Muller code. In [5.12, Chap. 13] it is shown that all Reed–Muller codes can be built in this way. □

In the $|u|u + v|$ construction, $\mathcal{U}$ is the union of M cosets of $\mathcal{V}$ in $\mathcal{U}$:

$$\mathcal{U} = \mathcal{V} \cup (\mathcal{V} + \mathbf{u}_1) \cup \cdots \cup (\mathcal{V} + \mathbf{u}_{M-1})$$

The trellis of $|u|u + v|$ consists of two sections joined at M intermediate states, as shown in Figure 5.19. The branches in each section correspond to cosets of $\mathcal{V}$ in $\mathcal{U}$, and branches in the past and in the future section labeled by the same coset are joined at a common intermediate state.

The distance properties of the construction can be derived directly from the trellis. Two code words that determine the same path through the trellis must differ in at least $d(\mathcal{V})$ positions, and two code words that determine two different paths must differ in at least $d(\mathcal{U})$ positions in each of the two branches.

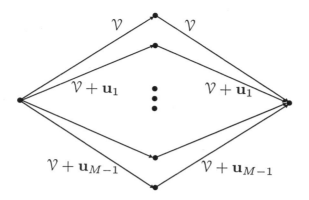

Figure 5.19: *Trellis of the* $|u|u + v|$ *construction.*

Figure 5.20: *A dog biting its tail, a Celtic symbol for renewal and immortality.*

5.9 Tail-biting code trellises

The "conventional" trellises used so far were defined on an ordered time axis $(0, 1, \ldots, n)$. A *tail-biting trellis* (Figure 5.20) is defined on a circular time axis $(0, 1, \ldots, n-1)$, corresponding to index arithmetics performed modulo n. Graphically, this can be represented as a trellis wrapped around a cylinder, with the states at time n coinciding with those at time 0. The valid paths in the trellis are those starting and ending at the same state. A conventional trellis may be regarded as a special case of a tail-biting trellis with a single starting and ending state. Conversely, a tail-biting trellis that has a single state at some time can be viewed as a conventional trellis.

Figure 5.21 shows conventional and tail-biting trellises of the quaternary repetition code $\{00, 11, 22, 33\}$. It can be seen that one of the versions of the tail-biting

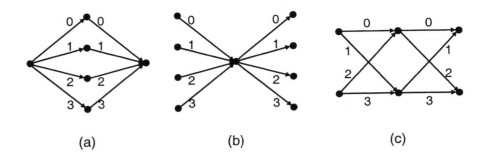

Figure 5.21: *Trellises of the quaternary repetition code. (a) Conventional trellis.*
(b) Tail-biting trellis. (c) Another tail-biting trellis with only two states.

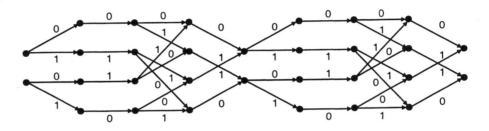

Figure 5.22: *Minimal tail-biting trellis of the $(8, 4, 4)$ Reed–Muller code.*

trellis has fewer states than the conventional one. Another example is shown in Figure 5.22. The interest for tail-biting trellises comes from the fact that in some cases the trellis complexity is lower than that of the conventional trellis (Figure 5.21 provides a simple example). Actually, the number of states can be as low as the square root of the number of states in a conventional trellis. In addition, tail-biting trellises can be viewed as the simplest form of a factor graph with cycles (see Chapter 8).

An exact maximum-likelihood algorithm for decoding a code on a tail-biting trellis has been derived [5.15]. It consists of running the VA under the assumption that the code starts and ends in each starting state in turn. The complexity is $|\Sigma_0|$ times that of a single VA, where $|\Sigma_0|$ denotes the number of possible starting states. Thus, this algorithm shows no complexity advantage over decoding a conventional trellis for the same code. An efficient iterative approximate decoding method on a tail-biting trellis is as follows. Initialize all the metrics at time 0 to zero, use the Viterbi algorithm going around and around in the trellis, and stop the iterations according to a preset stopping rule. If the preferred path starts and ends in the same state, it is chosen as the decoded code word; otherwise, an error is detected.

5.10 Bibliographical notes

The Viterbi algorithm was introduced in [5.17] as a method for decoding convolutional codes (Chapter 6). A survey of applications of the Viterbi algorithm, as well as a number of details on its implementation, can be found in [5.6]. Sequential algorithms and the M-algorithm are described in [5.18, Chap. 6] and [5.2].

The BCJR algorithm was proposed in [5.3]. It received limited attention because of its complexity until recently, where it was applied to iterative decoding techniques (Chapter 8).

The theory of trellis representation of linear block codes, after attracting some interest in the late 1970s [5.15, 5.20], has been rekindled in recent years, when it was recognized that, in addition to illuminating the code structure, it may lead the way to efficient decoding algorithms [5.5, 5.9–5.11, 5.13, 5.14, 5.16]. Tail-biting trellises were also introduced in the late 1970s [5.15]. Interest in them was recently enhanced after the discovery that the number of states in a tail-biting trellis can be as low as the square root of the number of states in the conventional trellis for the same code [5.5, 5.8, 5.19]. Iterative decoding of tail-biting trellises is now better understood [5.1, 5.7] than iterative decoding for more general graphs, that we shall discuss in Chapter 8.

5.11 Problems

1. Consider the linear binary code with parity-check matrix

$$\mathbf{H} = \begin{bmatrix} 0 & 1 & 1 & 1 \\ 1 & 0 & 1 & 1 \end{bmatrix}$$

   (a) Find a four-state trellis describing the code.

   (b) Decode the received vector $\mathbf{y} = (0.7, 1.2, -0.3, -0.5)$ by using the Viterbi algorithm (assume an AWGN channel and an elementary constellation $\mathcal{X} = \{\pm 1\}$ with the correspondence $0 \to +1, 1 \to -1$).

   (c) Decode as above by using the BCJR algorithm.

2. Derive the minimal trellis of the linear code whose parity-check matrix is

$$\mathbf{H} = \begin{bmatrix} 1 & 1 & 1 & 0 & 0 & 0 & 0 \\ 1 & 0 & 0 & 1 & 1 & 0 & 0 \\ 1 & 0 & 0 & 0 & 0 & 1 & 1 \end{bmatrix}$$

   Compare the resulting complexity with that of the trellis obtained by using the construction based on the parity-check matrix (Section 5.2).

3. Derive a version of the BCJR algorithm obtained by factoring the joint pdf $p(\mathbf{y}, \sigma_{i-1}, \sigma_i)$ in the reverse direction (this is equivalent to defining a reverse time axis).

References

[5.1] S. Aji, G. Horn, R. J. McEliece, and M. Xu, "Iterative min-sum decoding of tail-biting codes," in *Proc. IEEE Workshop on Information Theory*, Killarney, Ireland, pp. 68–69, June 1998.

[5.2] J. B. Anderson and S. Mohan, *Source and Channel Coding: An Algorithmic Approach*. Boston, MA: Kluwer Academic, 1991.

[5.3] L. R. Bahl, J. Cocke, F. Jelinek, and J. Raviv, "Optimal decoding of linear codes for minimizing symbol error rate," *IEEE Trans. Inform. Theory*, Vol. IT-20, No. 2, pp. 284–287, March 1974.

[5.4] G. Battail, M. C. Decouvelaere, and Ph. Godlewski, "Replication decoding," *IEEE Trans. Inform. Theory*, Vol. 25, No. 3, pp. 332–345, May 1979.

[5.5] A. R. Calderbank, G. D. Forney, Jr., and A. Vardy, "Minimal tail-biting trellises: The Golay code and more," *IEEE Trans. Inform. Theory*, Vol. 45, No. 5, pp. 1435–1455, July 1999.

[5.6] G. D. Forney, Jr., "The Viterbi algorithm," *IEEE Proceedings*, Vol. 61, No. 3, pp. 268–278, March 1973.

[5.7] G. D. Forney, Jr., F. R. Kschischang, B. Marcus, and S. Tuncel, "Iterative decoding of tail-biting trellises and connections with symbolic dynamics," in B. Marcus and J. Rosenthal, Eds., *Codes, Systems, and Graphical Models*. IMA Volumes in Mathematics and Its Applications, New York: Springer-Verlag, March 2001.

[5.8] R. Koetter and A. Vardy, "The structure of tail-biting trellises: Minimality and basic principles," *IEEE Trans. Inform. Theory*, Vol. 49, No. 9, pp. 2081–2105, September 2003.

[5.9] F. R. Kschischang, "The trellis structure of maximal fixed-cost codes," *IEEE Trans. Inform. Theory*, Vol. 42, No. 6, pp. 1828–1838, November 1996.

[5.10] F. R. Kschischang and V. Sorokine, "On the trellis structure of block codes," *IEEE Trans. Inform. Theory*, Vol. 41, No. 6, pp. 1924–1937, November 1995.

[5.11] A. Lafourcade and A. Vardy, "Lower bounds on trellis complexity of block codes," *IEEE Trans. Inform. Theory*, Vol. 41, No. 6, pp. 1938–1954, November 1995.

[5.12] F. J. MacWilliams and N. J. A. Sloane, *The Theory of Error-Correcting Codes*. Amsterdam: North-Holland, 1977.

[5.13] R. J. McEliece, "On the BCJR trellis for linear block codes," *IEEE Trans. Inform. Theory*, Vol. 42, No. 4, pp. 1072–1092, July 1996.

[5.14] D. J. Muder, "Minimal trellises for block codes," *IEEE Trans. Inform. Theory*, Vol. 34, No. 5, pp. 1049–1053, September 1988.

[5.15] G. Solomon and H. C. A. Van Tilborg, "A connection between block and convolutional codes," *SIAM J. Appl. Math.*, Vol. 37, pp. 358–369, October 1979.

[5.16] A. Vardy, "Trellis structure of codes," in V. S. Pless and W. C. Huffman, Eds., *Handbook of Coding Theory*. Amsterdam, Elsevier Science, pp. 1989–2118, 1998.

[5.17] A. J. Viterbi, "Error bounds for convolutional codes and an asymptotically optimum decoding algorithm," *IEEE Trans. Inform. Theory*, Vol. 13, No. 2, pp. 260–269, April 1967.

[5.18] A. J. Viterbi and J. K. Omura, *Principles of Digital Communication and Coding*. New York: McGraw-Hill, 1979.

[5.19] N. Wiberg, H.-A. Loeliger, and R. Kötter, "Codes and iterative decoding on general graphs," *Euro. Trans. Telecommun.*, Vol. 6, pp. 513–525, September/October 1995.

[5.20] J. K. Wolf, "Efficient maximum-likelihood decoding of linear block codes using a trellis," *IEEE Trans. Inform. Theory*, Vol. IT-24, No. 1, pp. 76–80, January 1978.

6

With his rent in his rears. /Give him six years

Coding on a trellis: Convolutional codes

In the previous chapter we have seen how a given block code can be represented by using a trellis. We now examine the problem of designing a binary code directly on a trellis. This can be done by first choosing a trellis with a preassigned complexity, then labeling its branches. The trellis is generated by using one or more binary shift registers. The choice of a periodic trellis, which simplifies the Viterbi algorithm, and of symbols generated as linear combinations of the contents of the shift registers, leads to the definition of convolutional codes. Invented in 1954, these codes have been very successful because they can be decoded in a simple way, have a good performance, and are well adapted to the transmission of continuous streams of data. In this chapter, we present the rudiments of an algebraic theory of convolutional codes, and show how code performance can be evaluated.

6.1   Introduction

In the previous chapter we have seen how a given code can be represented by a trellis. We now examine the problem of designing a code directly on a trellis. Specifically, we first choose a trellis (the number of its states being constrained by the decoding complexity we are willing to accept) and then we label the trellis branches so as to design a code that is optimum under a specified criterion. To simplify decoding, here the trellis will be assumed to have a *periodic* structure (apart from the initial and terminal transients caused by the finite length of the source sequence), so that the Viterbi-algorithm operations will be the same for every state-transition interval.

In this framework, we examine two design criteria. The first is based on binary symbols and linear encoders and leads to *convolutional codes*, which will be treated in the present chapter. In the next chapter, a more general set of elementary signals and encoders will be used, which gives rise to *trellis-coded modulation*.

A simple way of constructing a periodic trellis with a given number of states is by using a memory-ν binary shift register, as shown in Figure 6.1. This contains $\nu+1$ cells, or *stages*, and each binary symbol entering the register shifts its contents to the right by one place. Positions 1 to ν determine the *state* of the register, so there are 2^ν states. Position 0 contains the source symbol that is emitted at a given time: this forces the transition from one state to another. A trellis describes graphically the transitions among states. Notice that the resulting trellis structure is determined by the number of cells alone.

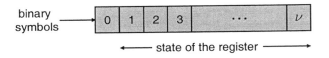

Figure 6.1: *A binary shift register with 2^ν states.*

Example 6.1

Take $\nu = 2$. We have four possible states, labeled by the contents of cells 1 and 2: these are 00, 01, 10, and 11. From state yz ($y \in \{0,1\}$ and $z \in \{0,1\}$) we can only move to state xy, where x denotes the symbol emitted by the source and

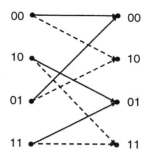

Figure 6.2: *A section of the trellis generated by a shift-register sequence with four states.*

forcing the transition. The resulting trellis section is shown in Figure 6.2. Here a dashed arrow denotes a transition driven by the binary source symbol "1," while a continuous arrow corresponds to a "0." □

6.2 Convolutional codes: A first look

A convolutional encoder (in a simplified definition that we shall generalize and make rigorous later on) combines linearly (with respect to modulo-2 operations) the contents of a binary $(\nu + 1)$-stage shift-register (nonbinary convolutional encoders can also be defined, although we shall not consider them here). We say that the code has *constraint length* $\nu + 1$, or *memory* ν.[1] If for every binary source symbol these linear combinations generate n_0 binary channel symbols, the resulting code has rate $1/n_0$. Figure 6.3 shows two constraint-length-3 convolutional encoders, one with rate $1/2$ and one with rate $1/3$.

The trellis diagram of the code has branches labeled by symbols generated by the linear combinations of the shift-register contents (n_0 binary symbols per branch). Moreover, it is customary to represent by a dashed line the branches corresponding to transitions between states forced by a source symbol "1" entering the shift register, and by a continuous line those forced by a "0". Figure 6.4 shows the initial sections of the trellis diagram, complete with its labels, corresponding to the rate-1/3 convolutional code of Figure 6.3.

Another representation of a convolutional encoder, which provides a useful tool for performance evaluation, is its *state diagram*, which describes the transitions

---

[1]Note that the definitions of constraint length and of memory are not consistent throughout the convolutional-code literature.

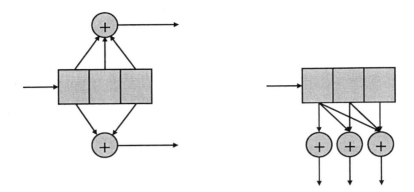

Figure 6.3: *Two convolutional encoders with constraint length* 3, *having rates* 1/2 *(left) and* 1/3 *(right).*

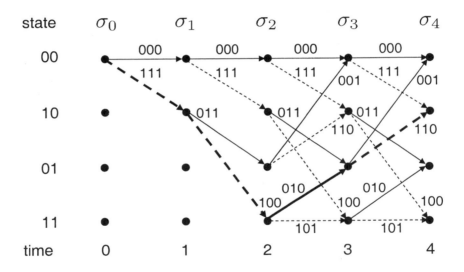

Figure 6.4: *Trellis diagram for the rate-1/3 convolutional code of Figure 6.3. The boldface path corresponds to the input sequence 1101. The initial state is chosen conventionally as the all-zero state.*

among states without explicitly including the time axis. The state diagram corresponding to the rate-1/3 convolutional code of Figure 6.3 is shown in Figure 6.5.

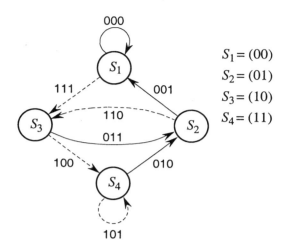

$$S_1 = (00)$$
$$S_2 = (01)$$
$$S_3 = (10)$$
$$S_4 = (11)$$

Figure 6.5: *State diagram for the rate-1/3 convolutional code of Figure 6.3.*

6.2.1 Rate-k_0/n_0 convolutional codes

More generally, we may define rate-k_0/n_0 convolutional codes. For these, n_0 binary symbols are sent to the channel for every k_0 input binary symbols. An example of a rate-2/3 encoder is shown in Figure 6.6. In general, in a rate-k_0/n_0 convolutional code we have k_0 shift registers with memories $\nu_1, \ldots, \nu_{k_0}$, and n_0 binary adders. An overall constraint length may also be defined, but its definitions are not consistent throughout the literature (see [6.6, pp. 12–13]).

6.3 Theoretical foundations

Contrary to what we did with linear block codes, here we must distinguish between a code (the set of all possible encoded sequences) and the encoder chosen for that code. As we shall see, the same convolutional code can be generated by several different encoders, whose properties may be helpful (minimum overall constraint length, and hence minimum decoding complexity) or harmful (catastrophicity— see *infra*).

We start by characterizing a rate-k_0/n_0 binary convolutional encoder as a linear causal time-invariant system over the binary field $\mathbb{F}_2$ having k_0 inputs and n_0 outputs (some additional properties are needed, as we shall see). In general, a single-

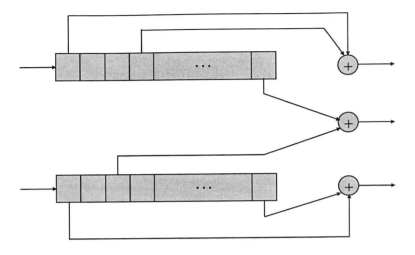

Figure 6.6: *A rate-2/3 convolutional encoder.*

input, single-output linear causal time-invariant system is characterized by its impulse response $\mathbf{g} \triangleq (g_i)_{i=0}^{\infty}$, with the system's output sequence $\mathbf{x} \triangleq (x_i)_{i=-\infty}^{\infty}$ being related to its input sequence $\mathbf{u} \triangleq (u_i)_{i=-\infty}^{\infty}$ by the convolution $\mathbf{x} = \mathbf{g} * \mathbf{u}$. Explicitly,

$$x_i = \sum_{j=0}^{\infty} g_j u_{i-j} \tag{6.1}$$

where sums and products are in $\mathbb{F}_2$, that is, are computed modulo-2.

It is convenient to associate with the sequences $\mathbf{g}$, $\mathbf{x}$, and $\mathbf{u}$ their D-transforms, i.e., the functions of the indeterminate D (the *delay operator* equivalent to the indeterminate z^{-1} of the z-transform) defined as

$$g(D) \triangleq \sum_{j=0}^{\infty} g_j D^j$$

$$x(D) \triangleq \sum_{j=-\infty}^{\infty} x_j D^j$$

and

$$u(D) \triangleq \sum_{j=-\infty}^{\infty} u_j D^j$$

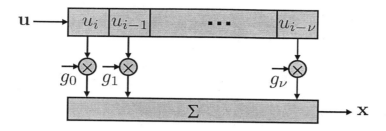

Figure 6.7: *An implementation of a causal linear time-invariant system with polynomial transfer function $g(D)$ having degree ν.*

These are related, through (6.1), by

$$x(D) = u(D)g(D) \tag{6.2}$$

Note that the causality constraint ($g_i = 0$ for $i < 0$) allows sequences to be represented by the sum of their D-series without potential ambiguities: thus, for example, $1/(1+D)$ represents the sequence $1+D+D^2+\ldots$ rather than $D^{-1}+D^{-2}+\ldots$. The function $g(D)$ is called the *transfer function* of the system. If $g(D)$ is a polynomial with degree ν (we write $\nu = \deg g(D)$), then the system whose input-output relationship is (6.1) can be implemented by using a shift register, as shown in Figure 6.7. Additionally, we say that a polynomial $g(D)$ is *delay free* if $g(0) = 1$.

The structure of Figure 6.7, while adequate for implementing polynomial transfer functions, cannot be used if $g(D)$ includes an infinite number of terms, i.e., if its corresponding sequence has no end. This may occur, for example, if $g(D)$ is *rational*, i.e., if it has the form of a ratio between two polynomials. If this is the case, we do not need an infinite number of stages in the shift-register implementation of the encoder: in fact, we can verify that the system of Figure 6.8 has the rational transfer function $g(D) = p(D)/q(D)$, where

$$p(D) \triangleq \sum_{i=0}^{\nu} p_i D^i \qquad q(D) \triangleq 1 + \sum_{i=1}^{\nu} q_i D^i$$

(i.e., $q(D)$ is delay free). If $p(D)$, $q(D)$ are relatively prime, then this realization has a feedback connection unless $q(D) = 1$. Conversely, every rational transfer function $p(D)/q(D)$, with a delay-free $q(D)$, can be realized in the "controller form" of Figure 6.8. For this reason, such a transfer function is called *realizable*.

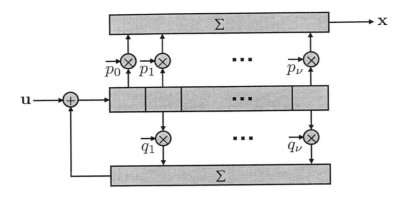

Figure 6.8: *Implementation of a system in controller form.*

Example 6.2

The causal transfer function

$$g(D) = (1 + D) \sum_{k=0}^{\infty} D^{3k} \qquad (6.3)$$

includes an infinite number of terms. From the equality

$$1 + D^3 = (1 + D)(1 + D + D^2)$$

we obtain, summing the series (6.3),

$$g(D) = \frac{1 + D}{1 + D^3} = \frac{1}{1 + D + D^2}$$

Thus, the system whose transfer function is $g(D)$ can be realized in the controller form of Figure 6.9. $\qquad \square$

Based on these definitions, we can now describe a rate-k_0/n_0 binary convolutional encoder by giving its $k_0 n_0$ impulse responses $g_{ij}(D)$, $1 \leq i \leq k_0$, $1 \leq j \leq n_0$, conveniently organized as the entries of a $k_0 \times n_0$ *generator matrix* $\mathbf{G}(D)$. If the k_0-input sequence and the n_0-output sequence are represented by the vectors $\mathbf{u}(D) \triangleq (u_1(D), \ldots, u_{k_0}(D))$ and $\mathbf{x}(D) \triangleq (x_1(D), \ldots, x_{n_0}(D))$, respectively, then we may write

$$\mathbf{x}(D) = \mathbf{u}(D)\mathbf{G}(D) \qquad (6.4)$$

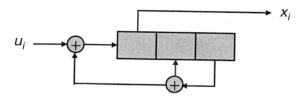

Figure 6.9: *An implementation of a causal linear time-invariant system with transfer function $g(D) = 1/(1 + D + D^2)$.*

Example 6.3

The rate-$1/2$ convolutional encoder of Figure 6.3 has

$$g_{11}(D) = 1 + D + D^2 \qquad g_{12}(D) = 1 + D^2$$

and is described by the generator matrix $\mathbf{G}(D) = (1 + D + D^2, 1 + D^2)$. □

6.3.1 Defining convolutional codes

We define now, in a natural way, a rate-k_0/n_0 convolutional code as the set of all possible sequences one can observe at the output of a convolutional encoder, which is a k_0-input, n_0-output system with the $k_0 \times n_0$ transfer function $\mathbf{G}(D)$. We require the encoder to be realizable and delay free (that is, at least one of its entries $p(D)/q(D)$ has $p(0) = 1$). Moreover, since the source sequence $\mathbf{u}(D)$ must be uniquely reconstructed from the observation of $\mathbf{x}(D)$, the matrix $\mathbf{G}(D)$ must have rank k_0.

The point here is that the same convolutional code can be generated by more than one encoder (encoders generating the same set of output sequences are called *equivalent*): each encoder defines a mapping between information sequences $\mathbf{u}(D)$ and code words $\mathbf{x}(D)$, but the set of code words does not depend on the mapping chosen. Observe in fact that if $\mathbf{Q}(D)$ denotes an invertible matrix with appropriate dimensions, (6.4) yields

$$\mathbf{x}(D) = \mathbf{u}(D)\mathbf{Q}(D)\mathbf{Q}^{-1}(D)\mathbf{G}(D) = \mathbf{u}'(D)\mathbf{G}'(D)$$

where $\mathbf{u}'(D) \triangleq \mathbf{u}(D)\mathbf{Q}(D)$ and $\mathbf{G}'(D) \triangleq \mathbf{Q}^{-1}(D)\mathbf{G}(D)$. Since $\mathbf{u}'(D)$ runs through all possible information sequences, the encoders with generators $\mathbf{G}(D)$ and $\mathbf{G}'(D)$ are equivalent. That said, it makes sense to look for encoders with certain useful properties. One of them is minimum number of states, which has a direct bearing on the complexity of the corresponding Viterbi decoder.

Example 6.4

The convolutional encoder of Figure 6.10 has the transfer function

$$\mathbf{G}(D) = \begin{bmatrix} 1 & D^2 & D \\ D & 1 & 0 \end{bmatrix} \tag{6.5}$$

Now, observe that

$$\begin{bmatrix} 1 & D^2 & D \\ D & 1 & 0 \end{bmatrix} = \begin{bmatrix} 1 & D^2 \\ D & 1 \end{bmatrix} \begin{bmatrix} 1 & 0 & \frac{D}{1+D^3} \\ 0 & 1 & \frac{D^2}{1+D^3} \end{bmatrix}$$

and that

$$\begin{bmatrix} 1 & D^2 \\ D & 1 \end{bmatrix}^{-1} = \frac{1}{1+D^3} \begin{bmatrix} 1 & D^2 \\ D & 1 \end{bmatrix}$$

which shows that the left-hand side of the last equation is an invertible matrix. Thus, we can write

$$
\begin{aligned}
\mathbf{x}(D) &= \mathbf{u}(D)\mathbf{G}(D) \\
&= \mathbf{u}(D) \begin{bmatrix} 1 & D^2 \\ D & 1 \end{bmatrix} \begin{bmatrix} 1 & 0 & \frac{D}{1+D^3} \\ 0 & 1 & \frac{D^2}{1+D^3} \end{bmatrix} \\
&= \mathbf{u}'(D)\mathbf{G}'(D) \tag{6.6}
\end{aligned}
$$

so that the encoder can also be implemented in an equivalent form as shown in Figure 6.11. This encoder is *systematic*: in fact, its input $(u_i^{(1)}, u_i^{(2)})$ yields the output $(x_i^{(1)}, x_i^{(2)}, x_i^{(3)})$ with $x_i^{(1)} = u_i^{(1)}$ and $x_i^{(2)} = u_i^{(2)}$. It can be seen that this encoder has the same number of states as that of Figure 6.10. Yet another equivalent encoder (but with more states) has the generator matrix

$$\mathbf{G}''(D) = \begin{bmatrix} 1+D & 1+D^2 & D \\ D+D^2 & 1+D & 0 \end{bmatrix} = \begin{bmatrix} 1 & 1 \\ 0 & 1+D \end{bmatrix} \mathbf{G}(D) \tag{6.7}$$

For future reference, observe that with this encoder the input sequence

$$\mathbf{u}(D) = \begin{bmatrix} 0 & \frac{1}{1+D} \end{bmatrix}$$

whose Hamming weight is ∞, generates the output sequence

$$\mathbf{x}(D) = \mathbf{u}(D)\mathbf{G}''(D) = [D \ 1 \ 0]$$

whose Hamming weight is 2.   □

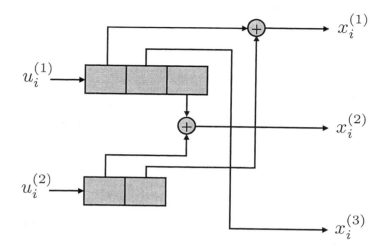

Figure 6.10: *A rate-2/3 convolutional encoder corresponding to the generator matrix (6.5).*

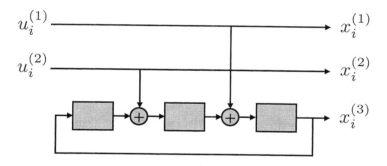

Figure 6.11: *A rate-2/3 convolutional encoder corresponding to the generator matrix $\mathbf{G}'(D)$ in (6.6).*

6.3.2 Polynomial encoders

If $q(D)$ denotes the least common multiple of all the denominators of the entries of $\mathbf{G}(D)$, the matrix

$$\mathbf{G}'(D) \triangleq q(D)\mathbf{G}(D)$$

yields an encoder equivalent to $\mathbf{G}(D)$ that is polynomial. Thus, every convolutional code admits a polynomial (i.e., feedback-free) encoder.

6.3.3   Catastrophic encoders

Example 6.4 shows how an encoder may map an infinite-weight input sequence $\mathbf{u}(D)$ to a finite-weight output sequence. We call this encoder *catastrophic* because a finite number of errors in the decoded word can lead to an infinite number of errors in the corresponding information sequence $\mathbf{u}(D)$. In Example 6.4, if the transmitted word is $\mathbf{x}(D) = [D\ 1\ 0]$ and the decoded word is $\hat{\mathbf{x}}(D) = [0\ 0\ 0]$, then the corresponding source sequence is $\hat{\mathbf{u}}(D) = [0\ 0]$, which entails an infinite number of errors.

Catastrophicity is an encoder property, not a code property: every convolutional code admits noncatastrophic as well as catastrophic generator matrices. For rate-$1/n_0$ codes with polynomial generators, a simple necessary and sufficient condition for a noncatastrophic encoder is that no two polynomials have a factor in common (see Problem 2 at the end of this chapter). More generally, it can be proved that if the entries of $\mathbf{G}(D)$ are rational, then a unique polynomial $q(D)$ exists with $q(0) = 1$ such that we can write, for $i = 1, \ldots, n_0$,

$$g_i(D) = \frac{p_i(D)}{q(D)}$$

A condition for this $\mathbf{G}(D)$ to be noncatastrophic is that no two polynomials $p_i(D)$ have a common factor other than D.

6.3.4   Minimal encoders

It can be shown that among all equivalent encoder matrices there exists one corresponding to the minimum number of trellis states: specifically, its realization in controller form requires the minimum number of memory elements [6.8, Section 2.6].

6.3.5   Systematic encoders

Every encoder can be transformed into an equivalent systematic rational one. It suffices to transform the generator matrix into the form

$$\mathbf{G}(D) = \begin{bmatrix} \mathbf{I}_{k_0} & \vdots & \mathbf{P} \end{bmatrix}$$

where $\mathbf{P}$ is a $k_0 \times (n_0 - k_0)$ matrix with rational entries. An interesting property of systematic encoding matrices is that they are minimal [6.8, Section 2.10].

Example 6.5

Consider the following generator matrix of a rate-2/3 code:

$$\mathbf{G}(D) = \begin{bmatrix} 1+D & D & 1 \\ D^2 & 1 & 1+D+D^2 \end{bmatrix}$$

The matrix consisting of the first two columns of $\mathbf{G}(D)$

$$\mathbf{T}(D) \triangleq \begin{bmatrix} 1+D & D \\ D^2 & 1 \end{bmatrix}$$

is invertible: in fact

$$\mathbf{T}^{-1}(D) = \frac{1}{1+D+D^3} \begin{bmatrix} 1 & D \\ D^2 & 1+D \end{bmatrix}$$

Premultiplication of $\mathbf{G}(D)$ by $\mathbf{T}^{-1}(D)$ yields a systematic generator matrix equivalent to $\mathbf{G}(D)$:

$$\mathbf{G}'(D) = \begin{bmatrix} 1 & 0 & \dfrac{1+D+D^2+D^3}{1+D+D^3} \\ 0 & 1 & \dfrac{1+D^2+D^3}{1+D+D^3} \end{bmatrix}$$

$\square$

6.4 Performance evaluation

The definition of error probabilities of convolutional codes, as contrasted with block codes, requires some extra care because in principle we are examining code words with infinite length. Consider transmission of the all-zero code word $\mathbf{x}$ and the competing code word $\hat{\mathbf{x}} \neq \mathbf{x}$. We decompose the error, made by choosing $\hat{\mathbf{x}}$, as a set of *error events*, each consisting in a single subpath diverging from the all-zero path and merging later into it, never to split again (see Figure 6.12). Note that, before and after this mergence, the correct path and the erroneous path accumulate the same metric increments.

The probability of error can then be written by using pairwise error probabilities and the union bound:

$$P(e \mid \mathbf{x}) \leq \sum P(\mathbf{x} \rightarrow \hat{\mathbf{x}}) \tag{6.8}$$

where the sum is extended to all possible error events (whose starting time is irrelevant, since the code trellis is periodic if we disregard, as we do, the initial and final transients). Of course, the actual computation of $P(\mathbf{x} \rightarrow \hat{\mathbf{x}})$ depends on the channel used for transmission.

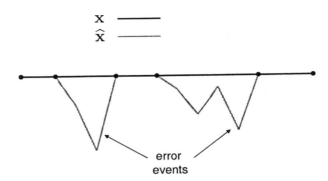

Figure 6.12: *Trellis paths associated with code words x and $\widehat{x}$, and two error events.*

6.4.1   AWGN channel

For this channel we can write, from (3.18):

$$P(\mathbf{x} \to \widehat{\mathbf{x}}) = Q\left(\frac{\|\mathbf{x} - \widehat{\mathbf{x}}\|}{\sqrt{2N_0}}\right)$$

Here, under the assumption of binary antipodal signaling with $\mathcal{X} = \{\pm\sqrt{\mathcal{E}}\}$, we have $\|\mathbf{x} - \widehat{\mathbf{x}}\|^2 = 4\mathcal{E}d_H(\mathbf{x}, \widehat{\mathbf{x}})$, where $d_H(\mathbf{x}, \widehat{\mathbf{x}})$ denotes the Hamming distance of $\mathbf{x}$ and $\widehat{\mathbf{x}}$. Notice that, for all code words $\mathbf{x}'$, we have

$$d_H(\mathbf{x}, \widehat{\mathbf{x}}) = d_H(\mathbf{x} + \mathbf{x}', \widehat{\mathbf{x}} + \mathbf{x}')$$

and hence, by choosing $\mathbf{x}' = \mathbf{x}$,

$$d_H(\mathbf{x}, \widehat{\mathbf{x}}) = d_H(\mathbf{0}, \widehat{\mathbf{x}} + \mathbf{x}) = w_H(\widehat{\mathbf{x}} + \mathbf{x}) \tag{6.9}$$

Since convolutional codes are linear, $\widehat{\mathbf{x}} + \mathbf{x}$ is itself a code word, and hence in (6.8) we may assume, without loss of generality, that $\mathbf{x}$ is the all-zero word.

In conclusion, we have

$$P(e) = P(e \mid \mathbf{0}) \leq \sum P(\mathbf{0} \to \widehat{\mathbf{x}}) \tag{6.10}$$

where

$$P(\mathbf{0} \to \widehat{\mathbf{x}}) = Q\left(\sqrt{2w_H(\widehat{\mathbf{x}})\frac{\mathcal{E}}{N_0}}\right) \tag{6.11}$$

This shows that the pairwise error probability depends on $\widehat{\mathbf{x}}$ only through its Hamming weight: thus, the union bound (6.10) can be evaluated by enumerating the set $\mathcal{W}$ of Hamming weights of all the nonzero words of the code. Specifically, we have the union bound

$$P(e) \leq \sum_{w \in \mathcal{W}} \nu(w) Q \left(\sqrt{2w \frac{\mathcal{E}}{N_0}} \right) \tag{6.12}$$

where $\nu(w)$ is the number of nonzero code words whose Hamming weight is w.

Weight enumerator function

The evaluation of bound (6.12) can be considerably simplified by using bound (3.19), which yields the union-Bhattacharyya bound

$$P(e) \leq \sum_{w \in \mathcal{W}} \nu(w) e^{-w\mathcal{E}/N_0} \tag{6.13}$$

To proceed further, we construct a *weight enumerator function* of the convolutional code: this lists the weights of all the words $\hat{x}$ that split from the all-zero-word trellis path and remerge later into the same path. We do the following: we start from the state diagram of the code and examine all state sequences splitting from the all-zero state and ending in it. Every edge of the diagram is labeled X^ω, where X is an indeterminate and ω is the weight of the code word segment associated with the edge in the original state diagram. Formally, we have $w = \sum \omega$, where the sum is extended to the trellis branches forming the code word whose weight is w. Figure 6.13 shows this modified state diagram for our usual rate-1/3 convolutional code (see Figure 6.5). The transfer function of the graph is the enumerator of path weights:

$$T(X) = \nu(a) X^a + \nu(b) X^b + \nu(c) X^c + \ldots \tag{6.14}$$

If $T(X)$ is known, then we have, from (6.13)

$$P(e) \leq T \left(e^{-\mathcal{E}/N_0} \right) \tag{6.15}$$

where we may want to write

$$\mathcal{E}/N_0 = \rho \mathcal{E}_b / N_0 \tag{6.16}$$

with $\rho = k_0/n_0$ the convolutional code rate (1/3 in our example). The minimum exponent of $T(X)$ is called the *free (Hamming) distance* of the code, denoted d_{free}.

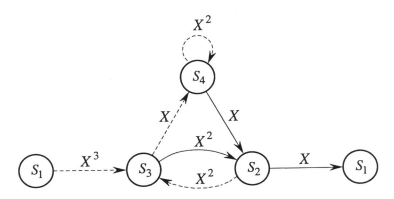

Figure 6.13: *State diagram for the rate-1/3 convolutional code of Figure 6.5. The labels allow the computation of the weight enumerator function $T(X)$.*

This is the smallest free distance of the error events and determines the asymptotic (for large $\mathcal{E}_b/N_0$) error probability of the code:

$$P(e) \overset{\sim}{\leq} \nu(d_{\text{free}})Q\left(\sqrt{2\rho d_{\text{free}}\frac{\mathcal{E}_b}{N_0}}\right) \qquad (6.17)$$

This also shows that ρd_{free} is the asymptotic coding gain of the code.

Computing the transfer function of a graph

The transfer function between a pair of vertices of a directed graph is defined as the sum of the labels of all paths of any length connecting the two vertices. The label of a path is defined as the product of the labels of its edges. This transfer function can be computed in several ways. One consists of the progressive reduction of the original graph to a single edge, obtained by repeated application of elementary rules like those summarized in Figure 6.14. These show how a graph with more than one edge can be replaced by a new graph with a single equivalent edge.

As an example, consider the graph of Figure 6.15 (a), with labels $A, B, \ldots, G$ and vertices $\alpha, \beta, \ldots, \varepsilon$. By using reduction rules (c) and (b) of Figure (6.14), this graph can first be replaced with that of Figure 6.15 (b) and then with that of Figure 6.15 (c). Final application of rule (a) of Figure 6.14 yields the transfer function: between α and ε

$$T(\alpha \to \varepsilon) = \frac{ACG(1-E) + ABFG}{1 - E - CD + CDE - BDF} \qquad (6.18)$$

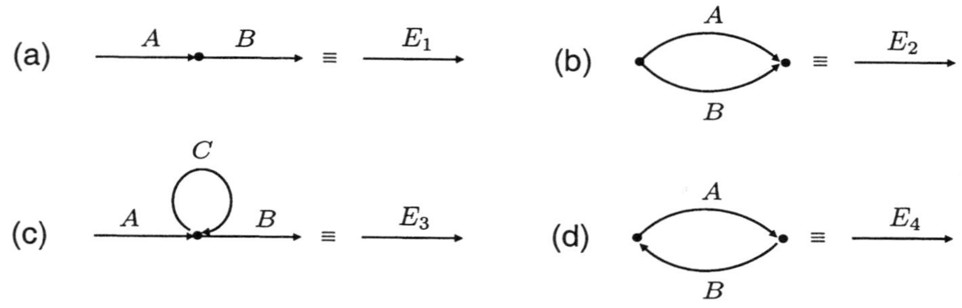

(a) $\xrightarrow{\quad A \quad B \quad} \equiv \xrightarrow{\quad E_1 \quad}$

(b) $\equiv \xrightarrow{\quad E_2 \quad}$

(c) $\xrightarrow{\quad A \quad B \quad} \equiv \xrightarrow{\quad E_3 \quad}$

(d) $\equiv \xrightarrow{\quad E_4 \quad}$

$E_1 = AB$

$E_2 = A + B$

$E_3 = AB + ACB + AC^2B + \ldots = \dfrac{AB}{1-C}$

$E_4 = A + ABA + ABABA + \ldots = \dfrac{A}{1-AB}$

Figure 6.14: *Reduction rules for the computation of the transfer function of a directed graph.*

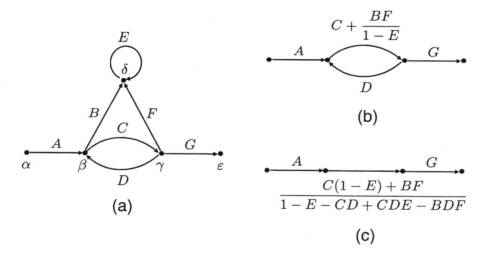

Figure 6.15: *(a) Directed graph with five vertices and seven edges; (b) first reduction step; (c) second reduction step.*

When the number of vertices is large, the reduction technique previously explained becomes too complex, and we must resort to a different technique for evaluating the transfer function. This technique is based on a matrix description of the graph: we shall explain it with reference to the graph of Figure 6.15.

Let us define by x_ι the value of the accumulated path labels from any vertex directed towards ι. The state equations for the graph of Figure 6.15 (a) are

$$
\begin{cases}
x_\beta = A + Dx_\gamma \\
x_\gamma = Cx_\beta + Fx_\delta \\
x_\delta = Bx_\beta + Ex_\delta \\
x_\varepsilon = Gx_\gamma
\end{cases}
\tag{6.19}
$$

With this approach, we obviously have $T(\alpha \rightarrow \varepsilon) = x_\varepsilon$, and therefore we can solve the system (6.19) and verify that x_ε is given again by (6.18).

The system of equations (6.19) can be given a more general and formal expression. Define the two column vectors

$$
\mathbf{x} \triangleq [x_\beta \ x_\gamma \ x_\delta \ x_\varepsilon]' \qquad \mathbf{x}_0 \triangleq [A \ 0 \ 0 \ 0]'
\tag{6.20}
$$

and the *state transition* matrix $\mathbf{T}$

$$
\mathbf{T} \triangleq \begin{bmatrix}
0 & D & 0 & 0 \\
C & 0 & F & 0 \\
B & 0 & E & 0 \\
0 & G & 0 & 0
\end{bmatrix}
\tag{6.21}
$$

Using (6.20) and (6.21), system (6.19) can be rewritten in matrix form as

$$
\mathbf{x} = \mathbf{Tx} + \mathbf{x}_0
\tag{6.22}
$$

whose formal solution is

$$
\mathbf{x} = (\mathbf{I} - \mathbf{T})^{-1}\mathbf{x}_0
\tag{6.23}
$$

or, equivalently, the matrix power series

$$
\mathbf{x} = (\mathbf{I} + \mathbf{T} + \mathbf{T}^2 + \cdots)\mathbf{x}_0
\tag{6.24}
$$

Example 6.6

From (6.18), the transfer function of the graph of Figure 6.13 is

$$T(X) = \frac{2X^6 - X^8}{1 - X^2 - 2X^4 + X^6} = 2X^6 + X^8 + 5X^{10} + \dots$$

which yields $d_{\text{free}} = 6$, and the bound

$$P(e) \leq T\left(e^{-\mathcal{E}_b/3N_0}\right)$$

$\square$

6.4.2 Independent Rayleigh fading channel

Assume now a fading channel with interleaving deep enough to validate the independent-fading model, and with CSI known at the receiver. Here we have

$$P(\mathbf{x} \to \widehat{\mathbf{x}}) = \mathbb{E}\, Q\left(\frac{\|\mathbf{R}(\mathbf{x} - \widehat{\mathbf{x}})\|}{\sqrt{2N_0}}\right) \tag{6.25}$$

with $\mathbf{R}$ the diagonal matrix of the fading gains. Now,

$$\|\mathbf{R}(\mathbf{x} - \widehat{\mathbf{x}})\|^2 = 4\mathcal{E} \sum_{i \in \mathcal{I}} R_i^2 \tag{6.26}$$

where $\mathcal{I}$ is the set of indices where the two vectors $\mathbf{x}$ and $\widehat{\mathbf{x}}$ differ. If the fading gains have a common Rayleigh pdf, the sum above is a chi-square-distributed random variable with $2|\mathcal{I}| = 2d_{\text{H}}(\mathbf{x}, \widehat{\mathbf{x}})$ degrees of freedom. As in the AWGN case, the PEP depends only on the Hamming distance between $\mathbf{x}$ and $\widehat{\mathbf{x}}$, so we can assume without loss of generality that $\mathbf{x}$ is the all-zero sequence and write

$$P(e) \leq \sum_{w \in \mathcal{W}} \nu(w)\, \mathbb{E}\, Q\left(\sqrt{2X_{2w}^2 \frac{\mathcal{E}}{N_0}}\right) \tag{6.27}$$

where X_{2w}^2 is chi-square with $2w$ degrees of freedom. Exact computation of the expectations in (6.27) can be done with the aid of (4.45)–(4.46). The asymptotic expression (4.48) can also be used to obtain

$$\mathbb{E}\, Q\left(\sqrt{2X_{2w}^2 \frac{\mathcal{E}}{N_0}}\right) \sim \frac{1}{(4\mathcal{E}/N_0)^w} \binom{2w-1}{w} \tag{6.28}$$

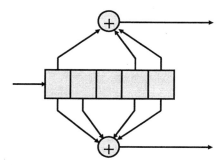

Figure 6.16: *The rate-1/2 convolutional code used in GSM.*

and hence, if there are $\nu(d_{\text{free}})$ words with Hamming weight d_{free}, then, for high SNR,

$$P(e) \stackrel{\sim}{\leq} \nu(d_{\text{free}}) \frac{1}{(4\mathcal{E}/N_0)^{d_{\text{free}}}} \binom{2d_{\text{free}} - 1}{d_{\text{free}}} \qquad (6.29)$$

The reader can verify that, if the results of Section 4.2.1 are directly applied to this case, a weaker bound on $P(e)$ is obtained.

6.4.3   Block-fading channel

From the results of Section 4.3, we can easily infer that the asymptotic performance of convolutional codes over block-fading channels depends on D_{free}, defined as the minimum Hamming block distance of error events.

Example 6.7

As an example of a convolutional code used on a block-fading channel, we describe the code used in GSM. Its performance is evaluated by determining its Hamming block distance with $F = 8$.

Consider the rate-1/2 binary convolutional code whose encoder is depicted in Figure 6.16. This is employed in full-rate GSM. Its Hamming free distance is 7, and the coded sequence with Hamming weight 7 is

$$\cdots 0, 0, 0, 1, 1, 0, 1, 0, 0, 1, 1, 1, 1, 0, 0, \cdots$$

The coded bits are interleaved over $F = 8$ blocks and transmitted over channels whose fading gains can be assumed to be independent, so the block-fading model applies. The above sequence is distributed among the eight blocks as follows:

$$\begin{array}{ccccccc}
\cdots & 0 & 1 & 1 & 0 & \cdots \\
\cdots & 0 & 1 & 1 & 0 & \cdots \\
\cdots & 0 & 0 & 0 & 0 & \cdots \\
\cdots & 0 & 1 & 0 & 0 & \cdots \\
\cdots & 0 & 0 & 0 & 0 & \cdots \\
\cdots & 0 & 0 & 0 & 0 & \cdots \\
\cdots & 0 & 1 & 0 & 0 & \cdots \\
\cdots & 0 & 1 & 0 & 0 & \cdots
\end{array}$$

which shows that its Hamming block-distance is 5. This turns out to be also D_{free} (see [6.9]), and use of the Singleton bound (4.35) shows that no binary rate-1/2 code can yield a higher distance. Hence, this code is optimum over the block-fading channel with $F = 8$. □

6.5 Best known short-constraint-length codes

Computer search methods have been used to find convolutional codes optimum in the sense that, for a given rate and a given constraint length, they have the largest possible free distance. The best rate-1/2 and rate-1/3 codes are listed in part in Tables 6.1 and 6.2 [6.3, 6.10]. The codes are identified by their generators, represented as octal numbers. By transforming an octal number into a binary number, we have a sequence of 1s and 0s, each of which represents the presence (or the absence, respectively) of the connection of a cell of a shift register with one modulo-2 adder.

6.6 Punctured convolutional codes

Puncturing is a procedure for obtaining, in an easy way, a convolutional code with a higher rate from one with a lower rate k_0/n_0. If a fraction ϵ of symbols are eliminated (*punctured*) from each encoded sequence, the resulting code has rate $(k_0/n_0)/(1 - \epsilon)$. For example, if 1/4 of the symbols of a rate-1/2 code are punctured, a new code with rate $(1/2)(4/3) = 2/3$ is obtained. An example will show how this can be done.

Example 6.8

Consider the four-state convolutional encoder of Figure 6.17(a). For each input bit

Memory ν	Generators in octal notation		d_{free}
1	1	3	3
2	5	7	5
3	15	17	6
4	23	35	7
5	53	75	8
6	133	171	10
7	247	371	10
8	561	753	12
9	1131	1537	12
10	2473	3217	14
11	4325	6747	15
12	10627	16765	16
13	27251	37363	16

Table 6.1: *Maximum-free-distance convolutional codes of rate $1/2$ and memory ν.*

Memory ν	Generators in octal notation			d_{free}
1	1	3	3	5
2	5	7	7	8
3	13	15	17	10
4	25	33	37	12
5	47	53	75	13
6	117	127	155	15
7	225	331	367	16
8	575	623	727	18
9	1167	1375	1545	20
10	2325	2731	3747	22
11	5745	6471	7553	24
12	10533	10675	17661	24
13	21645	35661	37133	26

Table 6.2: *Maximum-free-distance convolutional codes of rate $1/3$ and memory ν.*

entering the encoder, two bits are sent through the channel, so the code generated
has rate $1/2$. Its trellis is also shown, in Figure 6.17(b). Suppose now that, for every
four bits generated by the encoder, one (the last) is punctured, i.e., not transmit-

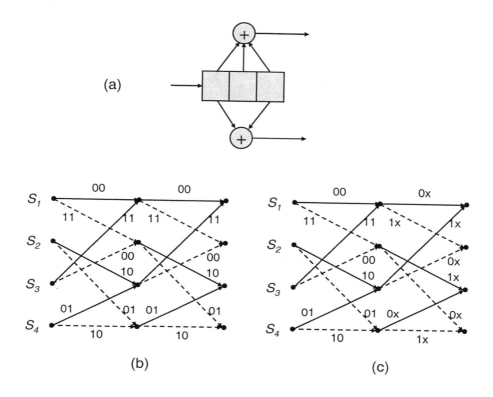

Figure 6.17: *Encoder (a) and trellis (b) for a rate-1/2 convolutional code. The trellis (c) refers to the rate-2/3 punctured code described in Example 6.8.*

ted. In this case, for every two input bits three bits are generated by the encoder, thus producing a rate-2/3 code. The trellis for the punctured code is shown in Figure 6.17 (c), and the letter x denotes a punctured output bit. As an example, the input sequence $\mathbf{u} = 101101\ldots$ would yield $\mathbf{x} = 111000010100$ for the rate-1/2 code and $\mathbf{x} = 111000010$ for the punctured rate-2/3 code. $\qquad\square$

In decoding, the samples corresponding to punctured bit locations are provided to the decoder after assigning them the value of 0.0, since there is no channel information about these bits.

The major upside of puncturing is that several rates can be obtained from the same "mother code," thus simplifying the implementation through a sort of "universal" encoder and decoder, a fact widely exploited in circuit implementations.

As for the downsides to the punctured solution, there are at least two of them. First, punctured codes are normally worse (albeit slightly), in terms of distance spectrum, than the best unpunctured codes with the same rate. Second, since the trellis of a punctured (n_0, k_0) code is time-varying with period k_0, the decoder needs to acquire frame synchronization in addition to symbol synchronization.

6.7   Block codes from convolutional codes

Since in practice a convolutional code is used to transmit a *finite* sequence of information symbols, its trellis must be terminated at a certain time. This operation yields a block code. Here we study how this termination can be done, focusing our attention to rate-$1/n_0$ convolutional codes for simplicity (the general case of rate-k_0/n_0 codes is left as an exercise to the willing reader).

Let us first derive the generator matrix $\mathbf{G}$ of the infinite-length block code obtained as the output of a rate-$1/n_0$ polynomial (i.e., nonrecursive) encoder. At each time $t > 0$, the n_0 output symbols are a linear combination of the $\nu + 1$ binary digits contained in the shift register: we write

$$\mathbf{x}_t = u_t \mathbf{g}_1 + u_{t-1} \mathbf{g}_2 + \ldots + u_{t-\nu} \mathbf{g}_{\nu+1} \tag{6.30}$$

where $\mathbf{g}_i$, $1 \le i \le \nu + 1$, is a "generator" row vector whose n_0 components describe the connections between the adders and the shift register. Specifically, the jth component of $\mathbf{g}_i$ is 1 if adder j is connected to the cell i of the shift register, and 0 otherwise. Equation (6.30) can be written in a matrix form involving the input sequence $\mathbf{u} = (u_0 \ u_1 \ \ldots)$ and the output sequence $\mathbf{x}$ as

$$\mathbf{x} = \mathbf{u} \mathbf{G}_\infty \tag{6.31}$$

where

$$\mathbf{G}_\infty \triangleq \begin{bmatrix} \mathbf{g}_1 & \mathbf{g}_2 & \cdots & \mathbf{g}_{\nu+1} & & & \\ & \mathbf{g}_1 & \mathbf{g}_2 & \cdots & \mathbf{g}_{\nu+1} & & \\ & & \mathbf{g}_1 & \mathbf{g}_2 & \cdots & \mathbf{g}_{\nu+1} & \\ & & & \mathbf{g}_1 & \mathbf{g}_2 & \cdots & \mathbf{g}_{\nu+1} \\ & & & & \cdots & \cdots & \cdots & \cdots \end{bmatrix} \tag{6.32}$$

(the blank entries in $\mathbf{G}_\infty$ are zero).

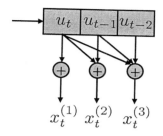

Figure 6.18: *A rate-1/3 polynomial convolutional encoder.*

Example 6.9

With the rate-1/3 encoder of Figure 6.18 we have

$$x_t^{(1)} = 1 \cdot u_t + 0 \cdot u_{t-1} + 0 \cdot u_{t-2}$$
$$x_t^{(2)} = 1 \cdot u_t + 1 \cdot u_{t-1} + 0 \cdot u_{t-2}$$
$$x_t^{(3)} = 1 \cdot u_t + 1 \cdot u_{t-1} + 1 \cdot u_{t-2}$$

so that

$$\mathbf{x}_t \triangleq [x_t^{(1)} \ x_t^{(2)} \ x_t^{(3)}] = u_t \mathbf{g}_1 + u_{t-1}\mathbf{g}_2 + u_{t-2}\mathbf{g}_3$$

with

$$\mathbf{g}_1 \triangleq [1\,1\,1] \qquad \mathbf{g}_2 \triangleq [0\,1\,1] \qquad \mathbf{g}_3 \triangleq [0\,0\,1]$$

which yields

$$\mathbf{G}_\infty = \begin{bmatrix} 111 & 011 & 001 & & \\ & 111 & 011 & 001 & \\ & & 111 & 011 & 001 \\ & & \cdots & \cdots & \cdots \end{bmatrix}$$

$\square$

6.7.1 Direct termination

Consider now an input sequence with finite length N. The first N output symbols can be computed from (6.30) by taking $0 \le t < N$, which is equivalent to writing

$$\mathbf{x} = \mathbf{u}\mathbf{G}_N \tag{6.33}$$

where

$$
\mathbf{G}_N \triangleq
\begin{bmatrix}
g_1 & g_2 & \cdots & \cdots & g_{\nu+1} \\
 & g_1 & g_2 & \cdots & \cdots & g_{\nu+1} \\
 & & \ddots & \ddots & & & \ddots \\
 & & & g_1 & g_2 & \cdots & \cdots & g_{\nu+1} \\
 & & & & g_1 & g_2 & \cdots & g_\nu \\
 & & & & & \ddots & \ddots & \vdots \\
 & & & & & & \ddots & g_2 \\
 & & & & & & & g_1
\end{bmatrix}
\qquad (6.34)
$$

is a matrix with N rows and $n_0 N$ columns. Hence, the resulting block code has the same rate as the original convolutional code. This is a very natural code construction, but it has a downside. In fact, it turns out that the last coded symbols are less protected from the noisy-channel effects: the BER as a function of the bit position in a block would typically be about constant for most bits, except for those at the beginning and at the end of the block. This occurs because the Viterbi algorithm leaves from a known starting state when it decodes the first bits, and hence yields a lower BER for them. The opposite occurs with the last bits in the block.

6.7.2 Zero-tailing

To avoid the poor protection at the end of the block caused by direct termination, the most common way (the *zero-tailing method*) of terminating a convolutional code consists of having the encoder end in a predefined state (typically, the all-zero state). To be able to do this, the encoder appends a deterministic sequence, at the end of the information sequence, that fills with zeros the entire shift register of the encoder. This sequence need not be any longer than the memory length of the encoder (multiplied by k_0 if the code has rate k_0/n_0). Its components generally depend on the encoder state at the end of the information sequence: however, for a polynomial generator matrix of memory ν, the zero-tailing sequence is the all-zero ν-tuple. This sequence causes a code-rate loss that may be substantial for short blocks: in fact, for rate-$1/n_0$ codes with N information symbols, the resulting block length is $(N + \nu)n_0$, so the resulting code rate is $N/((N + \nu)n_0) < 1/n_0$.

Example 6.10

Figure 6.19 shows that, to terminate the trellis by bringing the encoder back to state 00, the last two bits fed into the shift register must be zero. If the rate of the original convolutional code is $1/2$, that of the terminated block code is $5/14$. $\qquad \square$

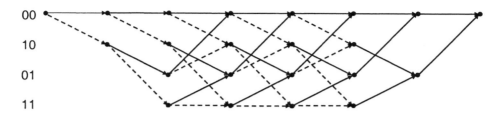

Figure 6.19: *Trellis termination.*

6.7.3 Tail-biting

If a system transmits short data frames, the loss due to trellis termination using the zero-tailing method may be unacceptable. In this case, one can resort to a transformation of the code trellis into a tail-biting one (see Section 5.9). With a tail-biting trellis the encoder starts and ends in the same state, and thus the code-rate loss is eliminated at the price of increased decoder complexity, due to the fact that the starting and ending states of the encoder are unknown to the decoder.

Consider a polynomial encoder for a memory-ν, rate-$1/n_0$ code (the generalization to rate-k_0/n_0 is straightforward). After N input symbols, the encoder state is $(u_N \, u_{N-1} \ldots u_{N-\nu+1})$, which we want to coincide with the initial state. If the all-zero sequence is sent to an encoder with this initial state (that is, preloaded with these bit values from the end of the block), then the output sequence has the form

$$\mathbf{x}^{(0)} = \mathbf{u}^{(0)} \mathbf{G}_N^{(0)}$$

where

$$\mathbf{u}^{(0)} \triangleq [0\,0\,0 \ldots 0\, u_{N-\nu+1} \, u_{N-\nu+2} \, \ldots \, u_N]$$

and

$$\mathbf{G}_N^{(0)} \triangleq \begin{bmatrix} \mathbf{g}_{\nu+1} & & & \\ \mathbf{g}_\nu & \mathbf{g}_{\nu+1} & & \\ \vdots & \vdots & \ddots & \\ \mathbf{g}_2 & \mathbf{g}_3 & \cdots & \mathbf{g}_{\nu+1} \end{bmatrix}$$

is a matrix with N rows and $n_0 N$ columns. For a general input sequence with length N, due to the linearity of the code we can write the code words obtained with the tail-biting method in the form $\mathbf{x} = \mathbf{u}\mathbf{G}_N^{\mathrm{TB}}$, where

$$
\mathbf{G}_N^{\mathrm{TB}} \triangleq \mathbf{G}_N + \mathbf{G}_N^{(0)}
$$

$$
= \begin{bmatrix}
g_1 & g_2 & \cdots & \cdots & g_{\nu+1} & & & & \\
 & g_1 & g_2 & \cdots & & \cdots & g_{\nu+1} & & \\
 & & \ddots & \ddots & & & & \ddots & \\
 & & & g_1 & g_2 & \cdots & & \cdots & g_{\nu+1} \\
g_{\nu+1} & & & & g_1 & g_2 & \cdots & & g_\nu \\
g_\nu & g_{\nu+1} & & & & \ddots & \ddots & & \vdots \\
\vdots & \vdots & \ddots & & & & & \ddots & g_2 \\
g_2 & g_3 & \cdots & g_{\nu+1} & & & & & g_1
\end{bmatrix} \quad (6.35)
$$

Observe how the matrix (6.35) can be obtained from (6.34) by wrapping around the last ν columns. Based on this observation, we may construct a convolutional code from a block code whose generator matrix has the form (6.35): to obtain $\mathbf{G}_\infty$ it suffices to unwrap it and to extend it to a semi-infinite matrix.

6.8   Bibliographical notes

A thorough treatment of the algebraic aspects of convolutional codes can be found in [6.8] (see also papers [6.4, 6.7]). Some information-theoretical aspects of these codes are examined in [6.13]. Tables of good codes can be found in [6.3, 6.10].

In this chapter we have not covered the calculation of bit error probability. This can be found, for example, in [6.1, Section 11.1]. The effect of trellis truncation on the error rate of convolutional codes is examined in [6.12]. Tail-biting convolutional codes are treated in [6.8, 6.11].

Tables of punctured codes can be found in [6.2, 6.14].

6.9   Problems

1. Find a parity-check matrix for the code of Example 6.3, i.e., a matrix $\mathbf{H}(D)$ satisfying

$$
\mathbf{x}(D)\mathbf{H}'(D) = \mathbf{0}
$$

2. Consider a rate-$1/n_0$ convolutional code with polynomial generators. Prove that its encoder is catastrophic if and only if all generator polynomials have a common polynomial factor of degree at least one.

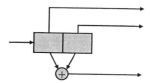

Figure 6.20: *Encoder of a rate-1/3 convolutional code.*

3. Prove that, for a rate-$1/n_0$ convolutional encoder,

    (a) Two generators $\mathbf{G}(D)$, $\mathbf{G}'(D)$ are equivalent if and only if $\mathbf{G}(D) = u(D)\mathbf{G}'(D)$ for some nonzero rational function $u(D)$.

    (b) Every generator $\mathbf{G}(D)$ is equivalent to a polynomial generator.

4. Prove that a systematic encoder for a rate-$1/n_0$ convolutional code with generator $\mathbf{G}(D) = (g_1(D), g_2(D), \ldots, g_{n_0}(D))$ may be obtained in the form $(1, g_2(D)/g_1(D), \ldots, g_{n_0}(D)/g_1(D))$.

5. Consider the rate-1/3 convolutional code shown in Figure 6.20.

    (a) Draw a section of the trellis diagram of the code, including the branch labels describing the coded symbols.

    (b) Derive the weight enumerator function $T(D)$ of the code.

    (c) Compute its free Hamming distance.

6. Use (6.28) and the inequality

$$\frac{1}{4^w}\binom{2w-1}{w} < 1$$

    to derive a transfer-function bound for the error probability $P(e)$ of a convolutional code used over the independent Rayleigh fading channel with perfect channel state information at the receiver.

7. Use (3.66) to derive a bound tighter than the union-Bhattacharyya bound (6.13).

8. Consider a convolutional encoder with feedback. Show that a finite zero-sequence at its input may not terminate the trellis in the null state. Examine the special case of systematic encoders with feedback.

9. Generalize the transfer-function bound to error probability derived in this chapter to the computation of $P(e)$ for block codes defined on a (generally nonperiodic) trellis.

References

[6.1] S. Benedetto and E. Biglieri, *Digital Transmission Principles with Wireless Applications*. New York: Kluwer/Plenum, 1999.

[6.2] J. B. Cain, G. C. Clark, Jr., and J. M. Geist, "Punctured convolutional codes of rate $(n-1)/n$ and simplified maximum likelihood decoding," *IEEE Trans. Inform. Theory*, Vol. 25, No. 1, pp. 97–100, January 1979.

[6.3] J.-J. Chang, D.-J. Hwang, and M.-C. Lin, "Some extended results on the search for good convolutional codes," *IEEE Trans. Inform. Theory*, Vol. 43, No. 5, pp. 1682–1697, September 1997.

[6.4] G. D. Forney, Jr., "Convolutional codes. I: Algebraic structure," *IEEE Trans. Inform. Theory*, Vol. IT-16, No. 6, pp. 720–738, November 1970.

[6.5] J. Hagenauer, "Rate-compatible punctured convolutional codes (RCPC codes) and their applications," *IEEE Trans. Commun.*, Vol. 36, No. 4, pp. 389–400, April 1988.

[6.6] C. Heegard and S. B. Wicker, *Turbo Coding*. Boston, MA: Kluwer Academic, 1999.

[6.7] R. Johannesson and Z.-X. Wan, "A linear algebra approach to minimal convolutional encoders," *IEEE Trans. Inform. Theory*, Vol. 39, No. 4, pp. 1219–1233, July 1993.

[6.8] R. Johannesson and K. Sh. Zigangirov, *Fundamentals of Convolutional Coding*. Piscataway, NJ: IEEE Press, 1999.

[6.9] R. Knopp, *Coding and Multiple Access over Fading Channels*, Ph.D. Thesis, École Polytechnique Fédérale de Lausanne, Lausanne, Switzerland, 1997.

[6.10] K. J. Larsen, "Short convolutional codes with maximal free distance for rates $1/2$, $1/3$ and $1/4$," *IEEE Trans. Inform. Theory*, Vol. 19, No. 3, pp. 371–372, May 1973.

[6.11] H. H. Ma and J. K. Wolf, "On tail biting convolutional codes," *IEEE Trans. Commun.*, Vol. 34, No. 2, pp. 104–111, February 1986.

[6.12] H. Moon, "Improved upper bound on bit error probability for truncated convolutional codes," *IEE Electronics Letters*, Vol. 34, No. 1, pp. 65–66, 8th January 1998.

[6.13] A. J. Viterbi and J. K. Omura, *Principles of Digital Communication and Coding*. New York: McGraw-Hill, 1979.

[6.14] Y. Yasuda, K. Kashiki, and Y. Hirata, "High-rate punctured convolutional codes for soft-decision Viterbi decoding," *IEEE Trans. Commun.*, Vol. 32, No. 3, pp. 315–319, March 1984.

and barnacled up to the eyes when he repented after seven

Trellis-coded modulation

In this chapter we introduce codes designed on a periodic trellis, and whose coded symbols are not restricted to the binary set $\{0, 1\}$ but rather chosen from a general elementary constellation $\mathcal{X}$. The basic idea underlying Trellis-Coded Modulation (TCM) is to use a convolutional code to generate the redundancy necessary to achieve a coding gain, while preventing bandwidth expansion by increasing the size of the constellation rather than the number of transmitted symbols. Specifically, a rate-k_0/n_0 convolutional code transforms a binary k_0-tuple into a binary n_0-tuple. The latter is used as the binary label of a constellation $\mathcal{X}$ with $|\mathcal{X}| = 2^{n_0}$ signals. Without coding, the constellation needed would be $\mathcal{X}'$, with $|\mathcal{X}'| = 2^{k_0}$ signals. Thus, coding does not entail any bandwidth expansion if $\mathcal{X}$, $\mathcal{X}'$ have the same dimensionality, while the convolutional code provides a coding gain. In practice, TCM uses $n_0 = k_0 + 1$, so $|\mathcal{X}| = 2|\mathcal{X}'|$ and data are transmitted at a rate of $\log |\mathcal{X}| - 1$ bits per signal.

7.1 Generalities

If a convolutional code is designed by properly choosing a periodic trellis and the binary symbols associated with its branches, its coded signals carry information at a rate lower than 1 bit/signal. To introduce the redundancy needed to decrease the rate (and hence to obtain a coding gain), we are forced to increase the number of transmitted binary symbols, which entails decreasing their duration, and consequently increasing the occupied bandwidth. Since bandwidth expansion is obviously an unwelcome effect, we may consider introducing redundancy in a different way. Trellis-coded modulation (TCM), described in this chapter, uses a redundant signal set, i.e., an elementary constellation with more signals than we would need if coding were not used. Specifically, TCM uses $2M$ signals to transmit at a rate of $\log M$ bit/symbol. As a result, a coding gain is obtained without any sacrifice in bandwidth. To understand how TCM works, consider Figure 3.4 or Figure 4.5. Here we see that, to transmit 1 bit per dimension pair (say), one could use 2-PSK and a high-enough SNR; if a lower SNR is available, then the rate decreases below 1 (this is the "standard" coding solution). Now, observe that 1 bit per dimension pair could also be transmitted at a lower SNR (and hence by achieving a coding gain) if 4-PSK is used. The challenge here is to design coding schemes yielding the coding gain promised by these considerations. TCM is one such solution.

Consider the transmission of uncoded signals chosen from $\mathcal{X}'$, and let $\mathcal{E}'$ denote the average transmitted energy. We have seen in Chapter 3 that the error probability of this uncoded system on the AWGN channel depends, asymptotically for high signal-to-noise ratios, on the energy per bit $\mathcal{E}'_b$ and on $\delta_{\min}$, the minimum Euclidean distance of $\mathcal{X}$. This dependence is summarized by the definition of asymptotic power efficiency, whose value is $\gamma' = \delta_{\min}^2 / 4\mathcal{E}'_b$ here. With a coding scheme using the constellation $\mathcal{X} \supset \mathcal{X}'$, the asymptotic error probability depends on the average transmitted energy per bit $\mathcal{E}_b$ and on δ_{free}, the smallest Euclidean distance among code words (since we are considering trellis codes here, it is appropriate to call it *free* Euclidean distance). The quality of the transmission solution can be expressed by its asymptotic power efficiency $\gamma = \delta_{\text{free}}^2 / 4\mathcal{E}_b$. Since the amount of bits per symbol carried by uncoded and coded constellations must be the same, $\mathcal{E}'_b / \mathcal{E}_b = \mathcal{E}' / \mathcal{E}$, and we can write the coding gain in the form

$$\eta \triangleq \frac{\gamma}{\gamma'} = \frac{\delta_{\text{free}}^2 / \mathcal{E}_b}{\delta_{\min}^2 / \mathcal{E}'_b} = \frac{\delta_{\text{free}}^2 / \mathcal{E}}{\delta_{\min}^2 / \mathcal{E}'} \tag{7.1}$$

Typically, TCM has $|\mathcal{X}| = 2|\mathcal{X}'|$. A bigger increase is indeed possible if $|\mathcal{X}| > 2|\mathcal{X}'|$ (look again at Figure 3.4 or Figure 4.5), but the performance improvement

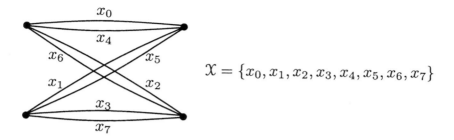

Figure 7.1: *A TCM trellis section with two states and parallel transitions.*

thus obtainable does not seem to compensate for the additional complexity introduced.

A TCM scheme is given by choosing a trellis and assigning to every transition
between pairs of states an element from constellation $\mathcal{X}$. For generality, unlike convolutional codes, in which only one branch can conjoin two states, TCM accepts
parallel transitions, i.e., the presence of two or more branches emanating from,
and merging into, the same state. Figure 7.1 illustrates a trellis section exhibiting
parallel branches.

7.2   Some simple TCM schemes

Before delving into the theory, we examine a few simple examples of TCM schemes
and evaluate their asymptotic coding gains. Consider first the transmission of 2 bits
per signal. Without coding, a constellation with four signals would suffice. Instead,
consider TCM schemes with eight signals.

With 4-PSK we have

$$\frac{\delta_{min}^2}{\mathcal{E}'} = 2$$

a value that provides a baseline for computing the coding gains. Consider next
8-PSK, with signals labeled $\{0, 1, 2, \ldots, 7\}$ as in Figure 7.2. Here,

$$\delta^2 = 4\mathcal{E}\sin^2 \pi/8$$

Two states.   Consider first a trellis with two states (Figure 7.3). If the encoder is
in state S_1, it picks its signals from the subconstellation $\{0, 2, 4, 6\}$. If it is in state
S_2, it picks them from $\{1, 3, 5, 7\}$. The free distance of this TCM scheme may

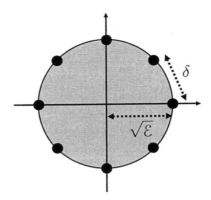

Figure 7.2: *8-PSK as used in a TCM scheme.*

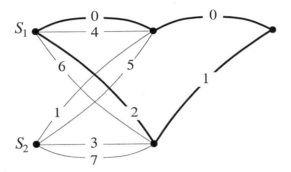

Figure 7.3: *TCM scheme with a two-state trellis. The pair of paths yielding the free distance is shown in bold.*

be equal to the smallest distance between pairs of signals associated with parallel transitions (competing paths with length 1), or to the smallest distance between pairs of paths diverging from a node and remerging after some instants (competing paths with length > 1). Through the use of techniques to be presented later in this chapter, it can be proved that the free distance is given by the pair of paths shown in bold in Figure 7.3. If $d_\mathrm{E}(i, j)$ denotes the Euclidean distance between signals i and j, we have

$$\frac{\delta_\mathrm{free}^2}{\mathcal{E}} = \frac{1}{\mathcal{E}}[d_\mathrm{E}^2(0, 2) + d_\mathrm{E}^2(0, 1)] = 2 + 4\sin^2\frac{\pi}{8} = 2.586$$

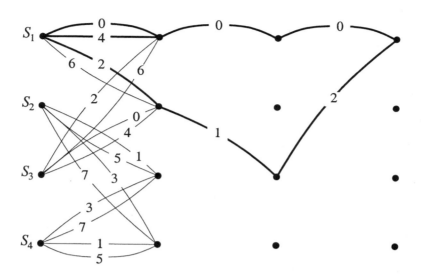

Figure 7.4: *Four-state TCM scheme.*

It follows that the asymptotic coding gain with respect to 4-PSK is

$$\eta = \frac{2.586}{2} = 1.293 \Rightarrow 1.1 \text{ dB}$$

Four states. More states in the TCM trellis yield a larger coding gain. Using again the constellation of Figure 7.2, we now consider the four-state trellis of Figure 7.4. Here we associate the subconstellation $\{0, 2, 4, 6\}$ with states S_1 and S_3, and $\{1, 3, 5, 7\}$ with states S_2 and S_4. In this case the error event generating the free distance δ_{free} has length 1 (parallel transition). This is shown in bold in Figure 7.4, along with another pair of competing paths with length 3 yielding a larger distance. We obtain

$$\frac{\delta_{\text{free}}^2}{\mathcal{E}} = d_{\text{E}}^2(0, 4) = 4$$

which entails that the asymptotic coding gain is

$$\eta = \frac{4}{2} = 2 \Rightarrow 3 \text{ dB}$$

Eight states. We can further increase the asymptotic coding gain by choosing an eight-state trellis as in Figure 7.5. To simplify Figure 7.5, the four symbols

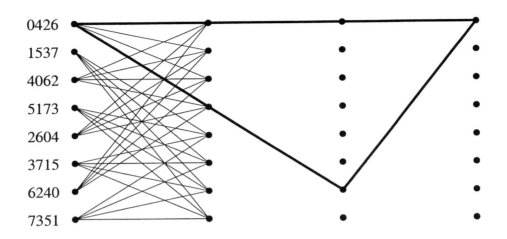

Figure 7.5: *Eight-state TCM scheme with 8-PSK.*

associated with the branches emanating from a node are indicated near the node. The first among the four symbols is associated with the uppermost transition, the second with the second transition from the top, etc. The pair of paths at δ_{free} is also shown. We obtain

$$\frac{\delta_{\text{free}}^2}{\mathcal{E}} = \frac{1}{\mathcal{E}}[d_{\text{E}}^2(0,6) + d_{\text{E}}^2(0,7) + d_{\text{E}}^2(0,6)] = 2 + 4\sin^2\frac{\pi}{8} + 2 = 4.586$$

and hence

$$\eta = \frac{4.586}{2} = 2.293 \Rightarrow 3.6 \text{ dB}$$

Consideration of QAM.   Consider now the transmission of 3 bits per signal and quadrature amplitude modulation (QAM) schemes. The octonary constellation of Figure 7.6 (black dots) will be used as the baseline uncoded scheme. It has

$$\frac{\delta_{\text{min}}^2}{\mathcal{E}'} = 0.8$$

A TCM scheme with eight states and based on this constellation is shown in Figure 7.7. The subconstellations used are

$$\{0, 2, 5, 7, 8, 10, 13, 15\}$$

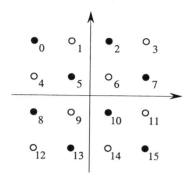

Figure 7.6: *The octonary QAM constellation* $\{0, 2, 5, 7, 8, 10, 13, 15\}$ *and the 16-ary QAM constellation* $\{0, 1, \ldots, 15\}$.

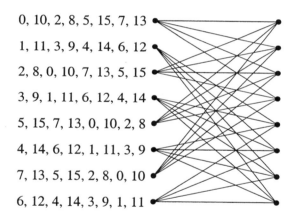

Figure 7.7: *A TCM scheme based on an eight-state trellis,* $M' = 8$, *and* $M = 16$.

and

$$\{1, 3, 4, 6, 9, 11, 12, 14\}$$

The free distance is obtained from

$$\frac{\delta_{\text{free}}^2}{\mathcal{E}} = \frac{1}{\mathcal{E}}[d_{\text{E}}^2(10,13) + d_{\text{E}}^2(0,1) + d_{\text{E}}^2(0,5)]$$
$$= \frac{1}{\mathcal{E}}[0.8\mathcal{E} + 0.4\mathcal{E} + 0.8\mathcal{E}]$$
$$= 2$$

so that

$$\eta = \frac{2}{0.8} = 2.5 \Rightarrow 3.98 \text{ dB}$$

7.2.1 Coding gain of TCM

The values of δ_{free} that can be obtained with TCM schemes based on two-dimensional, unit-energy elementary constellations (PSK and QAM) are summarized in Figure 7.8. Their free distances are expressed in dB and referred to the baseline value $\delta_{\text{min}}^2 = 2$, corresponding to uncoded unit-energy 4-PSK. The abscissa shows R_b/W, the spectral efficiency in bit/dimension pair (or in bit/s/Hz if W is the Shannon bandwidth, as defined in Chapter 3). We can see how relatively large coding gains can be achieved by using TCM schemes with a small number of states: 4, 8, and 16. Convolutional codes are the solution of choice if bandwidth efficiency is not at a premium; otherwise, TCM provides higher values of R_b/W, albeit at the price of a smaller coding gain for a comparable complexity.

7.3 Designing TCM schemes

As mentioned before, to describe a TCM scheme we give its trellis and the map associating each branch with an elementary signal. This map must be chosen in order to maximize the free Euclidean distance, and hence the asymptotic coding gain. In the following, we elaborate on how this maximization can be achieved.

7.3.1 Set partitioning

Consider the evaluation of the free distance δ_{free}, i.e., the Euclidean distance between a pair of trellis paths diverging from a node and remerging after L instants (Figure 7.9).

Let us first examine the case in which the free distance is determined by parallel transitions, viz., $L = 1$. In this case the free distance δ_{free} is equal to the smallest distance between any two signals associated with the parallel transitions. Next, consider $L > 1$; if A, B, C, D denote the subsets of signals associated with each

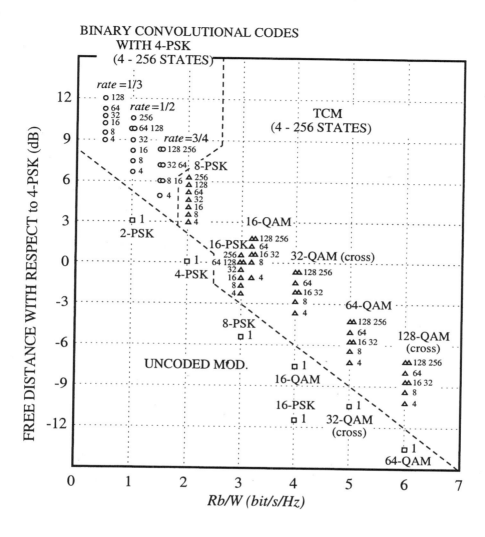

Figure 7.8: *Free distance vs. spectral efficiency of some TCM schemes using two-dimensional unit-energy constellations.*

branch, and $d_E(X,Y)$ is the minimum Euclidean distance between one signal in X and one signal in Y, then δ^2_{free} can be written in the form

$$\delta^2_{\text{free}} = d^2_E(A, B) + \cdots + d^2_E(C, D)$$

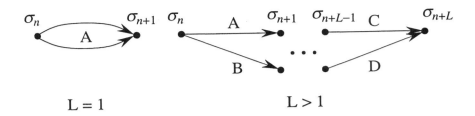

Figure 7.9: *A pair of paths diverging and remerging with $L = 1$ (parallel transitions) and $L > 1$.*

This implies that, for a good TCM scheme, the subsets associated with the same originating state (A and B in Figure 7.9) or with the same terminating state (C and D in Figure 7.9) must be separated by the largest possible distance. These observations form the basis for a technique suggested by Ungerboeck [7.6] and called *set partitioning*.

A constellation with M signals is successively partitioned into $2, 4, 8, \ldots$, subconstellations with size $M/2, M/4, M/8, \ldots$, respectively, having progressively larger minimum Euclidean distances: $\delta_{min}^{(1)}, \delta_{min}^{(2)}, \delta_{min}^{(3)}, \ldots$ (see Figure 7.10).

Three rules are applied, deemed to give rise to the best TCM schemes and called *Ungerboeck rules*:

U1 Parallel transitions are associated with signals belonging to the same subconstellation.

U2 Branches diverging from the same node or merging into the same node are associated with the same subconstellation at the level above that corresponding to rule **U1**.

U3 All signals are used equally often.

All the TCM schemes examined so far in this chapter satisfy U1–U3 (except the one in Figure 7.3: why?).

7.4 Encoders for TCM

In the previous chapter, we saw how the encoder of a code on a trellis can be generated by assigning one or more binary shift registers and a function mapping their contents to the elements of the signal constellation. With TCM, an encoder is

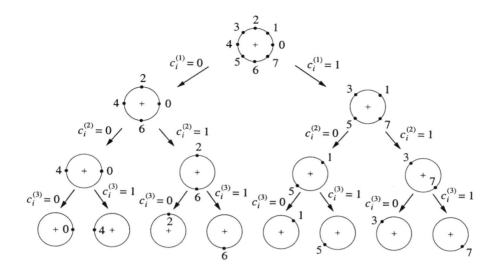

Figure 7.10: *Set-partitioning 8-PSK.*

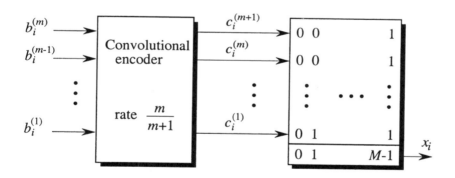

Figure 7.11: *A TCM encoder.*

slightly more complex, as two blocks are present. In the first block, m-tuples of source bits $b_i^{(1)}, \cdots b_i^{(m)}$ are presented simultaneously to a convolutional encoder with rate $m/(m+1)$. The second block consists of a memoryless modulator mapping the convolutionally encoded binary $(m+1)$-tuples onto a signal constellation (see Figure 7.11).

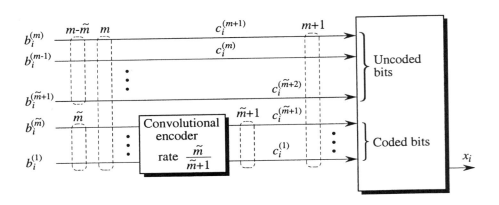

Figure 7.12: *TCM encoder. Here the uncoded bits are explicitly shown.*

It is convenient to modify the encoder scheme of Figure 7.11 as shown in Figure 7.12, which makes it explicit that some input binary digits may be left uncoded, hence generating parallel transitions (in fact, these bits, which do not pass through the memory elements of the convolutional encoder, cannot cause a state change). The convolutional code now has a rate $\tilde{m}/(\tilde{m}+1)$, and each trellis branch is associated with $2^{m-\tilde{m}}$ signals. The correspondence between convolutionally encoded bits and subconstellation signals is shown, for 8-PSK, in Figure 7.10.

Example 7.1

Figure 7.13 shows a TCM encoder and the trellis associated with it. Here $m = 2$ and $\tilde{m} = 1$, so that the nodes of the trellis (corresponding as usual to the states of the encoder) are connected by parallel transitions, each being associated with two signals.   □

7.5   TCM with multidimensional constellations

We have seen that, with a given constellation, the performance of TCM can be improved by increasing the number of trellis states. Nonetheless, when this number exceeds a certain (small) value, the coding gain increases little, in accordance with

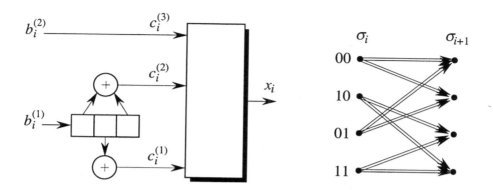

Figure 7.13: *A TCM encoder with* $m = 2$ *and* $\tilde{m} = 1$ *corresponding to the trellis at right.*

the principle of diminishing returns. In this situation, we may choose the solution of changing the constellation $\mathfrak{X}$: an option is to move from two-dimensional constellations to multidimensional constellations.

Consider, in particular, signals generated by taking the N-fold Cartesian product of two-dimensional constellations (such as PSK or QAM). These can be obtained, for example, by transmitting sequentially N two-dimensional signals, each with a duration T/N. The resulting signals can be interpreted as having a duration T and $2N$ dimensions.

Example 7.2

A TCM scheme with four dimensions can be obtained by concatenating two unit-energy 4-PSK signals. The constellation obtained is called 2×4-PSK. With the labels shown in Figure 7.14, the $4^2 = 16$ four-dimensional signals are

$$\{00, 01, 02, 03, 10, 11, 12, 13, 20, 21, 22, 23, 30, 31, 32, 33\}$$

This constellation has the same minimum Euclidean distance as 4-PSK:

$$\delta_{\min}^2 = d_E^2(00, 01) = d_E^2(0, 1) = 2$$

The subconstellation that follows has eight signals:

$$\mathfrak{X} = \{00, 02, 11, 13, 20, 22, 31, 33\}$$

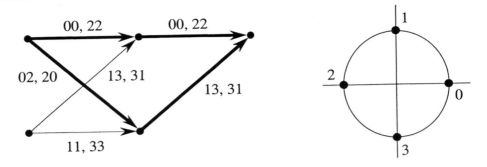

Figure 7.14: *A two-state TCM scheme with a* 2×4*-PSK constellation. The error event giving rise to the free distance is also shown.*

and a minimum squared distance equal to 4. If $\mathfrak{X}$ is partitioned into four subsets

$$\{00, 22\} \qquad \{02, 20\} \qquad \{13, 31\} \qquad \{11, 33\}$$

then the two-state TCM scheme of Figure 7.14 has a squared Euclidean distance equal to 8. $\square$

7.6   TCM transparent to rotations

We now examine channels affected by a phase offset, and the design of TCM for these channels. Consider coherent demodulation. This requires the estimate of the carrier phase before demodulation. Several techniques for this estimate exist, based on the removal of the phase shifts caused by the data transmitted (the *data noise*). This removal generates a carrier whose phase value is affected by an ambiguity that depends on the rotational symmetries of the constellation used. For example, QAM is left invariant by a rotation by multiples of $\pi/2$, and M-ary PSK by a rotation by multiples of $2\pi/M$. The presence of any such rotation cannot be directly detected by the carrier-phase recovery circuit. We model this effect by adding to the received signal a random, data-independent phase shift ϕ taking on values in the set $\{2\pi k/M\}_{k=0}^{M-1}$. To remove this phase shift, differential encoding and decoding are often used.

7.6.1   Differential encoding/decoding

With differential encoding, the information to be transmitted is associated with phase differences rather than with absolute phases. Consider differential encoding of PSK. The model of a channel introducing the phase ambiguity ϕ can be easily constructed by assuming that, when the phase information sequence $(\theta_n)_{n=0}^\infty$ is transmitted, the corresponding received phases are $(\theta_n + \phi)_{n=0}^\infty$. (We neglect noise for simplicity.) While the received phases differ from those transmitted, we may observe that the *phase differences* between adjacent bits are left invariant. Thus, if the information bits are associated with these differences rather than with absolute phases, the value of ϕ has no effect on the received information. We illustrate this by a simple example.

Example 7.3

Consider transmission of binary PSK. If the channel is affected by the phase offset $\phi = \pi$, we receive $(\theta_n + \pi)_{n=0}^\infty$, and hence all received bits differ from those transmitted. To avoid this, transform, before modulation, the information-carrying phases $(\theta_n)_{n=0}^\infty$ into the *differentially encoded* phase sequence $(\theta_n^*)_{n=0}^\infty$ according to the rule

$$\theta_n^* = \theta_n + \theta_{n-1}^* \qquad \text{mod } 2\pi \tag{7.2}$$

where it is assumed that $\theta_{-1}^* = 0$. Next, *differentially decode* the received phase sequence by inverting (7.2):

$$\hat{\theta}_n = \hat{\theta}_n^* - \hat{\theta}_{n-1}^* \qquad \text{mod } 2\pi \tag{7.3}$$

where a hat ˆ denotes phase estimates. We have the situation illustrated in Table 7.1.
□

Information digits	b_n		0	1	1	1	0	1	0	1	1
Corresponding phases	θ_n		0	π	π	π	0	π	0	π	π
Received phases	$\theta_n + \pi$		π	0	0	0	π	0	π	0	0
Detected info. digits	$\hat{b}_n$		1	0	0	0	1	0	1	0	0
Diff. encoded phases	θ_n^*	0	0	π	0	π	π	0	0	π	0
Received phases	$\theta_n^* + \pi$	π	π	0	π	0	0	π	π	0	π
Diff. decoded phases	$\hat{\theta}_n$		0	π	π	π	0	π	0	π	π
Decoded info. digits	$\hat{b}_n$		0	1	1	1	0	1	0	1	1

Table 7.1: *Effect of a rotation by π on uncoded and differentially encoded binary PSK.*

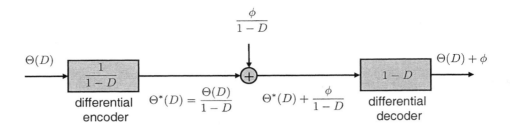

Figure 7.15: *Differential encoding and decoding.*

To explain in general terms the differential coding/encoding procedure, we use D-transform notations, which correspond to describing a semi-infinite sequence $(a_n)_{n=0}^{\infty}$ through the power series $A(D) \triangleq \sum_{n=0}^{\infty} a_n D^n$. With this notation, we write the transmitted phase sequence as

$$\Theta(D) = \theta_0 + \theta_1 D + \theta_2 D^2 + \cdots$$

and the received sequence as

$$
\begin{aligned}
\Theta(D) + \Phi(D) &= (\theta_0 + \phi) + (\theta_1 + \phi)D + (\theta_2 + \phi)D^2 + \cdots \\
&= \Theta(D) + \phi(1 + D + D^2 + \cdots) \\
&= \Theta(D) + \frac{\phi}{1 - D}
\end{aligned}
\tag{7.4}
$$

(we are still neglecting the effect of the additive Gaussian noise).

To get rid of the ambiguity term $\phi/(1-D)$, we multiply the received signal (7.4) by $(1 - D)$. This is accomplished by the *differential decoder*. In the time domain, this circuit subtracts, from the phase received at any instant, the phase that was received in the preceding symbol period: since both phases are affected by the same ambiguity ϕ, this is removed by the difference (except for the phase at time 0). The received sequence is now $(1 - D)\Theta(D) + \phi$, which shows that the ambiguity is now removed (except at the initial time $n = 0$, as reflected by the term ϕ multiplying D^0). Now, the information term $\Theta(D)$ is multiplied by $(1 - D)$: to recover it exactly we must divide $\Theta(D)$, before transmission, by $(1 - D)$. This operation, corresponding to (7.2), is called *differential encoding*. The whole process is summarized in Figure 7.15.

7.6.2   TCM schemes coping with phase ambiguities

If TCM is employed on a channel affected by a phase offset, we must make sure that signal sequences affected by phase rotations are still valid sequences, i.e., cor-

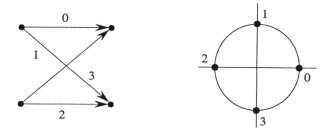

Figure 7.16: *TCM transparent to a phase rotation by* π *but not to a rotation by* $\pi/2$.

respond to paths traversing the trellis; otherwise, a single phase rotation might generate a long error sequence, because even in the absence of noise the TCM decoder cannot recognize what it receives as a valid TCM sequence.

Example 7.4

With the trellis of Figure 7.16, suppose that the all-zero sequence has been transmitted. A rotation by π introduced by the channel causes the reception of the all-2 sequence, which is valid. On the contrary, a rotation by $\pi/2$ generates the all-1 sequence, which is not a valid sequence, and hence will not be recognized as such by the Viterbi algorithm (see the trellis). □

The receiver can solve this ambiguity problem in several ways. One of these consists of transmitting a known "training" sequence that allows one to estimate, and hence to compensate for, the phase rotation introduced by the channel. Another one uses a code that is not transparent to rotations. A phase error generates a sequence that, not being recognized as valid by the decoder, triggers a phase-compensation circuit. The third solution, which we shall examine in some detail, is based on the design of a TCM scheme based on M-PSK and transparent to rotations: for it, every rotation by $2\pi/M$ of a TCM sequence generates another valid sequence so that the decoder will not be affected.

For a TCM scheme to be transparent to a certain set of rotations, we require the following:

1. Any rotated TCM sequence is a TCM sequence.

2. Any rotated TCM sequence is decoded into the same source-symbol sequence.

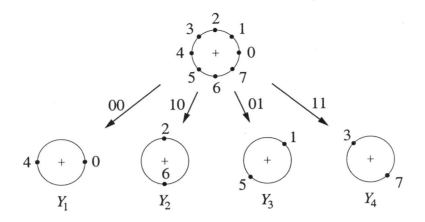

Figure 7.17: *Partition of 8-PSK transparent to rotations.*

The first property above is geometric: the coded sequences, which can be interpreted as a set of points in a Euclidean space with an infinite number of dimensions, must be invariant with respect to a given finite set of rotations. The second property is rather a structural property of the encoder: in fact, it is related to the input–output relationship of the encoder.

Partitions transparent to rotations.   The first fundamental principle in the design of a code transparent to rotations is the construction of a transparent partition.

Let $\mathcal{X}$ be a $2N$-dimensional signal constellation, $\{Y_1, \cdots, Y_K\}$ its partition into K subconstellations, and consider the rotations in the two-dimensional Euclidean space. The rotations of the $2N$-dimensional space are obtained by considering a separate rotation in each two-dimensional subspace. Next, consider the set of rotations that leave $\mathcal{X}$ invariant, and denote it by $R(\mathcal{X})$. If $R(\mathcal{X})$ leaves the partition invariant, that is, if the effect of every element of $R(\mathcal{X})$ on the partition is simply a permutation of its elements, then the partition is called *transparent to rotations*.

Example 7.5

Consider an 8-PSK constellation and its partition into four subsets of signals as in Figure 7.17. Let the elements of $R(\mathcal{X})$, the set (group) of rotations by multiples of $\pi/4$, be denoted by $\rho_0, \rho_{\pi/4}, \rho_{\pi/2}$, etc. This partition is transparent to rotations. For example, $\rho_{\pi/4}$ corresponds to the permutation $(Y_1 Y_3 Y_2 Y_4)$, $\rho_{\pi/2}$ to the permutation $(Y_1 Y_2)(Y_3 Y_4)$, ρ_π to the identity permutation, etc.   □

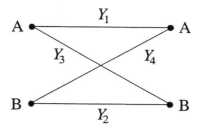

Figure 7.18: *Section of a two-state trellis.*

Example 7.6

Consider the four-dimensional signal set of Example 7.2 and its partition into the four subconstellations

$$Y = \{Y_1, Y_2, Y_3, Y_4, Y_5, Y_6, Y_7, Y_8\} \tag{7.5}$$
$$= \{\{00, 22\}, \{11, 33\}, \{02, 20\}, \{13, 31\},$$
$$\{01, 23\}, \{12, 30\}, \{03, 21\}, \{10, 32\}\} \tag{7.6}$$

The elements of $R(\mathcal{X})$ are the pairs of rotations, each a multiple of $\pi/2$, denoted ρ_0, $\rho_{\pi/2}$, ρ_π, and $\rho_{3\pi/2}$. We can see, for instance, that the effect of $\rho_{\pi/2}$ on signal xy is to change it into a signal $(x + 1)(y + 1)$, with sums taken mod 4. This partition is transparent to rotations. □

Trellises transparent to rotations.

Consider now the effect of a phase rotation on coded TCM sequences. If the partition Y of $\mathcal{X}$ is transparent to rotations, the TCM scheme becomes transparent to rotations if every rotation $\rho \in R(\mathcal{X})$ transforms a valid subconstellation sequence into a valid subconstellation sequence.

Examine a trellis section. If all the subconstellations labeling the trellis branches are affected by the same rotation ρ, we generally obtain a different trellis section. Now, for the TCM scheme to be transparent to rotations, this transformed trellis section must be equal to the unrotated section, apart from a possible permutation of its states.

Example 7.7

Consider a section of a two-state trellis (Figure 7.18) based on the partition

$$Y = \{Y_1, Y_2, Y_3, Y_4\}$$

(we use the same notations of Figure 7.17). This partition is transparent to rotations. We describe this trellis section by giving the set of its branches (S_i, Y_j, S_k), where Y_j is the subconstellation labeling the branch joining state S_j to state S_k. This trellis is described by the set

$$\mathcal{T} = \{(A, Y_1, A), (A, Y_3, B), (B, Y_4, A), (B, Y_2, B)\}$$

The rotations $\rho_{\pi/2}$ and $\rho_{3\pi/2}$ transform $\mathcal{T}$ into

$$\{(A, Y_2, A), (A, Y_4, B), (B, Y_3, A), (B, Y_1, B)\}$$

which corresponds to the permutation (A, B) of the states of $\mathcal{T}$. Similarly, ρ_0 and ρ_π correspond to the identity permutation. In conclusion, this TCM scheme is transparent to rotations.   □

It may happen that a TCM scheme satisfies the conditions of rotational invariance only for a subset of $R(\mathcal{X})$, rather than for the whole of $R(\mathcal{X})$. In this case we say that $\mathcal{X}$ is *partially* transparent to rotations.

Example 7.8

Consider the TCM scheme of Figure 7.19. This has eight states, and its subconstellations correspond to the partition $Y = \{Y_1, Y_2, Y_3, Y_4\}$ of Figure 7.17. This partition, as we know, is transparent to rotations. However, the TCM scheme is not fully transparent. If we consider the effect of a rotation by $\pi/4$ (Table 7.2), we can see that the effect of $\rho_{\pi/4}$ is not a simple permutation of the trellis states. In fact, take the branch (S_1, Y_3, S_1); in the initial trellis, there is no branch of the type (S_i, Y_3, S_i). This TCM scheme is partially transparent: in fact, it is transparent to rotations by multiples of $\pi/2$. For instance, the effect of $\rho_{\pi/2}$ is described in Table 7.2: it generates the following permutation of its states: $(S_1 S_8)(S_2 S_7)(S_3 S_6)(S_4 S_5)$.   □

Transparent encoder.   Consider finally the transparency of the encoder. We require every rotation of a TCM sequence to correspond to the same information sequence. If **u** denotes a sequence of source symbols, and **y** the corresponding sequence of subconstellations, then every rotation $\rho(\mathbf{y})$ to which the TCM scheme is transparent must correspond to the same sequence **u**.

We observe that for this condition to be satisfied it is sometimes necessary to introduce a differential encoder. This point is illustrated by the example that follows.

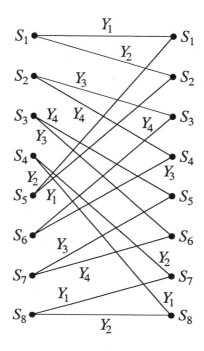

Figure 7.19: *A TCM scheme not transparent to rotations.*

Example 7.9

Consider again the eight-state TCM and the 8-PSK constellation of Example 7.7. We have seen in Example 7.5 that rotations by $\pi/2$ and $3\pi/2$ generate the permutation $(Y_1 Y_2)(Y_3 Y_4)$. If the encoder associates source symbols to constellations according to the rule:

$$00 \Rightarrow Y_1 \qquad 10 \Rightarrow Y_2 \qquad 01 \Rightarrow Y_3 \qquad 11 \Rightarrow Y_4$$

then the only effect of $\rho_{\pi/2}$ and of $\rho_{3\pi/2}$ is to change the first bit of the pair of binary source symbols, while ρ_0 and ρ_π change no bit. Thus, if the first bit is differentially encoded, the resulting TCM scheme will be transparent to rotations by multiples of $\pi/2$. $\qquad\qquad\square$

General considerations

We can generally say that the constraints of transparency to rotations may entail a reduction of the coding gain of two-dimensional TCM. Generating an encoder

ρ_0	$\rho_{\pi/4}$	$\rho_{\pi/2}$
(S_1, Y_1, S_1)	(S_1, Y_3, S_1)	(S_1, Y_2, S_1)
(S_1, Y_2, S_2)	(S_1, Y_4, S_2)	(S_1, Y_1, S_2)
(S_2, Y_3, S_3)	(S_2, Y_2, S_3)	(S_2, Y_4, S_3)
(S_2, Y_4, S_4)	(S_2, Y_1, S_4)	(S_2, Y_3, S_4)
(S_3, Y_4, S_5)	(S_3, Y_1, S_5)	(S_3, Y_3, S_5)
(S_3, Y_3, S_6)	(S_3, Y_2, S_6)	(S_3, Y_4, S_6)
(S_4, Y_2, S_7)	(S_4, Y_4, S_7)	(S_4, Y_1, S_7)
(S_4, Y_1, S_8)	(S_4, Y_3, S_8)	(S_4, Y_2, S_8)
(S_5, Y_2, S_1)	(S_5, Y_4, S_1)	(S_5, Y_1, S_1)
(S_5, Y_1, S_2)	(S_5, Y_3, S_2)	(S_5, Y_2, S_2)
(S_6, Y_4, S_3)	(S_6, Y_1, S_3)	(S_6, Y_3, S_3)
(S_6, Y_3, S_4)	(S_6, Y_2, S_4)	(S_6, Y_4, S_4)
(S_7, Y_3, S_5)	(S_7, Y_2, S_5)	(S_7, Y_4, S_5)
(S_7, Y_4, S_6)	(S_7, Y_1, S_6)	(S_7, Y_3, S_6)
(S_8, Y_1, S_7)	(S_8, Y_3, S_7)	(S_8, Y_2, S_7)
(S_8, Y_2, S_8)	(S_8, Y_4, S_8)	(S_8, Y_1, S_8)

Table 7.2: *Effect of rotations by $\pi/4$ and $\pi/2$ on the TCM scheme of Example 7.8.*

transparent to rotations may require a nonlinear convolutional code. The loss in performance caused by the invariance constraints is generally lower for multidimensional constellations.

7.7   Decoding TCM

As discussed in Section 5.3, due to the one-to-one correspondence between signal sequences and paths traversing the trellis, maximum-likelihood (ML) decoding over a stationary memoryless channel consists of searching for the trellis path with the maximum metric $m(\mathbf{y} \mid \mathbf{x})$. This is done by using the Viterbi algorithm. The branch metrics to be used are obtained as follows. If a branch of the trellis describing the code is labeled by signal x_i, then at discrete time i the metric associated with that branch is $m(y_i \mid x_i)$ if there are no parallel transitions. If a pair of nodes is connected by parallel transitions, with branches labeled x', x'', ..., in the set $\mathcal{X}^*$, then in the trellis used for decoding the same pair of nodes is connected by a single branch, whose metric is

$$\min_{x \in \mathcal{X}^*} m(y_i \mid x)$$

That is, in the presence of parallel transitions, the decoder first selects the signal with the maximum metric among x', x'', ..., (this is a *demodulation* operation), and then associates with a single branch the metric of the signal selected.

7.8 Error probability of TCM

This section is devoted to the calculation of error probability of TCM. We assume transmission over the AWGN channel, and maximum-likelihood detection. It should not come as a surprise that, asymptotically for large SNRs, upper and lower bounds to error probability decrease as δ_{free} increases. This finding shows that the free Euclidean distance is a sensible parameter for comparing TCM schemes used over the high-SNR AWGN channel.

No general constructive technique is available for the design of an optimal TCM scheme: hence, sensible designs should be based on a search among a set of schemes satisfying Ungerboeck rules. For this reason, it is important to ensure that a fast and accurate method is available for the evaluation of free distance and error probability.

7.8.1 Upper bound to the probability of an error event

A rate-$m/(m+1)$ convolutional code accepts blocks $\mathbf{b}_i$ of m binary source symbols each, and transforms them into blocks $\mathbf{c}_i$ of $m + 1$ binary symbols, each of these to be presented at the input of a nonlinear memoryless mapper (Figure 7.11). This mapper outputs the elementary signals x_i. From now on, the binary $(m + 1)$-tuple $\mathbf{c}_i$ will be called the *label* of signal x_i.

There is a one-to-one correspondence between each x_i and its label $\mathbf{c}_i$: thus, two sequences $\mathbf{x}_L$ and $\widehat{\mathbf{x}}_L$ of L signals each can be equivalently described by the sequences $\mathbf{C}_L$ and $\widehat{\mathbf{C}}_L$ of their labels, i.e,

$$\mathbf{C}_L = (\mathbf{c}_k, \mathbf{c}_{k+1}, \ldots, \mathbf{c}_{k+L-1})$$

and

$$\widehat{\mathbf{C}}_L = (\widehat{\mathbf{c}}_k, \widehat{\mathbf{c}}_{k+1}, \ldots, \widehat{\mathbf{c}}_{k+L-1})$$

where

$$\widehat{\mathbf{c}}_k = \mathbf{c}_k + \mathbf{e}_k, \quad \widehat{\mathbf{c}}_{k+1} = \mathbf{c}_{k+1} + \mathbf{e}_{k+1}, \quad \cdots \quad , \widehat{\mathbf{c}}_{k+L-1} = \mathbf{c}_{k+L-1} + \mathbf{e}_{k+L-1}$$

where $\mathbf{e}_i$, $i = k, \ldots, k+L-1$, is a sequence of binary vectors, called *error vectors*, and $+$ denotes mod-2 addition.

An *error event* of length L, as defined in Section 6.4, occurs when the demodulator chooses, in lieu of the transmitted sequence $\mathbf{x}_L$, the sequence $\widehat{\mathbf{x}}_L \neq \mathbf{x}_L$ corresponding to a path along the trellis that diverges from the correct path at a certain instant and remerges into it exactly L time instants later. The error probability (defined here, as we did in Section 6.4, as the probability of an error event) is obtained by summing over L, $L = 1, 2, \cdots$, the probabilities of the error events of length L, i.e., the joint probabilities that $\mathbf{x}_L$ is transmitted and $\widehat{\mathbf{x}}_L \neq \mathbf{x}_L$ decoded.

The union bound yields the following inequality for error probability:

$$P(e) \leq \frac{1}{|\Sigma|} \sum_{L=1}^{\infty} \sum_{\mathbf{x}_L} \sum_{\widehat{\mathbf{x}}_L \neq \mathbf{x}_L} P\{\mathbf{x}_L\} P\{\mathbf{x}_L \to \widehat{\mathbf{x}}_L\} \tag{7.7}$$

where $|\Sigma|$ is the number of trellis states.[1]

By exploiting once again the one-to-one correspondence between output signals and their labels, if $\mathbf{C}_L$ denotes a label sequence with length L, and $\mathbf{E}_L$ a sequence (with the same length L) of error vectors $\mathbf{e}_i$, we can rewrite (7.7) in the equivalent form

$$
\begin{aligned}
P(e) &\leq \frac{1}{|\Sigma|} \sum_{L=1}^{\infty} \sum_{\mathbf{C}_L} \sum_{\widehat{\mathbf{C}}_L \neq \mathbf{C}_L} P\{\mathbf{C}_L\} P\{\mathbf{C}_L \to \widehat{\mathbf{C}}_L\} \\
&= \frac{1}{|\Sigma|} \sum_{L=1}^{\infty} \sum_{\mathbf{C}_L} \sum_{\mathbf{E}_L \neq 0} P\{\mathbf{C}_L\} P\{\mathbf{C}_L \to \mathbf{C}_L + \mathbf{E}_L\} \\
&= \frac{1}{|\Sigma|} \sum_{L=1}^{\infty} \sum_{\mathbf{E}_L \neq 0} P\{\mathbf{E}_L\}
\end{aligned}
\tag{7.8}
$$

where

$$P\{\mathbf{E}_L\} \triangleq \sum_{\mathbf{C}_L} P\{\mathbf{C}_L\} P\{\mathbf{C}_L \to \mathbf{C}_L + \mathbf{E}_L\} \tag{7.9}$$

expresses the probability of a specific error event with length L generated by the error sequence $\mathbf{E}_L$. Although the PEP appearing in the last equation can be calculated exactly, we shall rather use a simple upper bound, which opens the way to a transfer-function approach to the calculation of $P(e)$.

---

[1]Division by $|\Sigma|$ does not take place for convolutional codes, since their linearity makes it possible to assume that all error events start and end in the same state.

Let $f(\mathbf{c})$ denote the signal whose label is $\mathbf{c}$, and $\mathbf{f}(\mathbf{C}_L)$ the signal sequence whose label is $\mathbf{C}_L$. Using the bound $Q(x) \leq \exp(-x^2/2)$, $x \geq 0$, we obtain

$$
\begin{aligned}
P\{\mathbf{C}_L \rightarrow \mathbf{C}_L + \mathbf{E}_L\} &= Q\left(\frac{\|\mathbf{f}(\mathbf{C}_L) - \mathbf{f}(\mathbf{C}_L + \mathbf{E}_L)\|}{\sqrt{2N_0}}\right) \\
&\leq \exp\left\{-\frac{1}{4N_0}\|\mathbf{f}(\mathbf{C}_L) - \mathbf{f}(\mathbf{C}_L + \mathbf{E}_L)\|^2\right\} \quad (7.10)
\end{aligned}
$$

Define now the function

$$
W(\mathbf{E}_L) \triangleq \sum_{\mathbf{C}_L} P\{\mathbf{C}_L\} e^{-\|\mathbf{f}(\mathbf{C}_L) - \mathbf{f}(\mathbf{C}_L + \mathbf{E}_L)\|^2/4N_0} \quad (7.11)
$$

This allows us to finally bound the error probability $P(e)$ in the form

$$
P(e) \leq \frac{1}{|\Sigma|} \sum_{L=1}^{\infty} \sum_{\mathbf{E}_L \neq \mathbf{0}} W(\mathbf{E}_L) \quad (7.12)
$$

Equality (7.12) shows that $P(e)$ is upper bounded by a sum, extended to all possible lengths of the error events, of functions of the vectors $\mathbf{E}_L$ generating them. We shall enumerate these vectors. Before proceeding further, we observe here that a technique often used to evaluate error probabilities (especially for TCM schemes with a large number of states, or being transmitted on non-AWGN channels) consists of including in (7.12) a finite number of terms, chosen among those with a small value of L. It is expected that these terms contribute to the smallest Euclidean distances, so they should provide the most relevant contribution to error probability. However, this technique should be used with the utmost care, because truncating the series does not necessarily yield an upper bound.

Enumerating the error events

We show now how $W(\mathbf{E}_L)$ can be computed. The error vectors can be enumerated by looking for the transfer function of an error state diagram, i.e., a graph whose branch labels are $|\Sigma| \times |\Sigma|$ matrices. Observe first that we can write

$$
\|\mathbf{f}(\mathbf{C}_L) - \mathbf{f}(\mathbf{C}_L + \mathbf{E}_L)\|^2 = \sum_{\ell=1}^{L} |f(\mathbf{c}_\ell) - f(\mathbf{c}_\ell + \mathbf{e}_\ell)|^2 \quad (7.13)
$$

Next, observe that, under our assumptions, all L-tuples of labels have the same probability 2^{-mL} to be transmitted, and let us define the $|\Sigma| \times |\Sigma|$ *error weight* matrix $\mathbf{G}(\mathbf{e}_\ell)$ as follows: The component i,j of $\mathbf{G}(\mathbf{e}_\ell)$ is equal to zero if no transition is allowed between trellis states S_i and S_j; otherwise, it is given by

$$[\mathbf{G}(\mathbf{e}_\ell)]_{i,j} = 2^{-m} \sum_{\mathbf{c}_{i \to j}} X^{|f(\mathbf{c}_{i \to j}) - f(\mathbf{c}_{i \to j} + \mathbf{e}_\ell)|^2} \tag{7.14}$$

where X is an indeterminate and $\mathbf{c}_{i \to j}$ are the vectors of the labels generated by the transition going from state S_i to state S_j. The sum in (7.14) accounts for the parallel transitions that may occur between these two states.

To every sequence $\mathbf{E}_L = (\mathbf{e}_1, \cdots, \mathbf{e}_L)$ of labels in the error state diagram, there corresponds a sequence of L matrices of error weights $\mathbf{G}(\mathbf{e}_1), \ldots, \mathbf{G}(\mathbf{e}_L)$, with

$$W(\mathbf{E}_L) = \mathbf{1}' \left[\prod_{\ell=1}^{L} \mathbf{G}(\mathbf{e}_\ell) \right] \mathbf{1} \Bigg|_{X = \exp(-1/4N_0)} \tag{7.15}$$

where $\mathbf{1}$ is the column vector all of whose components are 1. (Consequently, if $\mathbf{A}$ denotes a $|\Sigma| \times |\Sigma|$ matrix, then $\mathbf{1}'\mathbf{A}\mathbf{1}$ is the sum of all entries of $\mathbf{A}$.) The entry i,j of matrix $\prod_{n=1}^{L} \mathbf{G}(\mathbf{e}_\ell)$ enumerates the Euclidean distances generated by the transitions from state S_i to state S_j in exactly L steps.

Now, to compute $P(e)$ we should sum $W(\mathbf{E}_L)$ over all possible error sequences $\mathbf{E}_L$. To this purpose, we use (7.12).

The error state diagram. Consider now the enumeration of the error sequences $\mathbf{E}_L$. Assume first the simpler situation of no parallel transitions in the TCM trellis. The nonzero $\mathbf{E}_L$ correspond to error events in the convolutional code: in fact, since the latter is linear, and $\mathbf{C}_L$ and $\widehat{\mathbf{C}}_L = \mathbf{C}_L + \mathbf{E}_L$ are admissible label sequences in an error event, then also $\mathbf{E}_L = \widehat{\mathbf{C}}_L + \mathbf{C}_L$ is an admissible label sequence. Thus, the error sequences can be described by using the same trellis associated with the encoder and can be enumerated by using a state diagram that mimics that of the code. This diagram is called the *error state diagram*. Its structure is uniquely determined by the convolutional code underlying the TCM scheme, and differs from the code state diagram only in its branch labels, which are now the matrices $\mathbf{G}(\mathbf{e}_\ell)$.

If there are parallel transitions, each one of the $\mathbf{e}_\ell$ can be decomposed in the form $\mathbf{e}_\ell = (\mathbf{e}'_\ell, \mathbf{e}''_\ell)$, where $\mathbf{e}'_\ell$ contains $m - \tilde{m}$ "unconstrained" 0s and 1s generated by the uncoded bits, and $\mathbf{e}''_\ell$ contains the $\tilde{m} + 1$ components that are constrained by the structure of the convolutional code. Thus, the set of possible sequences $\mathbf{E}''_L \triangleq (\mathbf{e}''_1, \ldots, \mathbf{e}''_L)$ is the same as the set of convolutionally encoded sequences, and $\mathbf{E}_L = \widehat{\mathbf{C}}_L + \mathbf{C}_L$ is again an admissible label sequence. In this case, the branch

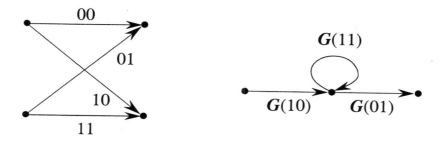

Figure 7.20: *Trellis diagram of a TCM scheme with two states and $m = 1$ (the branch labels are the components of* **c***) and error state diagram.*

labels of the error state diagram are the matrices $\sum \mathbf{G}(\mathbf{e}'_\ell, \mathbf{e}''_\ell)$, where the sum is taken with respect to all values of $\mathbf{e}'_\ell$.

Transfer-function bound.   Using (7.12) and (7.15), we can write

$$P(e) \leq \frac{1}{|\Sigma|} T(X) \Big|_{X=\exp(-1/4N_0)} \tag{7.16}$$

where

$$T(X) = \mathbf{1}'\mathbf{G}\mathbf{1} \tag{7.17}$$

and the matrix

$$\mathbf{G} \triangleq \sum_{L=1}^{\infty} \sum_{\mathbf{E}_L \neq \mathbf{0}} \prod_{\ell=1}^{L} \mathbf{G}(\mathbf{e}_\ell) \tag{7.18}$$

is the transfer function of the error state diagram. We call $T(X)$ the scalar transfer function of the error state diagram.

Example 7.10

Consider the TCM scheme whose trellis diagram is shown in Figure 7.20, where $m = 1$ and $M = 4$ (binary source, quaternary constellation). The error state diagram is also shown in the figure. Denoting by $\mathbf{e} = (e_2 e_1)$ the error vector and by $\bar{\mathbf{e}} = 1 + e$ ($\bar{e}$ the complement of e), we can write the general form of matrix $\mathbf{G}(\mathbf{e})$ as follows:

$$\mathbf{G}(e_2 e_1) = \frac{1}{2} \begin{bmatrix} X^{|f(00)-f(e_2 e_1)|^2} & X^{|f(10)-f(\bar{e}_2 e_1)|^2} \\ X^{|f(01)-f(e_2 \bar{e}_1)|^2} & X^{|f(11)-f(\bar{e}_2 \bar{e}_1)|^2} \end{bmatrix} \tag{7.19}$$

The transfer function of the error state diagram is

$$\mathbf{G} = \mathbf{G}(10)\,[\mathbf{I}_2 - \mathbf{G}(11)]^{-1}\,\mathbf{G}(01) \qquad (7.20)$$

with $\mathbf{I}_2$ the 2×2 identity matrix.

We can observe that (7.19) and (7.20) can be written without knowing the signals used for TCM. In fact, giving the constellation is tantamount to giving the four values of function $f(\cdot)$. In turn, these values provide those of the entries of $\mathbf{G}(e_2e_1)$ for which the transfer function $T(X)$ is computed.

Consider first 4-PAM, with the following correspondence between labels and signals:

$$f(00) = +3 \quad f(01) = +1 \quad f(10) = -1 \quad f(11) = -3$$

In this case we have

$$\mathbf{G}(00) = \frac{1}{2}\begin{bmatrix} 1 & 1 \\ 1 & 1 \end{bmatrix} \qquad (7.21)$$

$$\mathbf{G}(01) = \frac{1}{2}\begin{bmatrix} X^4 & X^4 \\ X^4 & X^4 \end{bmatrix} \qquad (7.22)$$

$$\mathbf{G}(10) = \frac{1}{2}\begin{bmatrix} X^{16} & X^{16} \\ X^{16} & X^{16} \end{bmatrix} \qquad (7.23)$$

and

$$\mathbf{G}(11) = \frac{1}{2}\begin{bmatrix} X^{36} & X^4 \\ X^4 & X^{36} \end{bmatrix} \qquad (7.24)$$

which allows us to obtain, from (7.20),

$$\mathbf{G} = \frac{X^{20}}{2 - X^4 - X^{36}}\begin{bmatrix} 1 & 1 \\ 1 & 1 \end{bmatrix} \qquad (7.25)$$

Finally, the transfer function has the form

$$T(X) = \mathbf{1}'\mathbf{G}\mathbf{1} = \frac{4X^{20}}{2 - X^4 - X^{36}} \qquad (7.26)$$

If we consider a unit-energy 4-PSK constellation as shown in Figure 7.21, we obtain

$$f(00) = 1 \quad f(01) = j \quad f(10) = -1 \quad f(11) = -j$$

and hence

$$\mathbf{G}(00) = \frac{1}{2}\begin{bmatrix} 1 & 1 \\ 1 & 1 \end{bmatrix} \qquad (7.27)$$

$$\mathbf{G}(01) = \frac{1}{2}\begin{bmatrix} X^2 & X^2 \\ X^2 & X^2 \end{bmatrix} \qquad (7.28)$$

$$\mathbf{G}(10) = \frac{1}{2}\begin{bmatrix} X^4 & X^4 \\ X^4 & X^4 \end{bmatrix} \qquad (7.29)$$

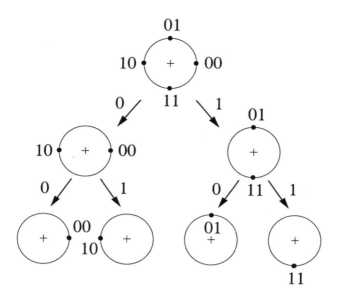

Figure 7.21: *4-PSK signal constellation and its partition.*

and

$$\mathbf{G}(11) = \frac{1}{2} \begin{bmatrix} X^2 & X^2 \\ X^2 & X^2 \end{bmatrix} \tag{7.30}$$

Finally,

$$\mathbf{G} = \frac{1}{2} \frac{X^6}{1 - X^2} \begin{bmatrix} 1 & 1 \\ 1 & 1 \end{bmatrix} \tag{7.31}$$

so the transfer function is

$$T(X) = \mathbf{1}'\mathbf{G1} = 2 \frac{X^6}{1 - X^2} \tag{7.32}$$

$\square$

Interpretation and symmetry considerations

Examining the matrix $\mathbf{G}$ defined in (7.18), we can observe that its entry i, j provides us with an upper bound to the probability that an error event starts at node S_i and ends at node S_j. Similarly, $\mathbf{G1}$ is a vector whose entry i is an upper bound to an error event starting at node S_i, and $\mathbf{1}'\mathbf{G}$ is a vector whose entry j is an upper bound to the probability of all error events ending at node S_j.

By inspection of matrix **G** we can observe that different levels of symmetry may occur in a TCM scheme. Matrix **G** may have all its elements equal, as in (7.31). We interpret this by saying that all paths in the trellis bring the same contribution to error probability (more precisely, to the *upper bound* to error probability). In the context of the analysis of TCM, we can choose a single reference path and compute the error probability under the assumption that this path corresponds to the coded sequence that has been transmitted. A sufficient condition for this to occur is that all matrices **G**(**e**) have equal components. However, this condition is not necessary, as we can observe by considering the 4-PAM case of Example 7.10: **G** has equal entries, although those of **G**(11) are not equal.

If all matrices **G**(**e**) have equal components, then, in the calculation of the transfer-function bound, the error state diagram branches can be simply labeled by the common entry of these matrices, which yields a scalar transfer function. However, for this to be possible, the code needs a high degree of symmetry. Actually, what is needed is only the following weaker form of symmetry: *the sum of all entries of a row (or a column) of **G** is the same for all rows (or columns)*. With this symmetry, since all states play the same role, a single state can be chosen as a reference, rather that a state pair. It suffices to consider only the error events starting from a certain state (when all rows have the same sum) or merging into the same state (when all columns have the same sum).

Algebraic conditions for a scalar transfer function. Here we derive simple conditions to have a graph whose labels are scalars rather than matrices.

If **A** is a square matrix, and **1** is the eigenvector of its transpose **A**′, that is,

$$\mathbf{1}'\mathbf{A} = \alpha\mathbf{1}'$$

where α is a constant, then the sum of the components of any column of **A** does not depend on the column index. We say that **A** is *column-uniform*. Similarly, if **1** is an eigenvalue of the square matrix **B**, that is, if

$$\mathbf{B}\mathbf{1} = \beta\mathbf{1}$$

where β is a constant, then the sum of the components of any row does not depend on the row index. In this case we say that **B** is *row-uniform*.

Now, the product and the sum of two (row- or column-) uniform matrices are uniform matrices. For example, if $\mathbf{B}_1$ and $\mathbf{B}_2$ are row-uniform with eigenvalues β_1 and β_2, then $\mathbf{B}_3 \triangleq \mathbf{B}_1 + \mathbf{B}_2$ and $\mathbf{B}_4 \triangleq \mathbf{B}_1\mathbf{B}_2$ satisfy the following relationships:

$$\mathbf{B}_3\mathbf{1} = (\beta_1 + \beta_2)\mathbf{1}$$

and

$$\mathbf{B}_4 \mathbf{1} = \beta_1 \beta_2 \mathbf{1}$$

which show that $\mathbf{B}_3$ and $\mathbf{B}_4$ are also row-uniform, with eigenvalues $\beta_1 + \beta_2$ and $\beta_1 \beta_2$. Moreover, for a (row- or column-) uniform matrix $\mathbf{A}$ of order N, we have

$$\mathbf{1}' \mathbf{A} \mathbf{1} = N\alpha$$

It follows from this discussion that, if all matrices $\mathbf{G}(\mathbf{e})$ are row-uniform or column-uniform, the transfer function (which is a sum of matrix products, as we can see from (7.18)) can be computed using only scalar labels in the error state diagram. These labels are sums of the row (or column) components (notice that, when scalar labels are used, the resulting transfer function should be multiplied by $|\Sigma|$ to be consistent with the original definition (7.17)). In this case we say that the TCM scheme is *uniform*.[2] From the definition of matrices $\mathbf{G}(\mathbf{e})$, we can observe that $\mathbf{G}(\mathbf{e})$ is row-uniform if the transitions splitting from any trellis node carry the same label set (the order of the transitions is not relevant). $\mathbf{G}(\mathbf{e})$ is column-uniform if the transitions leading to any trellis node carry the same label set.

Asymptotics

For large signal-to-noise ratios, i.e., when $N_0 \to 0$, the only elements of the matrix $\mathbf{G}$ that contribute significantly to error probability are those proportional to $X^{\delta_{\text{free}}^2}$. Hence, asymptotically,

$$P(e) \sim \nu(\delta_{\text{free}}) e^{-\delta_{\text{free}}^2 / 4N_0}$$

where $\nu(\delta_{\text{free}})$ is the average number of competing paths at distance δ_{free}.

A tighter upper bound

A better approximation to $P(e)$ can be obtained by using in (7.10) the tighter bound (3.66). Recall that we have, exactly,

$$P\{\mathbf{C}_L \to \mathbf{C}_L + \mathbf{E}_L\} = Q\left(\frac{\|\mathbf{f}(\mathbf{C}_L) - \mathbf{f}(\mathbf{C}_L + \mathbf{E}_L)\|}{\sqrt{2N_0}} \right) \qquad (7.33)$$

Since the minimum value of $\|\mathbf{f}(\mathbf{C}_L) - \mathbf{f}(\mathbf{C}_L + \mathbf{E}_L)\|$ equals δ_{free}, then, by using the inequality (3.66), we obtain

$$P\{\mathbf{C}_L \to \widehat{\mathbf{C}}_L\} \le Q\left(\frac{\delta_{\text{free}}}{\sqrt{2N_0}} \right) e^{\delta_{\text{free}}^2 / 4N_0} \cdot \exp\left\{ -\frac{1}{4N_0} \|\mathbf{f}(\mathbf{C}_L) - \mathbf{f}(\mathbf{C}_L + \mathbf{E}_L)\|^2 \right\}$$

$$(7.34)$$

---

[2]This uniformity is weaker than the "geometric uniformity" introduced in Section 3.6. See [7.5].

In conclusion, we obtain the following error-probability bound:

$$P(e) \leq Q\left(\frac{\delta_{\text{free}}}{\sqrt{2N_0}}\right) e^{\delta_{\text{free}}^2/4N_0} T(X)\bigg|_{X=e^{-1/4N_0}} \qquad (7.35)$$

Lower bound to error probability

A lower bound to the probability of an error event can also be computed. Our calculations rely upon the fact that the error probability of a real-life decoder is bigger than that of an ideal decoder using side information provided by a benevolent genie. The *genie-aided decoder* operates as follows. The genie observes a long sequence of transmitted symbols, or, equivalently, the label sequence

$$\mathbf{C} = (\mathbf{c}_i)_{i=0}^{K-1}$$

and informs the decoder that the transmitted sequence is either $\mathbf{C}$ *or* the sequence

$$\mathbf{C}' = (\mathbf{c}_i')_{i=0}^{K-1}$$

where $\mathbf{C}'$ is selected at random among the possible transmitted sequences at the smallest Euclidean distance from $\mathbf{C}$ (this is not necessarily δ_{free}, because $\mathbf{C}$ may not have a sequence $\mathbf{C}'$ at free distance).

The error probability of this genie-aided receiver is that of a binary transmission scheme whose only two transmitted sequences are $\mathbf{C}$ and $\mathbf{C}'$:

$$P_G(e \mid \mathbf{C}) = Q\left(\frac{\|\mathbf{f}(\mathbf{C}) - \mathbf{f}(\mathbf{C}')\|}{\sqrt{2N_0}}\right) \qquad (7.36)$$

Consider now the error probability of the genie-aided receiver, denoted $P_G(e)$. We have

$$\begin{aligned}
P_G(e) &= \sum_{\mathbf{C}} P(\mathbf{C}) Q\left(\frac{\|\mathbf{f}(\mathbf{C}) - \mathbf{f}(\mathbf{C}')\|}{\sqrt{2N_0}}\right) \\
&\geq \sum_{\mathbf{C}} I(\mathbf{C}) P(\mathbf{C}) Q\left(\frac{\delta_{\text{free}}}{\sqrt{2N_0}}\right) \qquad (7.37)
\end{aligned}$$

where $I(\mathbf{C}) = 1$ if $\mathbf{C}$ admits a sequence at distance δ_{free}:

$$\min_{\mathbf{C}'} \|\mathbf{f}(\mathbf{C}), \mathbf{f}(\mathbf{C}')\| = \delta_{\text{free}}$$

and $I(\mathbf{C}) = 0$ otherwise. In conclusion,

$$P(e) \geq \psi \, Q\left(\frac{\delta_{\text{free}}}{\sqrt{2N_0}}\right)$$

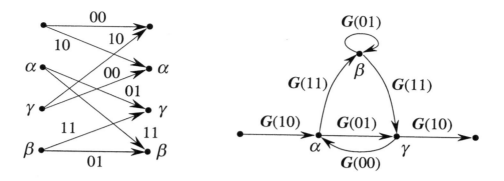

Figure 7.22: *Trellis diagram of a TCM scheme with four states and* $m = 1$, *and corresponding error state diagram.*

where

$$\psi \triangleq \sum_{\mathbf{C}} P(\mathbf{C})I(\mathbf{C}) \tag{7.38}$$

is the probability that, at any time instant, a path traversing the code trellis, chosen at random, has another path diverging from it at this instant, and remerging later, such that the Euclidean distance between the two is δ_{free}. If all paths have this property, we obtain the following lower bound:

$$P(e) \geq Q\left(\frac{\delta_{\text{free}}}{\sqrt{2N_0}}\right) \tag{7.39}$$

For the validity of (7.39) we need all the trellis paths to be equivalent and each of them to have a path at distance δ_{free}. This is obtained if every error matrix has equal entries.

Example 7.11

We develop here an example of calculation of error probability. From the theory above, this calculation is performed in two steps. First, we evaluate the transfer function of the error state diagram with formal branch labels. Next, we replace formal labels with actual labels and compute $T(X), P(e)$.

A four-state TCM is shown in Figure 7.22 along with its error state diagram. T_α, T_β, and T_γ denote the transfer functions of the error state diagram from the initial node to nodes α, β, and γ, respectively. We can write

$$
\begin{aligned}
T_\alpha &= \mathbf{G}(10) + T_\gamma \mathbf{G}(00) \\
T_\beta &= T_\alpha \mathbf{G}(11) + T_\beta \mathbf{G}(01) \\
T_\gamma &= T_\alpha \mathbf{G}(01) + T_\beta \mathbf{G}(11) \\
T(X) &= T_\gamma \mathbf{G}(10)
\end{aligned}
$$

To simplify, we examine here only the case of scalar labels, for which commutativity holds. Denoting by g_0, g_1, g_2, and g_3 the scalar labels associated with $\mathbf{G}(00)$, $\mathbf{G}(01)$, $\mathbf{G}(10)$, and $\mathbf{G}(11)$, respectively, we obtain the following result:

$$
T(X) = 4\frac{g_2^2(g_1 - g_1^2 + g_3^2)}{(1 - g_0 g_1)(1 - g_1) - g_0 g_3^2} \tag{7.40}
$$

Using (7.40), we can obtain an upper bound to the probability of an error event by replacing g_i with the values obtained from the calculation of the error matrices $\mathbf{G}(\cdot)$. This operation will be performed for a unit-energy 4-PSK with the map

$$
f(00) = 1 \quad f(01) = j \quad f(10) = -1 \quad f(11) = -j
$$

We have uniformity here, and we obtain matrices $\mathbf{G}$ whose associated scalar labels are

$$
g_0 = 1 \quad g_1 = g_3 = X^2 \quad g_2 = X^4
$$

so that from (7.40) the transfer function is

$$
T(X) = 4\frac{X^{10}}{1 - 2X^2} = 4(X^{10} + 2X^{12} + 4X^{14} + \ldots) \tag{7.41}
$$

We have $\delta_{\text{free}}^2 = 10$ (this value is obtained with $\mathcal{E} = 1$). A binary PSK constellation with antipodal signals ± 1 has distance $\delta_{\text{min}}^2 = 4$ and energy $\mathcal{E}' = 1$, which yield a coding gain

$$
\eta = \frac{10}{4} = 2.5 \Rightarrow 4 \text{ dB}
$$

If error probabilities are rewritten so as to show explicitly the ratio $\mathcal{E}_b/N_0$, then, observing that $\mathcal{E} = \mathcal{E}_b = 1$, we obtain the following from (7.41) and (7.16):

$$
P(e) \leq \frac{e^{-5\mathcal{E}_b/2N_0}}{1 - 2e^{-\mathcal{E}_b/2N_0}}
$$

The improved upper bound (7.35) yields

$$
P(e) \leq Q\left(\sqrt{5\frac{\mathcal{E}_b}{N_0}}\right) \cdot \frac{1}{1 - 2e^{-\mathcal{E}_b/2N_0}}
$$

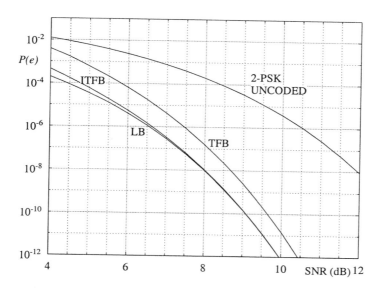

Figure 7.23: *Error probabilities of a four-state TCM scheme with 4-PSK. TFB: Transfer-function upper bound. ITFB: Improved transfer-function bound. LB: Lower bound. The error probability of uncoded 2-PSK is also shown. Here* $SNR == \mathcal{E}_b/N_0$.

The lower bound (7.39) yields

$$P(e) \geq Q\left(\sqrt{5\frac{\mathcal{E}_b}{N_0}}\right)$$

These error probabilities should be compared with those of uncoded 2-PSK, i.e.,

$$P(e) = Q\left(\sqrt{2\frac{\mathcal{E}_b}{N_0}}\right)$$

These four error probabilities are plotted in Figure 7.23. We observe from Figure 7.23 that lower bound and improved upper bound are very close, and hence approximate well the exact error probability. Unfortunately, this occurs only for TCM schemes built upon small constellations and for a small number of trellis states. Moreover, comparing the probability $P(e)$ for uncoded 2-PSK with the two TCM bounds, we can observe that the coding gain is very close to 4 dB, its asymptotic value. □

7.8.2 Computing δ_{free}

The results obtained for the upper and lower bounds to error probability show that δ_{free} plays a central role in assessing the performance of a TCM scheme. If a single parameter has to be chosen to evaluate the quality of a coded scheme, this is δ_{free}. For this reason, it is sensible to look for algorithms aimed at computing this quantity.

Using the error state diagram

The first technique we shall describe for the calculation of δ_{free} is based on the error state diagram. We have previously observed that the transfer function $T(X)$ contains information on the distance δ_{free}. In previous examples we have seen that the value of δ_{free}^2 can be obtained from the series expansion of this function: the smallest exponent of X in this series is δ_{free}^2. However, an exact expression for $T(X)$ may not be available.

For this reason, we describe here an algorithm for the numerical calculation of δ_{free}. The algorithm is based on the update of matrices $\mathbf{D}^{(n)} = (\delta_{ij}^{(n)})$ whose elements are the squared minimum distances between all pairs of paths diverging at the initial instant and merging at time n into states i and j (here and in the following we simply write i, j to denote states S_i, S_j). Two pairs of such paths are shown in Figure 7.24. We observe that the matrix $\mathbf{D}^{(n)}$ is symmetric and that its entries on the main diagonal are the distances between paths converging to a single state (the *error events*).

The algorithm goes as follows.

Step 1 For every state i, find the 2^m states (the *predecessors*) from which a transition to state i is possible, and store them in a matrix. Let $\delta_{ij} = -1$ for every i and $j \geq i$. If there are parallel transitions, for every i let δ_{ii} be equal to the smallest Euclidean distance between signals associated with the parallel transitions leading to state i.

Step 2 For every state pair (i, j), $j \geq i$, find the minimum Euclidean distance between pairs of paths diverging from the same initial state and merging into the state pair i, j in a single instant. Two such pairs are shown in Figure 7.25. This distance is $\delta_{ij}^{(1)}$.

Step 3 For the two states of the pair (i, j), $j > i$, find in the matrix defined at Step 1 the 2^m predecessors $i_1, \cdots, i_{2^m}$ and $j_1, \cdots, j_{2^m}$ (see Figure 7.26). In general we have 2^{2m} possible paths at time $n - 1$ passing by i and j at

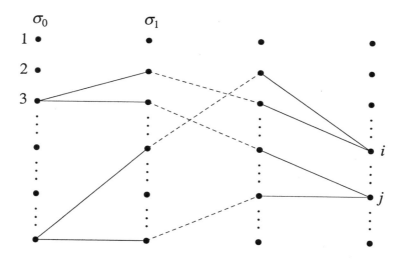

Figure 7.24: *Two pairs of paths diverging at time $n = 0$ and reaching states i, j at the same time.*

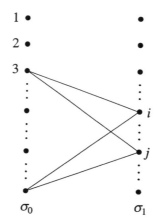

Figure 7.25: *Two pairs of paths leaving two different states and merging into the same pair of states in a single time instant.*

time n. They pass by the pairs

$$(i_1, j_1), (i_1, j_2), \quad \cdots \quad , (i_1, j_{2m})$$

$$\cdots$$

$$(i_{2m}, j_1), (i_{2m}, j_2), \quad \cdots \quad , (i_{2m}, j_{2m})$$

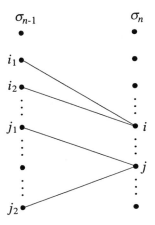

Figure 7.26: *Predecessors of states i, j.*

The minimum distance between pairs passing by (i, j) at time n is

$$
\delta_{ij}^{(n)} = \min \quad \left\{
\begin{aligned}
&\delta_{i_1 j_1}^{(n-1)} + \delta^2(i_1 \to i, j_1 \to j), \\
&\delta_{i_1 j_2}^{(n-1)} + \delta^2(i_1 \to i, j_2 \to j), \\
&\qquad \cdots \\
&\delta_{i_1 j_{2^m}}^{(n-1)} + \delta^2(i_1 \to i, j_M \to j), \\
&\qquad \cdots \\
&\delta_{i_{2^m} j_{2^m}}^{(n-1)} + \delta^2(i_{2^m} \to i, j_{2^m} \to j)
\end{aligned}
\right\}
\tag{7.42}
$$

In (7.42), the distances $\delta^{(n-1)}$ come from the calculations performed in Step 2, where for example $\delta(i_1 \to i, j_1 \to j)$ is the Euclidean distance between signals associated with transitions $i_1 \to i$ and $j_1 \to j$. These can be computed only once at the beginning. When one of the distances $\delta_{\ell m}^{(n-1)}$ already computed is equal to -1, the corresponding term in the right-hand side of (7.42) vanishes. In fact, the value $\delta_{\ell m}^{(n-1)} = -1$ tells us that there is no pair of paths passing by states ℓ and m at time $n-1$. When $i = j$, $\delta_{ii}^{(n)}$ represents the squared distance between two paths merging at step n and at state i. This is an error event. If $\delta_{ii}^{(n)} < \delta_{ii}^{(n-1)}$, then $\delta_{ii}^{(n)}$ takes the place of $\delta_{ii}^{(n-1)}$ in matrix $\mathbf{D}^{(n)}$.

Step 4 If

$$\delta_{ij}^{(n)} < \min_i \delta_{ii}^{(n)} \qquad (7.43)$$

for at least one pair (i, j), then n is changed into $n+1$, and we return to Step 3. Otherwise, the iterations are stopped, and we have

$$\delta_{\text{free}}^2 = \min_i \delta_{ii}^{(n)}$$

7.9 Bit-interleaved coded modulation

From results of Chapters 3 and 4 we can infer that, to perform well on the AWGN as well as on the independent Rayleigh fading channel with high SNR, a code should exhibit large Euclidean as well as Hamming distances. We may call such a code *robust*. Binary coding schemes are intrinsically robust, due to the proportionality between Hamming and Euclidean distance, but nonbinary codes may not be robust. In particular, TCM schemes that are optimum for the AWGN channel (in the sense that they maximize the free Euclidean distance) may not be optimum for the independent Rayleigh channel, whenever their free *Hamming* distance is not maximum. For example, many schemes that are optimum for the AWGN channel exhibit parallel transitions, and hence have free Hamming distance 1: consequently, while they perform well on the AWGN channel, they do poorly on the Rayleigh fading channel.

Now, if the channel model is nonstationary, in the sense that the propagation environment changes during transmission (think of a wireless telephone call initiated indoors, and ended while driving a car on the freeway), we are interested in robust codes, rather than in codes that are optimum only for a specific channel. One such robust scheme is provided by *bit-interleaved coded modulation* (BICM). BICM separates coding from modulation and hence cannot achieve optimum Euclidean distance: however, it can achieve a Hamming distance larger than TCM. The idea here is to transform the channel generated by the multilevel constellation $\mathcal{X}$ into parallel *and independent* binary channels. Any transmission of a multilevel signal from $\mathcal{X}$, with $|\mathcal{X}| = 2^m$, can actually be thought of as taking place over m parallel channels, each carrying one binary symbol from the signal label. However, these channels are generally not independent, due to the constellation structure. To make them independent, binary symbols are interleaved (with infinite depth) before being used as signal labels.

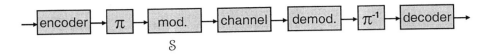

Figure 7.27: *Block diagram of a transmission system with TCM and with BICM. For TCM, π denotes a symbol interleaver, while for BICM it denotes a bit interleaver.*

Example 7.12

Consider 8-PSK: the three bits labeling each signal can be thought of as being transmitted over three nonindependent binary channels: in fact, erroneous reception of one bit influences the probability that the others are also received erroneously. In other terms, the probability that an 8-PSK symbol is received erroneously is not equal to the product of the probabilities of receiving each bit erroneously. The idea of bit interleaving is to remove the statistical connections that were created by the modulator among bits. ☐

The BICM block diagram is shown in Figure 7.27. In the decoder, the metrics must reflect the fact that we are separating bits. Suppose we transmit the code word

$$\mathbf{x} = (x_1,\ x_2,\ \ldots,\ x_n)$$

and we receive $\mathbf{y}$ at the output of a stationary memoryless channel. With TCM, we decode by maximizing the metric

$$\log p(\mathbf{y} \mid \mathbf{x}) = \sum_{i=1}^{n} \log(y_i \mid x_i) \tag{7.44}$$

with respect to $\mathbf{x}$, while with BICM we must consider, instead of the symbol metric $\log p(y_i \mid x_i)$, the *bit metric*

$$\log \sum_{x_i \in \mathfrak{X}(b,j)} p(y_i \mid x_i), \quad b = 0, 1, \quad j = 1, 2, \ldots, \log |\mathfrak{X}| \tag{7.45}$$

where $\mathfrak{X}(b, j)$ denotes the subset of $\mathfrak{X}$ having bit b in position j of its label.

As computation of this metric may be too complex for implementation, a convenient approximation here is based on

$$\log \sum_{j} z_j \approx \max_{j} \log z_j \tag{7.46}$$

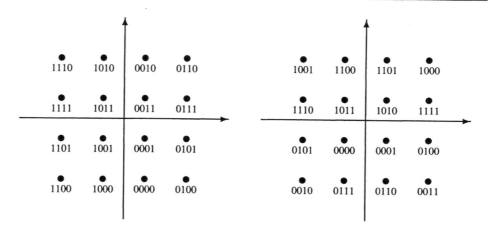

Figure 7.28: *16-QAM with Gray and Ungerboeck labeling.*

which yields the suboptimum metrics

$$\max_{x_i \in \mathcal{X}(b,j)} \log p(y_i \mid x_i) \tag{7.47}$$

Notice that the term $\log p(y_i \mid x_i)$ is the same as that appearing in TCM metric computations, and hence only the set of comparisons leading to the maximization (7.47) adds to the complexity of BICM.

It should be also observed that *labeling* plays a key role in BICM; in particular, empirical evidence suggests that Gray labeling should be preferred. This labeling is such that signals at minimum Euclidean distance differ in only one bit of their labels; as such, it is convenient in uncoded modulation, where one symbol error at high SNR causes only one bit to be in error. A formal definition of Gray labeling, useful for BICM, is the following: we say that $x \in \mathcal{X}(b,i)$ satisfies the Gray condition if it has at most one $x' \in \mathcal{X}(\bar{b},i)$ at distance $d_{\mathrm{E,min}}$. A Gray labeling is one in which every $x \in \mathcal{X}$ satisfies the Gray condition. The performance of BICM depends on the labeling used: in particular, Gray labeling performs better than Ungerboeck labeling, that is, the labeling generated by set partitioning (see Figure 7.28).

Since for some constellations Gray labeling does not exist, we may define *quasi-Gray* labeling as one that minimizes the number of signals for which the Gray condition is not satisfied.

Table 7.3 shows free Euclidean and Hamming distances of selected BICM and TCM schemes with the same state complexity. It can be seen that BICM increases the Hamming distance considerably, while reducing (often marginally) the Euclidean distance. This means that BICM will outperform TCM over independent Rayleigh fading channels, while suffering a moderate loss of performance over the AWGN channel.

Encoder	BICM		TCM	
memory	δ^2_{free}	d_{H}	δ^2_{free}	d_{H}
2	1.2	3	2.0	1
3	1.6	4	2.4	2
4	1.6	4	2.8	2
5	2.4	6	3.2	2
6	2.4	6	3.6	3
7	3.2	8	3.6	3
8	3.2	8	4.0	3

Table 7.3: *Euclidean and Hamming distances of some BICM and TCM schemes for a 16-QAM elementary constellation. Both schemes have a transmission rate of 3 bits per dimension pair (the average energy is normalized to 1).*

7.9.1   Capacity of BICM

An equivalent BICM channel model consists of $\log |\mathfrak{X}|$ parallel, independent, and memoryless binary-input channels. A *random switch*, whose position selects at random one of the label positions with which the coded symbol is associated, models ideal interleaving. Specifically, for every symbol c_i in the coded sequence **c**, this switch selects (independently from previous and future selections) a position index $i \in \{1, 2, \ldots, \log |\mathfrak{X}|\}$, and transmits c_i over the channel. The decoder knows the sequence of channels used for the transmission of **c**, and makes ML decisions accordingly.

Computations [7.3] show that with Gray labeling the capacity of BICM is very close to the capacity of coded modulation. Similar results hold for the independent Rayleigh fading channel with channel-state information known at the receiver.

7.10   Bibliographical notes

Before the introduction of TCM by Ungerboeck [7.6], it was commonly accepted that a coding scheme would necessarily expand the bandwidth. Once this belief was dispelled, TCM gained quick acceptance for applications not tolerating

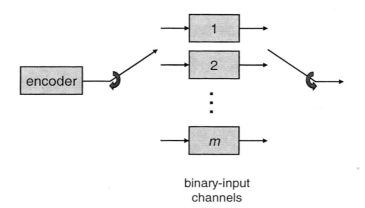

Figure 7.29: *Equivalent parallel channel model of BICM with ideal interleaving and $m = \log |\mathfrak{X}|$.*

a loss in spectral efficiency. An important application was in modem design for voice-grade telephone lines. Before the introduction of TCM, the International Telecommunication Union's ITU-T V.29 modem used uncoded 16-QAM to transmit 9,600 bit/s. This standard was introduced in 1976. Further improvement of the data rate would require using a bigger signal constellation, at the price of a worse error probability due to power constraints. New standards (V.32, V.33, and V.34), introduced after 1986, included TCM with two-dimensional or four-dimensional QAM [7.7, 7.8]. Further details on this, as well as a historical perspective of TCM, can be found in the review paper [7.4]. Further details on TCM can be found in the book [7.2]. A thorough analysis of the geometric uniformity of TCM can be found in [7.5]. BICM is covered in [7.3].

7.11 Problems

1. Consider the TCM encoder shown in Figure 7.30 and based on the 8-PSK constellation of Figure7.2. Draw one section of its trellis diagram, and label the trellis branches by the signals associated with them. Are the three Ungerboeck conditions satisfied?

2. Consider the TCM scheme whose encoder and signal constellation are shown in Figure 7.31. Here the block $\otimes$ denotes a bit multiplier. Notice the presence of a *nonlinear* trellis code in lieu of a convolutional code.

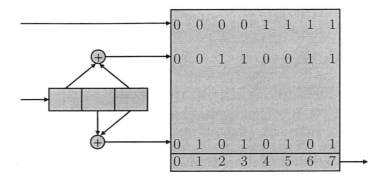

Figure 7.30: *A TCM encoder for Problem* 1.

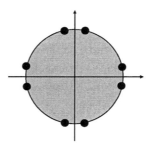

Figure 7.31: *TCM encoder and signal constellation for Problem 2.*

(a) Find a set partitioning of the constellation.

(b) Find the trellis of the encoder, and associate with each branch a source symbol.

(c) Is this a good TCM scheme?

3. A signal constellation ("asymmetric 8-PSK") is shown in Figure 7.32. Do "set partitioning" of the constellation.

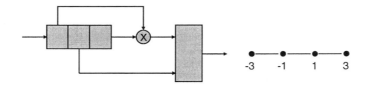

Figure 7.32: *Asymmetric 8-PSK for Problem 3.*

4. Asymmetric 4-PSK is shown in Figure 7.33. Design TCM schemes with 2 and 4 states based on this constellation and transmitting 1 bit per signal. Discuss how the asymptotic coding gain depends on the value of angle ϕ.

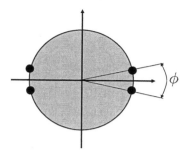

Figure 7.33: *Asymmetric 4-PSK for Problem 4.*

5. Partition the 32-QAM "cross" constellation of Figure 7.34 into eight subconstellations such that the minimum Euclidean distance within any of them is $2\sqrt{2}$ larger than the minimum Euclidean distance within the original constellation. (This partition is used in the V.32 telephone modem standard, which incorporates an eight-state rotationally invariant TCM scheme.)

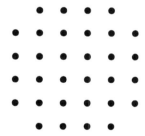

Figure 7.34: *32QAM "cross" constellation.*

6. Consider the TCM encoder shown in Figure 7.35. The signal constellation used is 16-QAM. Design the memoryless mapper, and compute the resulting Euclidean free distance.

7. Consider the 16-point constellation of Figure 7.36. Design good TCM schemes with four and eight states for transmitting 3 bit/signal with this constellation.

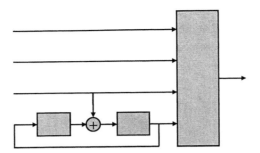

Figure 7.35: *A TCM encoder for* 16-*QAM.*

Compute the ratio between the square Euclidean free distance and the constellation energy.

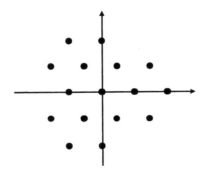

Figure 7.36: *A* 16-*point signal constellation.*

References

[7.1] E. Biglieri, "High-level modulation and coding for nonlinear satellite channels," *IEEE Trans. Commun.*, Vol. 32, No. 5, pp. 616–626, May 1984.

[7.2] E. Biglieri, D. Divsalar, P. J. McLane, and M. K. Simon. *Introduction to Trellis-Coded Modulation with Applications.* New York: Macmillan, 1991.

[7.3] G. Caire, G. Taricco, and E. Biglieri, "Bit-interleaved coded modulation," *IEEE Trans. Inform. Theory*, Vol. 44, No. 3, pp. 927–946, March 1998.

[7.4] D. J. Costello, J. Hagenauer, H. Imai, and S. B. Wicker, "Applications of error-control coding," *IEEE Trans. Inform. Theory*, Vol. 44, No. 6, pp. 2531–2560, October 1998.

[7.5] G. D. Forney, Jr., "Geometrically uniform codes," *IEEE Trans. Inform. Theory*, Vol. 37, No. 5, pp. 1241–1260, September 1991.

[7.6] G. Ungerboeck, "Channel coding with multilevel/phase signals," *IEEE Trans. Inform. Theory*, Vol. IT-28, pp. 55–67, January 1982.

[7.7] L.-F. Wei, "Rotationally invariant convolutional channel encoding with expanded signal space, Part II: Nonlinear codes," *IEEE J. Select. Areas Commun.*, Vol. 2, pp. 672–686, September 1984.

[7.8] L.-F. Wei, "Trellis-coded modulation using multidimensional constellations," *IEEE Trans. Inform. Theory*, Vol. IT-33, pp. 483–501, July 1982.

[7.9] L.-F. Wei, "Rotationally invariant trellis-coded modulations with multidimensional M-PSK," *IEEE J. Select. Areas Commun.*, Vol. 7, pp. 1285–1295, December 1989.

8

A full octavium below me!

Codes on graphs

This chapter introduces a new code description. We first develop a graphical representation for the factorization of a function of several variables into a product of functions of a lower number of variables. This representation allows us to derive efficient algorithms for computing the marginals of the original function with respect to any one of its variables. Given a function describing membership in a code $\mathcal{C}$, its marginalization leads to a method for decoding $\mathcal{C}$. Thus, graphical representations of codes provide a natural setting for the description of symbol-by-symbol decoding techniques, much as the code trellis is a natural setting for the description of maximum-likelihood decoding using the Viterbi algorithm. This chapter is centered on these representations of codes and on a general procedure (the sum–product algorithm) for their decoding. The importance of the theory presented here lies in the fact that all known codes that approach capacity and are practically decodable admit a simple graphical representation.

8.1   Factor graphs

To motivate the introduction of factor graphs in this Chapter, let us consider a specific problem, viz., *maximum a posteriori* (MAP) probability decoding of a given code. In previous chapters we have examined decoding of code $\mathcal{C}$ based on the maximum-likelihood (ML) rule, which consists of maximizing, over $\mathbf{x} \in \mathcal{C}$, the function $p(\mathbf{y} \mid \mathbf{x})$. We recall from Section 1.1 that, under the assumption that all code words are equally likely, this rule minimizes the word error probability.

Now, assume that we want to minimize the error probability of a single symbol of the code word (possibly at the price of a nonminimum word error probability). In this case, we must do symbol-MAP decoding, which consists of maximizing the *a posteriori probabilities* (APPs)

$$p(x_i \mid \mathbf{y}), \qquad i = 1, \ldots, n$$

Specifically, we decode x_i into 0 or 1 by comparing the probabilities that the ith bit in $\mathbf{x}$ is equal to 0 or 1, given the received vector $\mathbf{y}$ and the fact that $\mathbf{x}$ must satisfy the constraints describing the code $\mathcal{C}$. The APP $p(x_i \mid \mathbf{y})$ can be expressed as

$$p(x_i \mid \mathbf{y}) = \sum_{\mathbf{x} \in \mathcal{C}_i(x_i)} p(\mathbf{x} \mid \mathbf{y}) \tag{8.1}$$

and $\mathcal{C}_i(x_i)$ denotes the subset of code words whose ith component is x_i. Similarly, we may want to do symbol-ML decoding, which consists of maximizing the probabilities[1]

$$p(\mathbf{y} \mid x_i), \qquad i = 1, \ldots, n \tag{8.2}$$

In this chapter we develop efficient algorithms for decoding according to criteria (8.1) and (8.2) (see Table 8.1).

The central concept in our development is that of a *marginalization*, which consists of associating with a function $f(x_1, \ldots, x_n)$ of n variables its n *marginals*, defined as the functions

$$f_i(x_i) \triangleq \sum_{x_1} \cdots \sum_{x_{i-1}} \sum_{x_{i+1}} \cdots \sum_{x_n} f(x_1, \ldots, x_n) \tag{8.3}$$

For each value taken on by x_i, these are obtained by summing the function f over all of its arguments consistent with the value of x_i. It is convenient to introduce

---

[1]One should observe that, if the symbol-MAP (or symbol-ML) rule is used to decode all symbols of $\mathbf{x} \in \mathcal{C}$, theresulting word does not necessarily belong to $\mathcal{C}$, and, if it does, this is note necessarily the word one would obtain by using the block-MAP (or block-ML) rule. [8.2]

(block-)MAP	$\widehat{\mathbf{x}} = \max_{\mathbf{x}}^{-1} p(\mathbf{x} \mid \mathbf{y})$
(block-)ML	$\widehat{\mathbf{x}} = \max_{\mathbf{x}}^{-1} p(\mathbf{y} \mid \mathbf{x})$
symbol-MAP	$\hat{x}_i = \max_{x_i}^{-1} p(x_i \mid \mathbf{y})$
symbol-ML	$\hat{x}_i = \max_{x_i}^{-1} p(\mathbf{y} \mid x_i)$

Table 8.1: *Summary of different decoding rules. The maximizations are performed with respect to words and symbols compatible with code $\mathcal{C}$.*

the compact notation $\sim x_i$ to denote the set of indices $x_1, \ldots, x_{i-1}, x_{i+1}, \ldots, x_n$ to be summed over, so that we can write

$$f_i(x_i) \triangleq \sum_{\sim x_i} f(x_1, \ldots, x_n) \tag{8.4}$$

If $x_i \in \mathcal{X}$, $i = 1, \ldots, n$, then the complexity of this computation grows as $|\mathcal{X}|^{n-1}$. A simplification can be achieved when f can be factored as a product of functions, each with less than n arguments. Consider for example a function $f(x_1, x_2, x_3)$ that factors as follows:

$$f(x_1, x_2, x_3) = g_1(x_1, x_2) g_2(x_1, x_3) \tag{8.5}$$

Its marginal $f_1(x_1)$ can be computed as

$$\begin{aligned} f_1(x_1) &\triangleq \sum_{x_2} \sum_{x_3} f(x_1, x_2, x_3) \\ &= \sum_{x_2} \sum_{x_3} g_1(x_1, x_2) g_2(x_1, x_3) = \sum_{x_2} g_1(x_1, x_2) \cdot \sum_{x_3} g_2(x_1, x_3) \end{aligned}$$

where we see that this marginalization can be achieved by computing separately the two simpler marginals $\sum_{x_2} g_1(x_1, x_2)$ and $\sum_{x_3} g_2(x_1, x_3)$ and finally taking their product. This procedure can be represented in graphical form by defining a *factor graph* describing the fact that the function f factors in the form (8.5). The factor graph corresponding to the function f is shown in Figure 8.1. The nodes here can be viewed as processors that compute a function whose arguments label the incoming edges, and the edges as channels by which these processors exchange data. We see that the first sum $\sum_{x_2} g_1(x_1, x_2)$ can be computed locally at the g_1 node because x_1 and x_2 are available there; similarly, the second sum $\sum_{x_3} g_2(x_1, x_3)$ can be computed locally at the g_2 node because x_1 and x_3 are available there.

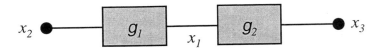

Figure 8.1: *Factor graph of the function* $f(x_1, x_2, x_3) = g_1(x_1, x_2)g_2(x_1, x_3)$.

Formally, we describe a ("normal") factor graph as a set of nodes, edges, and half-edges. Every factor corresponds to a unique node, and every variable to a unique edge or half-edge. The node representing the function g is connected to the edge or half-edge representing the variable x if and only if x is an argument of the factor g. Half-edges are connected to only one node and terminate in a filled circle ●. Edges represent *states* of the system, and filled circles represent *external variables*. In the example of Figure 8.1 we have two nodes representing the factors g_1 and g_2, one edge, and two half-edges. The factors g are called *local functions*, or *constraints*, and the function f the *global function*. An important feature of factor graphs is the presence or absence of *cycles*: we say that a factor graph has no cycles if removing any (regular) edge partitions the graph into two disconnected subgraphs. More specifically, a cycle of length ℓ is a path through the graph that includes ℓ edges and closes back on itself. The *girth* of a graph is the minimum cycle length of the graph.

The definition of normality assumes implicitly that no variable appears in more than two factors. For example, the graph of Figure 8.2 does not satisfy our definition: in fact, the variable x_1 appears as a factor of g_1, g_2, and g_3, and as a result it corresponds to more than one edge. To be able to include as well in our graphical description those functions that factor as in Figure 8.2, we need to "clone" the variables appearing in more than two factors. We shall explain below how this can be done.

8.1.1   The Iverson function

An important role in factor graphs is played by functions taking on values 0 or 1 as follows. Let P denote a proposition that may be either true or false; we denote by $[P]$ the *Iverson function*

$$[P] \triangleq \begin{cases} 1, & P \text{ is true} \\ 0, & P \text{ is false} \end{cases}$$

Clearly, if we have n propositions $P_1, \ldots, P_n$, we have the factorization

$$[P_1 \text{ and } P_2 \cdots \text{ and } P_n] = [P_1] [P_2] \cdots [P_n]$$

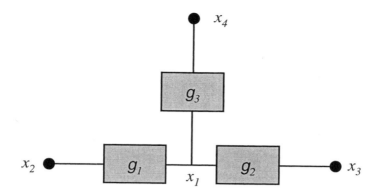

Figure 8.2: Factor graph of the function $f(x_1, x_2, x_3, x_4) = g_1(x_1, x_2)g_2(x_1, x_3)g_3(x_1, x_4)$.

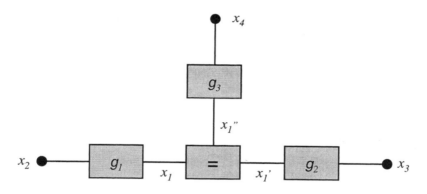

Figure 8.3: Normal factor graph of the function $f(x_1, x_2, x_3, x_4) = g_1(x_1, x_2)g_2(x_1, x_3)g_3(x_1, x_4) = g_1(x_1, x_2)g_2(x_1', x_3)g_3(x_1'', x_4)f_=(x_1, x_1', x_1'')$.

This function has several applications in our context. In particular, it allows the transformation of any graph into one satisfying normality. In fact, define the *repetition* function $f_=$ as

$$f_=(x_1, x_1', x_1'') \triangleq [x_1 = x_1' \text{ and } x_1 = x_1''] \tag{8.6}$$

This transforms the branching point of Figure 8.2 into a node representing a repetition function. Thus, the graph of Figure 8.2 is transformed into that of Figure 8.3, which satisfies the definition of a normal factor graph.

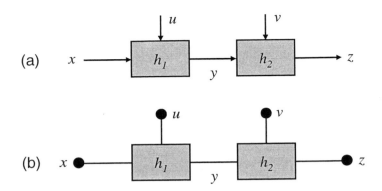

Figure 8.4: *A block diagram (a) and the corresponding normal graph (b).*

Another application of the Iverson function is the natural transformation of a block diagram into a normal factor graph. For example, an adder $x_1 + x_2 = x_3$ can be represented by the Iverson function $[x_1 + x_2 = x_3]$, which takes value 1 if the condition $x_1 + x_2 = x_3$ is satisfied, and 0 otherwise. Consider the block diagram of Figure 8.4. The block with input–output relationship $y = h_1(x, u)$ can be interpreted as representing the factor $[y - h_1(x, u)]$, and the block $z = h_2(y, v)$ as representing the factor $[z - h_2(y, v)]$. Thus, the global Iverson function $[y - h_2(x, u)][z - h_2(y, v)]$ takes on value 1 if and only if the values of the variables x, y, z, u, v are consistent with the input–output relationships of the block diagram. The resulting normal graph is shown in Figure 8.4(b).

8.1.2   Graph of a code

For our purposes, the most important application of the Iverson function is the representation of a block code described through the set of its parity-check equations, as summarized by the parity-check matrix (Section 3.7). For example, consider the parity-check matrix

$$\mathbf{H} = \begin{bmatrix} 1 & 1 & 1 & 0 & 0 & 0 & 0 \\ 1 & 0 & 0 & 1 & 1 & 0 & 0 \\ 1 & 0 & 0 & 0 & 0 & 1 & 1 \end{bmatrix} \qquad (8.7)$$

defining a linear binary block code $\mathcal{C}$. This consists of the 2^4 binary length-7 code words $\mathbf{x} = (x_1, x_2, \ldots, x_7)$ that satisfy the three parity checks

$$
\begin{aligned}
x_1 + x_2 + x_3 &= 0 \\
x_1 + x_4 + x_5 &= 0 \\
x_1 + x_6 + x_7 &= 0
\end{aligned}
$$

and membership in $\mathcal{C}$ is determined by verifying that each parity check is satisfied. Thus, the Iverson function that expresses the membership of an n-tuple $\mathbf{x}$ in $\mathcal{C}$ is

$$
[\mathbf{x} \in \mathcal{C}] = [\mathbf{Hx'} = \mathbf{0}] \tag{8.8}
$$

which, in our example, becomes

$$
[(x_1, \ldots, x_7) \in \mathcal{C}] = [x_1 + x_2 + x_3 = 0]\,[x_1 + x_4 + x_5 = 0]\,[x_1 + x_6 + x_7 = 0]
$$

so a linear block code can be described by a factor graph.

A *Tanner graph* is a graphical representation of a linear block code corresponding to the set of parity checks that specify the code. Each symbol is represented by a filled circle ●, and every parity check by a *check node* ⊕. Each check node specifies a set of symbols whose sum must be zero. Tanner graphs are *bipartite*: filled circles are connected only to check nodes and vice versa. For the code described by (8.7) we have the Tanner graph shown in Figure 8.5(a) and 8.5(b) (the bipartite structure of the graph is especially evident in the latter, where symbol nodes are on the left and check nodes are on the right).

The normal factor graph representing a general linear binary block code $\mathcal{C}$ can be obtained from its Tanner graph as shown in Figure 8.6. It has n variables x_1, $\ldots$, x_n, n repetition nodes, and $n - k$ parity-check nodes. A parity-check node corresponds to an Iverson function (for example, the uppermost node of Figure 8.6 corresponds to the function $[x_1 + x_2 + x_n = 0]$). Generally, in the normal graph of a linear binary code, each variable node corresponds to one bit of the code word, i.e., to one column of $\mathbf{H}$, and each check node to one parity-check equation, i.e., to one row of $\mathbf{H}$. The edges in the graph are in one-to-one correspondence with the nonzero entries of $\mathbf{H}$.

The Tanner graph of a code (and hence its normal factor graph) may have cycles: the $(7, 4, 3)$ '*Hamming code* described by the parity-check matrix

$$
\mathbf{H} = \begin{bmatrix} 1 & 1 & 0 & 0 & 1 & 1 & 0 \\ 1 & 0 & 0 & 1 & 0 & 1 & 1 \\ 1 & 0 & 1 & 0 & 1 & 0 & 1 \end{bmatrix}
$$

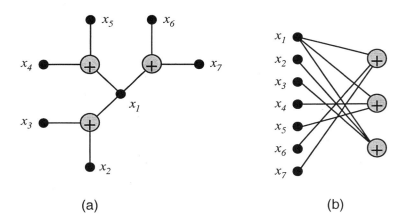

Figure 8.5: *Two equivalent forms of the Tanner graph of the* $(7, 4)$ *code.*

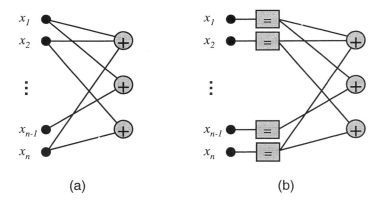

Figure 8.6: *(a) Tanner graph of a linear binary code and (b) its normal version.*

i.e., whose code words $\mathbf{x} = (x_1, x_2, \ldots, x_7)$ satisfy the three parity checks

$$\begin{aligned} x_1 + x_2 + x_5 + x_6 &= 0 \\ x_1 + x_4 + x_6 + x_7 &= 0 \\ x_1 + x_3 + x_5 + x_7 &= 0 \end{aligned}$$

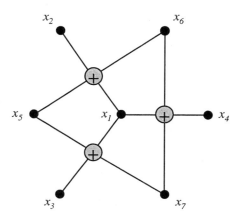

Figure 8.7: *A Tanner graph with cycles: the* $(7, 4, 3)$ *Hamming code.*

shows this fact (Figure 8.7).

Notice that, since a given code can be represented by several parity-check matrices, then the same code can be represented by several Tanner graphs. It is possible that some representations have cycles, while others are cycle free (see the Problem section at the end of this chapter).

Example 8.1 (LDPC codes)

Figure 8.8 shows the normal graph of a *regular low-density parity-check* (LDPC) code. This is a long linear binary block code such that every code symbol is checked by the same number w_c of parity checks, and every parity equation checks the same number w_r of code symbols. Equivalently, the parity-check matrix $\mathbf{H}$ of the code has the same number w_c of 1s in each column and the same number w_r of 1s in each row. (We have $w_c = 3$ and $w_r = 5$ in Figure 8.8.) An interleaver π, which applies a permutation to the input symbols before they enter the modulo-2 adders, describes the connections between symbols and parity checks. The term *low-density* refers to the fact that the number of 1s in $\mathbf{H}$ is small as compared to the number of entries. For large block lengths (say, above 1,000), LDPC codes rank among the best codes known. They will be studied in depth in the next chapter. □

Normal code graphs can now be generalized by considering codes originally described by a trellis (e.g., terminated convolutional codes). A trellis can be viewed as a set of triples $(\sigma_{i-1}, x_i, \sigma_i)$ describing which state transitions $\sigma_{i-1} \rightarrow \sigma_i$ at time

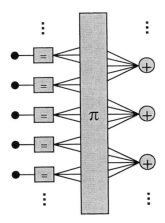

Figure 8.8: *Normal graph of a regular LDPC code with $w_c = 3$ and $w_r = 5$.*

$i - 1$, $i = 1, \ldots, n$, are driven by the channel symbol x_i. Let $\mathcal{T}_i$ denote the set of branches in the trellis at time $i-1$. Then the set of branch labels in $\mathcal{T}_i$ is the domain of a variable x_i, while the set of nodes at time $i - 1$ (i) is the domain of the state variable σ_{i-1} (σ_i). The initial and final state variables take on a single value. The local function corresponding to the ith trellis section is

$$[(\sigma_{i-1}, x_i, \sigma_i) \in \mathcal{T}_i] \qquad (8.9)$$

and the whole trellis corresponds to a product of Iverson functions (Figure 8.9):

$$[\mathbf{x} \in \mathcal{C}] = \prod_{i=1}^{n} [(\sigma_{i-1}, x_i, \sigma_i) \in \mathcal{T}_i] \qquad (8.10)$$

In some cases (see Example 8.5 *infra*) it is convenient to include, in the description of the trellis sections, also the information symbols u_i that drive the transitions between states. If this is the case, the local function corresponding to the ith trellis section becomes

$$[(\sigma_{i-1}, u_i, x_i, \sigma_i) \in \mathcal{T}_i]$$

Example 8.2

Figure 8.10 shows the trellis of an $(8, 4, 4)$ binary linear code and its normal factor graph. Here the filled circles correspond to two-bit inputs.   □

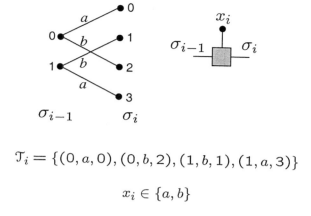

$$\mathcal{T}_i = \{(0, a, 0), (0, b, 2), (1, b, 1), (1, a, 3)\}$$

$$x_i \in \{a, b\}$$

Figure 8.9: *A section of a trellis and the node representing it in a normal graph.*

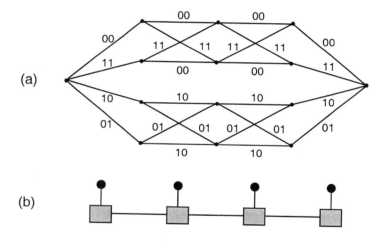

(a)

(b)

Figure 8.10: *An $(8, 4, 4)$ binary linear code: (a) Trellis diagram, and (b) Normal graph.*

Notice how in this representation the graph edges are associated with states, the filled dots with symbols, and the nodes with constraints. We may say that symbols represent *visible* variables, while states represent *hidden* variables: the latter are unobserved, as parts of the internal realization of the code. We may keep this interpretation even for codes described by normalized Tanner graphs, even though here the concept of state does not come naturally. In this latter case, the constraints are simply represented by adders.

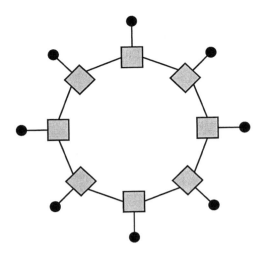

Figure 8.11: *Normal graph of a code defined through its tail-biting trellis.*

Since the factor graph of the trellis code consists of a chain of nodes, each corresponding to a function (8.9), it is cycle free. Since we know that all codes can be represented by a trellis, the above shows that any of them can also be represented by a cycle-free factor graph: however, the resulting complexity (number of states) might be so large that this kind of representation becomes useless (for a quantitative version of this statement, see [8.6]).

Example 8.3 (Tail-biting trellises)

Figure 8.11 shows the normal graph of a code described by a tail-biting trellis. It is seen that this trellis has a single cycle.   □

Example 8.4

This example illustrates the fact that the same code admits different graph representations. Figure 8.12 shows the Tanner graph of the $(4, 1, 4)$ binary repetition code, its trellis, and the graph derived from the trellis. Notice how the repetition blocks in the figure illustrate the fact that, in each trellis section, coded symbol, starting state, and ending state coincide. Figure 8.13 shows two graph representations for the $(4, 3, 2)$ binary single-parity-check code, its trellis, and the graph derived from

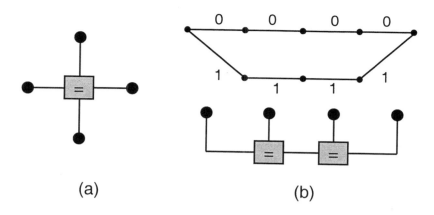

Figure 8.12: *Two graph representations of the binary* $(4, 1, 4)$ *repetition code.*

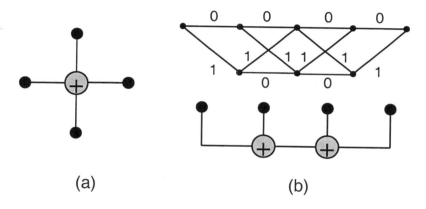

Figure 8.13: *Two graph representations of the binary* $(4, 3, 2)$ *single-parity-check code.*

the trellis. Notice that for both codes the two endmost states, rather than being represented in the graph by two dangling branches, correspond to two code symbols. This reflects the fact that the initial and final trellis states are fixed, and, consequently, in the functions $\mathcal{T}_1$ and $\mathcal{T}_4$ the arguments σ_0 and σ_4, respectively, are fixed. Observe finally how the normal graph of the repetition code can be obtained from the graph of the single-parity-check code (its dual code: see Section 3.7) by replacing the parity-check nodes with repetition nodes. This is a special case of a general result [8.6] connecting the graphs of dual codes. $\qquad\square$

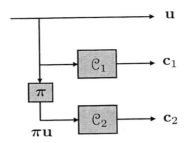

Figure 8.14: *Encoder of a turbo code.*

Example 8.5 (Turbo codes)

Figure 8.14 shows the encoder of a turbo code, the *parallel concatenation* of two terminated convolutional codes. Here $\mathcal{C}_1$ and $\mathcal{C}_2$ are systematic convolutional encoders. A block $\mathbf{u}$ of source data is fed to the encoder $\mathcal{C}_1$, which produces the block $\mathbf{c}_1$, and to an interleaver π, which produces the permuted block $\pi\mathbf{u}$. This block in turn is sent into the encoder $\mathcal{C}_2$ to produce the block $\mathbf{c}_2$. The blocks $\mathbf{c}_1$ and $\mathbf{c}_2$ contain only the parity-check part of the code words generated by the convolutional encoders. The three blocks $\mathbf{u}$, $\mathbf{c}_1$, and $\mathbf{c}_2$ are multiplexed and sent through the transmission channel. Observe that $\mathcal{C}_1$ and $\mathcal{C}_2$ may be (and often are) the same code. Turbo codes are among the best codes known and will be studied in the next chapter. Their normal graph is shown in Figure 8.15. We see how the two trellises generating $\mathbf{c}_1$ and $\mathbf{c}_2$ share, via the interleaver π, the common symbols $\mathbf{u}$.

□

Example 8.6 (Repeat–accumulate codes)

Figure 8.16 shows the encoder of a repeat–accumulate code. This is obtained by cascading two codes separated by an interleaver. The first one is an $(n, 1, n)$ binary repetition code. The second one is a rate-1 convolutional code, whose generator is $g(D) = 1/(1 + D)$, corresponding to the input–output *accumulation* relationship $x_i = u_i + x_{i-1}$. The trellis of the latter code has two states, corresponding to coded symbols. Originally introduced as tools for deriving coding theorems [8.3], these codes exhibit a surprisingly good performance on the additive white Gaussian noise channel. The structure of their normal graph is shown in Figure 8.17, where the trellis constraints $x_i + u_i + x_{i-1} = 0$ are shown explicitly.   □

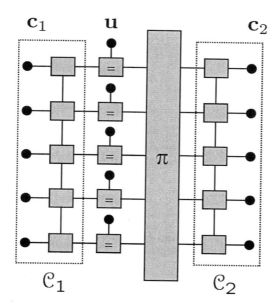

Figure 8.15: *Normal graph of a (truncated) turbo code.*

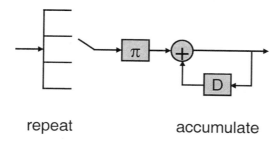

repeat   accumulate

Figure 8.16: *Encoder of a repeat–accumulate code.*

8.2 The sum–product algorithm

We now describe an algorithm for the efficient computation of the marginals of a function whose factors are described by a normal factor graph. This works when the graph is cycle free, and yields, after a finite number of steps, the marginal function corresponding to each variable associated with an edge. Initially, we limit ourselves to stating the algorithm; the principles on which it is based will be described later on, in the context of its application to symbol-MAP decoding.

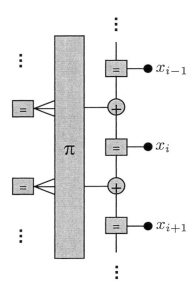

Figure 8.17: *Normal graph of a rate-1/3 repeat–accumulate code.*

In this algorithm, two *messages* are transmitted along each branch, one for each direction. Each message is a function of the variable associated with that branch, and depends on the direction. It is given in the form of a vector, whose components are all the values taken on by the corresponding variable. For messages that are probability distributions of binary variables, a convenient choice consists of representing each of them as the ratio between two probabilities, or as the logarithm of this ratio.

Consider the node representing the factor $g(x_1, \ldots, x_n)$ (see Figure 8.18). The message $\mu_{g \to x_i}(x_i)$ out of this function node along the edge x_i is the function

$$\mu_{g \to x_i}(x_i) = \sum_{\sim x_i} g(x_1, \ldots, x_n) \prod_{\ell \neq i} \mu_{x_\ell \to g}(x_\ell) \qquad (8.11)$$

where $\mu_{x_\ell \to g}(x_\ell)$ is the message incoming on edge x_ℓ, and the notation $\sum_{\sim x_i}$ indicates that all variables are summed over except x_i. In words, the message $\mu_{g \to x_i}(x_i)$ is the product of g and all messages towards g along all edges except x_i, summed over all variables except x_i. Half-edges, which are connected to a single node, transmit towards it a message with value 1.

Two important special cases are as follows:

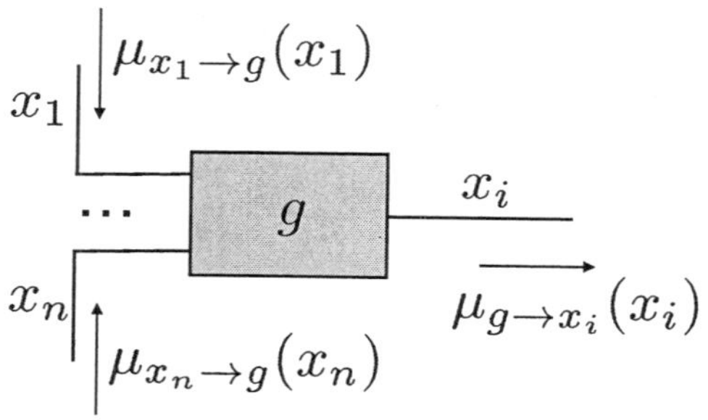

Figure 8.18: *The basic step of the sum–product algorithm.*

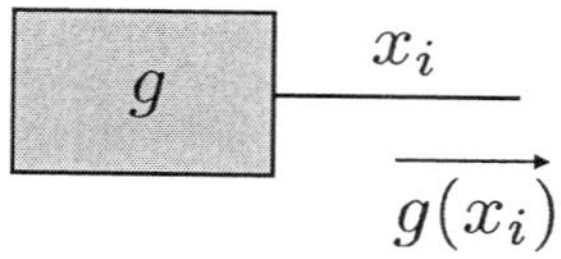

Figure 8.19: *The basic step of the sum–product algorithm when a node is a function of only one argument.*

1. If g is a function of only one argument x_i, then the product in (8.11) is empty, and we simply have (see Figure 8.19)

$$\mu_{g \to x_i}(x_i) = g(x_i)$$

2. If g is the repetition function $f_=$, then we have (see Figure 8.20)

$$\mu_{f_= \to x_i}(x_i) = \prod_{\ell \neq i} \mu_{x_\ell \to f_=}(x_i) \tag{8.12}$$

8.2.1 Scheduling

The messages in the graph must be computed in both directions for each edge. After all of them are computed according to some schedule, the product of the two messages associated with an edge yields the marginal function sought. It should be observed here that the choice of the computational schedule may affect the algorithm efficiency. A possible schedule consists of requiring all nodes to update their outgoing messages whenever their incoming messages are updated. In a factor graph without cycles, message computation may start from the leaves and proceed from node to node as the necessary terms in (8.11) become available. In the

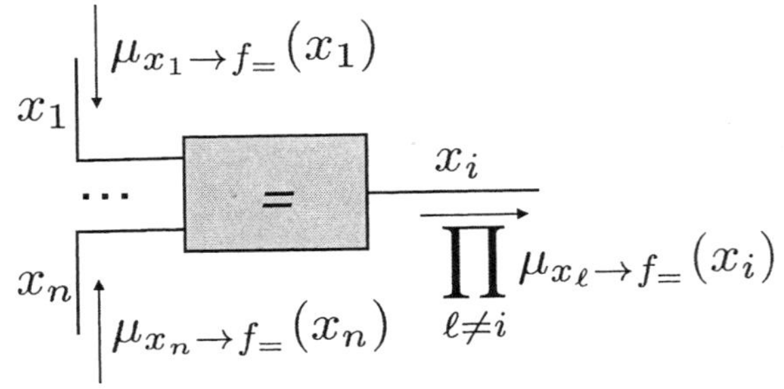

Figure 8.20: *The basic step of the sum–product algorithm when a node represents the repetition function $f_=$.*

"flooding" schedule, the messages are transmitted along all edges simultaneously. For linear graphs like that shown in Figure 8.10, a natural schedule consists of a single forward sweep and a single backward sweep (see *infra*, Section 8.3.2).

8.2.2 Two examples

Later on, we shall prove why the sum–product algorithm (SPA) solves the marginalization problem in the context of symbol-MAP decoding. Here we illustrate it in a simple case, followed by a numerical example of application.

Example 8.7

Consider the function

$$f(x_1, x_2, x_3, x_4, x_5) = g_1(x_1, x_2, x_5)g_2(x_3, x_4, x_5)$$

whose factor graph is shown in Figure 8.21. Its marginalization with respect to x_5 can be computed as follows:

$$\begin{aligned}
f_5(x_5) &= \sum_{x_1}\sum_{x_2}\sum_{x_3}\sum_{x_4} g_1(x_1, x_2, x_5)g_2(x_3, x_4, x_5) \\
&= \underbrace{\sum_{x_1}\sum_{x_2} g_1(x_1, x_2, x_5)}_{\mu_{g_1 \to x_5}(x_5)} \cdot \underbrace{\sum_{x_3}\sum_{x_4} g_2(x_3, x_4, x_5)}_{\mu_{g_2 \to x_5}(x_5)}
\end{aligned}$$

which corresponds to the product of the two messages along edge x_5 exchanged by the SPA.   □

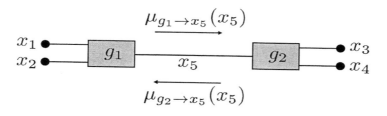

Figure 8.21: *Factor graph of the function* $g_1(x_1, x_2, x_5)g_2(x_3, x_4, x_5)$.

Figure 8.22: *Factor graph for the burglar alarm problem.*

Example 8.8

Consider the *burglar alarm* graph [8.8, 8.13], which describes a burglar alarm sensitive not only to burglary, but also to earthquakes. There are three binary variables: b (for "burglary"), e (for "earthquake"), and a (for "alarm"). A value of 0 for any of these variables indicates that the corresponding event has not occurred, whereas a value of 1 indicates that it has occurred. Suppose that the alarm went off. We want to infer the probability of the two possible causes, namely, derive $p(b \mid a = 1)$ and $p(e \mid a = 1)$. These can be computed by marginalizing $p(b, e \mid a = 1)$; since we have, assuming independence of e and b,

$$p(b, e \mid a = 1) = \frac{p(a = 1 \mid b, e)p(b)p(e)}{p(a = 1)} \propto p(a = 1 \mid b, e)p(b)p(e)$$

then the factor graph appropriate to the problem is shown in Figure 8.22, where

$$f(b, e) \triangleq p(a = 1 \mid b, e) \qquad f_b(b) \triangleq p(b) \qquad f_e(e) \triangleq p(e)$$

The data of the problem consist of the values taken on by these three functions. Let them be

$$f_b(0) = 0.9 \qquad f_b(1) = 0.1$$

$$f_e(0) = 0.9 \qquad f_e(1) = 0.1$$

and

$$f(0, 0) = .001 \qquad f(1, 0) = .368 \qquad f(0, 1) = .135 \qquad f(1, 1) = .607$$

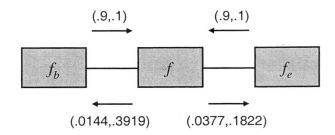

Figure 8.23: *Messages passed by the sum–product algorithm along the edges of the factor graph for the burglar alarm problem.*

By representing a function g of a binary argument as the two-component vector $(g(0), g(1))$, the SPA goes as follows:

$$\mu_{f_b \to b}(b) = (.9, .1)$$

$$\mu_{f_e \to e}(e) = (.9, .1)$$

$$\mu_{f \to e}(e) = \sum_b f(b, e)\mu_{b \to f}(b)$$

$$= (\underbrace{.001 \times .9}_{b=0} + \underbrace{.368 \times .1}_{b=1}, \underbrace{.135 \times .9}_{b=0} + \underbrace{.607 \times .1}_{b=1}) = (.0377, .1822)$$
$$\underbrace{\hspace{3.5cm}}_{e=0} \underbrace{\hspace{3.5cm}}_{e=1}$$

$$\mu_{f \to b}(b) = \sum_e f(b, e)\mu_{e \to f}(e)$$

$$= (\underbrace{.001 \times .9}_{e=0} + \underbrace{.135 \times .1}_{e=1}, \underbrace{.368 \times .9}_{e=0} + \underbrace{.607 \times .1}_{e=1}) = (.0144, .3919)$$
$$\underbrace{\hspace{3.5cm}}_{b=0} \underbrace{\hspace{3.5cm}}_{b=1}$$

The messages passed along the graph edges are summarized in Figure 8.23. Thus, we have

$$p(b \mid a = 1) \propto (.9 \times .0144, .1 \times .3919) = (.01296, .03919)$$

and

$$p(e \mid a = 1) \propto (.9 \times .0377, .1 \times .1822) = (.03393, .01822)$$

After proper rescaling of these vectors (we account for the fact that they are probability vectors) we obtain the final result:

$$p(b \mid a = 1) = (.249, .751) \qquad p(e \mid a = 1) = (.651, .349)$$

$\square$

8.3 ★ Decoding on a graph: Using the sum–product algorithm

Consider now the problem mentioned at the beginning of this chapter, viz., symbol-MAP decoding of a code $\mathcal{C}$ defined on a graph. Specifically, we transmit $\mathbf{x}$ and observe a sequence $\mathbf{y}$ at the output of a channel; the APP $p(\mathbf{x} \mid \mathbf{y})$ is proportional (see (1.1)) to the product $p(\mathbf{y} \mid \mathbf{x})p(\mathbf{x})$. Now, for a stationary memoryless channel, we have

$$p(\mathbf{y} \mid \mathbf{x}) = \prod_{i=1}^{n} p(y_i \mid x_i) \qquad (8.13)$$

while, assuming that the a priori distribution of the transmitted code words is uniform, we have

$$p(\mathbf{x}) = \frac{1}{|\mathcal{C}|}[\mathbf{x} \in \mathcal{C}] \qquad (8.14)$$

and $[\mathbf{x} \in \mathcal{C}]$ factors according to the graph of the code.

Thus, the APPs can be computed by marginalizing the function

$$p(\mathbf{x} \mid \mathbf{y}) \propto [\mathbf{x} \in \mathcal{C}] \prod_{i=1}^{n} p(y_i \mid x_i) \qquad (8.15)$$

This is done by applying the SPA to the graph of the code (which describes the factorization of $[(x_1, \ldots, x_n) \in \mathcal{C}]$) in which the filled dots are replaced by the function nodes $p(y_i \mid x_i)$ (each of these to be interpreted as a function of x_i with parameter y_i). The resulting graph describes the factorization of (8.15). As an example, Figure 8.24 shows the graph to be used for symbol-MAP decoding of a block code described by a normalized Tanner graph.

For systematic codes, whose code words we express in the form

$$\mathbf{x} = (u_1, u_2, \ldots, u_k, x_{k+1}, \ldots, x_n)$$

we can write, instead of (8.14),

$$p(\mathbf{x}) = [\mathbf{x} \in \mathcal{C}] \prod_{i=1}^{k} p(u_i)$$

so that the APPs are obtained by marginalizing the function

$$p(\mathbf{x} \mid \mathbf{y}) \propto [\mathbf{x} \in \mathcal{C}] \prod_{i=1}^{k} p(y_i \mid u_i)p(u_i) \prod_{i=k+1}^{n} p(y_i \mid x_i) \qquad (8.16)$$

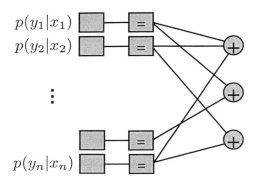

Figure 8.24: *Normal factor graph for the calculation of the APPs $p(x_i \mid \mathbf{y})$.*

and the graph of Figure 8.24 should be modified accordingly, by introducing a priori information about the source symbols u_i.

The above can be extended to the computation, for a state variable σ_j, of the APP

$$p(\sigma_j \mid \mathbf{y}) \propto \sum_{\mathbf{x} \in \mathcal{C}_j(\sigma_j)} \prod_{i=1}^{n} p(y_i \mid x_i) \tag{8.17}$$

where now $\mathcal{C}_j(\sigma_j)$ denotes the set of code words consistent with σ_j, i.e., whose jth state is σ_j.

8.3.1 Intrinsic and extrinsic messages

In a terminology that is often used, the message depending only on the channel observation generated by symbol x_i (and possibly on its probability distribution if the code is systematic and (8.16) is used) is called *intrinsic*. The message in the opposite direction, whose multiplication by the intrinsic message yields the APP of x_i, is called *extrinsic*. This depends on the code structure and on the observation of all components of $\mathbf{y}$ except y_i. Extrinsic information plays a central role in the "turbo algorithm," to be described in Section 9.2.1.

Example 8.9

The single-parity-check binary code with length 3 has words $\mathbf{x} = (x_1, x_2, x_3)$ with $x_3 = x_1 + x_2$. It can be decoded using the graph of Figure 8.25. Information on x_1 can be gathered from the observation of y_1 (intrinsic message), and also from the observation of y_2, y_3, because we have $x_1 = x_2 + x_3$ (extrinsic message). □

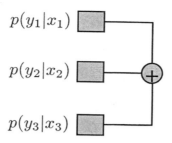

Figure 8.25: *Graph of the binary* $(3, 2, 2)$ *single-parity-check code for symbol-MAP decoding.*

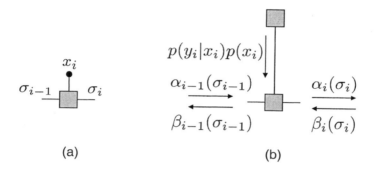

Figure 8.26: *Illustrating the BCJR algorithm on a linear graph. (a) A section of the graph. (b) Messages exchanged by the BCJR algorithm.*

8.3.2 The BCJR algorithm on a graph

As emphasized before, application of the sum–product algorithm depends on the choice of a computational schedule. Sometimes, this choice follows quite naturally from the graph structure, as is the case with linear graphs derived from a trellis (see the example of Figure 8.10). Here, the schedule consists of a single forward sweep and a single backward sweep, which makes the SPA equivalent to the BCJR algorithm introduced in Section 5.4. We now prove the latter statement.

Consider a segment of the graph as shown in Figure 8.26(a), and the corresponding messages exchanged in the application of SPA, with the notations indicated in Figure 8.26(b). Direct use of (8.11) yields the messages

$$\alpha_i(\sigma_i) = \sum_{\sigma_{i-1}} \sum_{x_i} [(\sigma_{i-1}, x_i, \sigma_i) \in \mathcal{T}_i] \, p(y_i|x_i) p(x_i) \alpha_{i-1}(\sigma_{i-1})$$

$$= \sum_{\sigma_{i-1}} \sum_{x_i} \gamma_{i-1,i}(\sigma_{i-1}, x_i, \sigma_i) \alpha_{i-1}(\sigma_{i-1}) \qquad (8.18)$$

and

$$\beta_{i-1}(\sigma_{i-1}) = \sum_{\sigma_i} \sum_{x_i} [(\sigma_{i-1}, x_i, \sigma_i) \in \mathcal{T}_i] \, p(y_i|x_i) p(x_i) \beta_i(\sigma_i)$$

$$= \sum_{\sigma_i} \sum_{x_i} \gamma_{i-1,i}(\sigma_{i-1}, x_i, \sigma_i) \beta_i(\sigma_i) \qquad (8.19)$$

where

$$\gamma_{i-1,i}(\sigma_{i-1}, x_i, \sigma_i) \triangleq [(\sigma_{i-1}, x_i, \sigma_i) \in \mathcal{T}_i] \, p(y_i|x_i) p(x_i) \qquad (8.20)$$

In conclusion, the APP of x_i is given by

$$p(x_i \mid \mathbf{y}) = \sum_{\sigma_{i-1}} \sum_{\sigma_i} \gamma_{i-1,i}(\sigma_{i-1}, x_i, \sigma_i) \alpha_{i-1}(\sigma_{i-1}) \beta_i(\sigma_i)$$

in agreement with the results of Section 5.4. Notice that the values of $p(x_i)$ may not be available, in which case they are substituted with a constant (see the next chapter, where the BCJR algorithm is applied to iterative decoding of turbo codes).

8.3.3 Why the sum–product algorithm works

We now prove that the sum–product algorithm achieves the marginalization of a function represented by a factor graph without cycles. We consider in particular the calculation of APPs in a decoding problem.

The algorithm is based on two principles, called the *past–future decomposition* and the *sum–product decomposition*. For the first principle, observe that, if an edge is cut in a graph without cycles, then the graph is partitioned into two disconnected subgraphs. Consider the edge corresponding to state σ_j, and the value S_j taken on by σ_j. Then, similarly to what we did in Chapter 5, a code $\mathcal{C}$ can be decomposed as the Cartesian product of two "past" and "future" projection codes, denoted here $\mathcal{P}_j^P(S_j)$ and $\mathcal{P}_j^F(S_j)$, respectively. If for a moment we think of the code words of $\mathcal{C}$ as paths traversing a trellis, then at time j the code $\mathcal{P}_j^P(S_j)$ corresponds to the set of subpaths merging into S_j, and $\mathcal{P}_j^F(S_j)$ to the set of subpaths emanating from S_j (Figure 8.27 illustrates this).

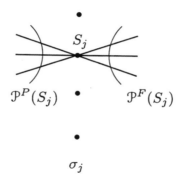

Figure 8.27: *Decomposing a code at state σ_j. The code words corresponding to paths through S_j are the Cartesian product of two projection codes, the "past" $\mathcal{P}_j^P(S_j)$ and the "future" $\mathcal{P}_j^F(S_j)$.*

We now use the *Cartesian-product distributive law*, which states the following: If $\mathcal{A}$ and $\mathcal{B}$ are disjoint discrete sets, and $\alpha(a)$, $\beta(b)$ are two functions defined on $\mathcal{A}$ and $\mathcal{B}$, respectively, then

$$\sum_{(a,b)\in\mathcal{A}\times\mathcal{B}} \alpha(a)\beta(b) = \left(\sum_{a\in\mathcal{A}}\alpha(a)\right)\left(\sum_{b\in\mathcal{B}}\beta(b)\right) \tag{8.21}$$

This law, on which several "fast algorithms" are based [8.1], says that, rather than computing the sum of $|\mathcal{A}\times\mathcal{B}| = |\mathcal{A}|\cdot|\mathcal{B}|$ products, we can (and it is faster to) compute a single product of separate sums over $\mathcal{A}$ and $\mathcal{B}$.

Now apply this law to (8.17). After defining $\mathbf{x}^P$ and $\mathbf{x}^F$ as code words of $\mathcal{P}_j^P(\sigma_j)$ and $\mathcal{P}_j^F(\sigma_j)$, $\mathcal{J}^P$, $\mathcal{J}^F$ as their index ranges, and $\mathbf{y}^P$, $\mathbf{y}^F$ as the projections of $\mathbf{y}$ to $\mathcal{J}^P$, $\mathcal{J}^F$, respectively, we obtain

$$p(\sigma_j\mid\mathbf{y}) \;\propto\; \left(\sum_{\mathbf{x}^P\in\mathcal{P}_j^P(\sigma_j)}\prod_{i\in\mathcal{J}^P}p(y_i\mid x_i)\right)\left(\sum_{\mathbf{x}^F\in\mathcal{P}_j^P(\sigma_j)}\prod_{i\in\mathcal{J}^F}p(y_i\mid x_i)\right)$$

$$\propto\; p(\sigma_j\mid\mathbf{y}^P)\,p(\sigma_j\mid\mathbf{y}^F) \tag{8.22}$$

The SPA computes the two functions in (8.22) separately, then multiplies them. In practice, each function will be represented as a vector whose components correspond to the values taken by S_j, with the multiplication occurring componentwise.

Application of the Cartesian-product distributive law to (8.15), which involves symbol variables rather than state variables, is simpler. In fact, one of the two

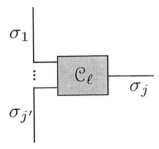

Figure 8.28: *Illustrating the sum–product decomposition principle.*

factors in the product is just $p(x_i \mid \mathbf{y})$, and we have

$$
\begin{aligned}
p(x_i \mid \mathbf{y}) \;&\propto\; p(x_i \mid y_i) \left(\sum_{\mathbf{x} \in \mathcal{C}_i(x_i)} \prod_{\ell \neq i} p(y_\ell \mid x_\ell) \right) \\
&\propto\; p(x_i \mid y_i) p(x_i \mid \mathbf{y}_{\sim i})
\end{aligned}
\tag{8.23}
$$

where $\mathbf{y}_{\sim i}$ is a shorthand notation to indicate the vector $\mathbf{y}$ with its ith component removed.

Consider now the second principle, the sum–product decomposition. This deals with the computation of $p(\sigma_j \mid \mathbf{y}^P)$ and $p(\sigma_j \mid \mathbf{y}^F)$ in (8.22) from analogous quantities that are one step further upstream. Consider the edge labeled σ_j, emanating from the node corresponding to constraint $\mathcal{C}_\ell$, that we interpret here as a code (for example, for normal graphs derived from Tanner graph, the nodes may only represent repetition functions or modulo-2 adders, which we interpret as repetition codes and single-parity-check codes, respectively). For notational simplicity, renumber the edges entering this node $\sigma_1, \dots, \sigma_{j'}$, as shown in Figure 8.28. Next, observe that, since the graph is cycle free, each of the edges entering the node has its own distinct past, whose union must be the projection code $\mathcal{P}_j^P(\sigma_j)$. For each word of code $\mathcal{C}_i$ corresponding to σ_j, the set of possible pasts is the Cartesian product of possible pasts of the other states, i.e., $\sigma_1, \dots, \sigma_{j'}$, and the total set of possible pasts is the disjoint union of these Cartesian products. Using once again the Cartesian-product distributive law, we have

$$
p(\sigma_j \mid \mathbf{y}^P) = \sum_{\mathcal{C}_i(\sigma_j)} \prod_{i=1}^{j'} p(\sigma_i \mid \mathbf{y}^{P_i})
\tag{8.24}
$$

where the factors in the product are known and hence can be used to compute $p(\sigma_j \mid \mathbf{y}^P)$.

8.3.4   The sum–product algorithm on graphs with cycles

If the code graph is finite and has no cycles, the sum–product algorithm, applied in conjunction with a suitable computational schedule that is usually easy to find, yields the required exact APP distribution of the code word symbols in a finite number of steps. On a graph with cycles, the sum–product algorithm can still be applied by implementing the sum–product step (8.11) at any node, using the incoming messages that are actually available and initializing all unknown messages to the constant function. However, the algorithm may not converge at all, or it may converge to an incorrect APP distribution: the derivation of conditions for convergence is a topic of current research. Fortunately, in many practical cases the algorithm *does* converge and yields the correct answer: for this reason it is commonly used to decode powerful codes such as the turbo codes and low-density parity-check codes to be studied in next chapter.

Regrettably, codes whose Tanner graphs have no cycles are rather poor: in [8.4] it is shown that, if C is an (n, k, d) cycle-free linear code with rate $\rho = k/n \geq 0.5$, then $d \leq 2$. On the other hand, if C has a rate lower than 0.5, then $d < 2/\rho$.

In a code whose graph is not cycle-free, the presence of short cycles should be avoided, as they hinder convergence [8.14]: if the girth of the factor graph is very large, the loop-free approximation can be made. In [8.14], it is proved that the assumption of a graph without cycles holds asymptotically, as n grows large, for LDPC codes, while for turbo codes it has only a heuristic justification. The presence of the interleaver should be exploited to maximize the girth of the factor graph.

8.4   Algorithms related to the sum–product

The formulation of the sum–product algorithm described above is based on the fact that two operations are available (sum, product) and that distributivity holds:

$$(a \cdot b) + (a \cdot c) = a \cdot (b + c)$$

An immediate generalization of the sum–product algorithm can be obtained whenever we can define two operations that are distributive. For example, assuming that the quantities we are operating on are nonnegative, the operators *max* and *product* are such that

$$\max\{ab, ac\} = a \max\{b, c\}$$

Similarly, the operators *min* and *sum* are such that

$$\min\{a + b, a + c\} = a + \min\{b, c\}$$

while the operators *max* and *sum* yield

$$\max\{a + b, a + c\} = a + \max\{b, c\}$$

For every specific distributive law we are using, a version of the sum–product algorithm can be derived. For example, from the latter property a *max–sum algorithm* can be obtained by changing sums into *max* and products into sums. This allows one to marginalize a function $f(x_1, \ldots, x_n)$ by computing the equivalent of (8.3) as

$$f_i(x_i) \triangleq \max_{\sim x_i} f(x_1, \ldots, x_n) \tag{8.25}$$

where the function f is "factored" as a sum of functions: (8.5) corresponds to

$$f(x_1, x_2, x_3) = g_1(x_1, x_2) + g_2(x_1, x_3)$$

The basic step of the max–sum algorithm then becomes, upon modification of (8.11),

$$\mu_{g \to x_i}(x_i) = \max_{\sim x_i} \left[g(x_1, \ldots, x_n) + \sum_{\ell \neq i} \mu_{x_\ell \to g}(x_\ell) \right] \tag{8.26}$$

Notice that the definition of the Iverson function must also be generalized: in general, we have a null element z for the "sum" and an identity u for the "product," such that $x + z = x$, $u \cdot x = x$, and $z \cdot x = z$ for all x. We define $[P]$ as taking the value u when P is true, and z otherwise. For example, in the max–sum context we have $z = -\infty$ and $u = 0$. The situation is summarized in Table 8.2.

"+"	"·"	"1"	"0"
sum	product	1	0
min	sum	0	∞
max	sum	0	$-\infty$
max	product	1	0

Table 8.2: *Multiple aspects of the sum–product algorithm.*

8.4.1 Decoding on a graph: Using the max–sum algorithm

We now show how the max–sum algorithm can be used to decode a code described by a factor graph by a variant of ML decoding. Observe that we have, for a station-

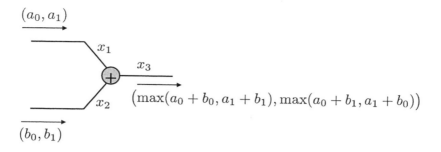

Figure 8.29: *A step of the max–sum algorithm applied to the node* $[x_1 + x_2 + x_3 = 0]$.

ary memoryless channel,

$$\log p(\mathbf{y} \mid \mathbf{x}) = \log \prod_{i=1}^{n} p(y_i \mid x_i)$$

$$= \sum_{i=1}^{n} \log p(y_i \mid x_i) \tag{8.27}$$

Denote, as before, by $\mathcal{C}_i(x_i)$ the subset of code words whose ith component is x_i. Now, ML decoding of $\mathcal{C}$ consists of finding the code word whose symbols $\hat{x}_i$ are solutions of

$$\hat{x}_i = \max_{\mathbf{x} \in \mathcal{C}_i(x_i)}{}^{-1} p(\mathbf{y} \mid \mathbf{x}) = \max_{\mathbf{x} \in \mathcal{C}_i(x_i)}{}^{-1} \log p(\mathbf{y} \mid \mathbf{x}) \tag{8.28}$$

Since $\log p(\mathbf{y} \mid \mathbf{x})$ "factors" as in (8.27), the max–sum algorithm can be applied to the factor graph of the code. In fact, observe that we can write, by appropriately defining the Iverson function,

$$\max_{\mathbf{x} \in \mathcal{C}_i(x_i)} \log p(\mathbf{y} \mid \mathbf{x}) = \max_{\sim x_i} \{ [\mathbf{x} \in \mathcal{C}] + \log p(\mathbf{y} \mid \mathbf{x}) \} \tag{8.29}$$

As an illustration, consider the basic step of the max–sum algorithm for a $\oplus$ node with three branches emanating from it, as shown in Figure 8.29 (the case with more than three branches can be derived as a simple exercise). We have

$$\mu_{\oplus \to x_3}(x_3) = \max_{x_1, x_2} \{ [x_1 + x_2 + x_3 = 0] + \mu_{x_1 \to \oplus}(x_1) + \mu_{x_2 \to \oplus}(x_2) \}$$

Again, since $[x_1 + x_2 + x_3 = 0]$ takes on value $-\infty$ when $x_1 + x_2 + x_3 \neq 0$, and 0 otherwise, the maximum value of the term in curly brackets occurs when the constraint $x_1 + x_2 = x_3$ is satisfied, so

$$\mu_{\oplus \to x_3}(x_3) = \max_{x_1, x_2 | x_1 + x_2 = x_3} \{\mu_{x_1 \to \oplus}(x_1) + \mu_{x_2 \to \oplus}(x_2)\}$$

In words, for a value of x_3 the outgoing message is the maximum of the sums of the incoming messages over the set of all incoming pairs x_1, x_2 consistent with x_3.[2] For example, with the incoming message from each edge x_i being represented by the two-component vector

$$\left(\mu_{x_i \to \oplus}(0), \ \mu_{x_i \to \oplus}(1)\right)$$

we have the situation of Figure 8.29.

Similarly, the outgoing messages from a repetition node are determined according to the rule, derived from (8.12),

$$\mu_{f_= \to x_i}(x_i) = \sum_{\ell \neq i} \mu_{x_\ell \to f_=}(x_i) \tag{8.30}$$

We should also observe the following. Consider decoding of $\mathcal{C}$. Taking the logarithm of the function marginalized by the SPA, we have the approximation (see Problem 6 below)

$$\log \sum_{\sim x_i} [\mathbf{x} \in \mathcal{C}] \, p(\mathbf{y} \mid \mathbf{x}) \approx \max_{\sim x_i} \{\log[\mathbf{x} \in \mathcal{C}] + \log p(\mathbf{y} \mid \mathbf{x})\}$$

which expresses how the max–sum algorithm turns out to be an approximation of the sum–product algorithm as far as decoding is concerned.

Example 8.10

We now provide an example of decoding a linear binary code using the max–sum algorithm. We choose the code whose Tanner graph is shown in Figure 8.5. The normal form of the graph used for decoding is shown in Figure 8.30. The values of the input variables are shown in Figure 8.31, in the form of the two-component vector $\left(\log p(y_i \mid x_i = 0), \ \log p(y_i \mid x_i = 1)\right)$. Since for binary messages only the difference between the two components is significant in this algorithm, the symbol with the lowest value of $\log p(y_i \mid x_i)$ is assigned zero value. The centripetal messages associated with each edge are also shown (notice how these messages can be computed in parallel). Figure 8.32 shows the centrifugal messages, and Figure 8.33

---

[2]It may be observed that this step is reminiscent of the ACS step in the Viterbi algorithm. This is not a coincidence: a variant of the Viterbi algorithm can be expressed as a max–sum algorithm on a factor graph derived from a code trellis. Not only the BCJR algorithm, but also the Fast-Fourier Transform (FFT) algorithm and the Kalman filtering algorithm turn out to be special cases of the SPA [8.1, 8.11].

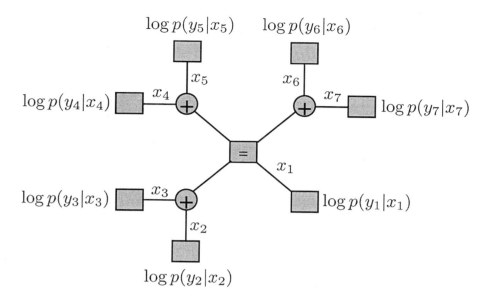

Figure 8.30: *Normal graph of a binary code, as transformed for the application of the max–sum algorithm.*

the final values obtained by summing the latter with the centripetal messages of Figure 8.31.

At this point, we can do ML decoding (see Figure 8.33): the most likely code word is 1010110 and has weight 15. However, the max–sum algorithm tells us more: it computes the relative weights of all possible symbol values. The magnitudes of the weight differences (the *soft information*)

$$1\ 1\ 3\ 2\ 1\ 1\ 3$$

tell us that the decisions on x_3 and x_7 are the most reliable, while those on x_1, x_2, x_5, and x_6 are the least reliable. This additional information may prove useful, for example to monitor the channel quality. □

8.5   Bibliographical notes

In his landmark 1981 paper [8.15], Tanner introduced the graphical-model description of codes and proved the optimality of the sum–product algorithm for cycle-free graphs. In Tanner's original formulation, all variables were code symbols. Wiberg

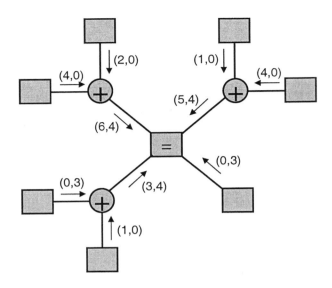

Figure 8.31: *Max–sum decoding of a binary code: initial values and centripetal messages, computed according to Figure 8.29.*

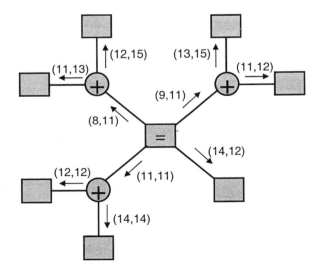

Figure 8.32: *Max–sum decoding of a binary code: centrifugal messages.*

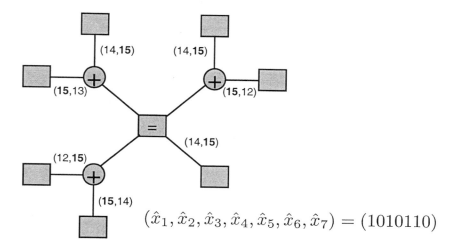

$$(\hat{x}_1, \hat{x}_2, \hat{x}_3, \hat{x}_4, \hat{x}_5, \hat{x}_6, \hat{x}_7) = (1010110)$$

Figure 8.33: *Max–sum decoding of a binary code: final values of the messages and ML decoding.*

et al. [8.16, 8.17] introduced state variables in the model. The sum–product algorithm was discovered by Gallager in [8.9] as a decoding algorithm for LDPC codes, but it took a long time (and the invention of turbo codes) for its full potential to be appreciated. Tanner [8.15] introduced the min-sum algorithm. In [8.17] it was observed how the Viterbi and BCJR algorithms can be reinterpreted in the message-passing context of the SPA. Reference [8.1] describes how the SPA can be described in the context of "belief propagation in Bayesian networks" [8.13], a theory developed with applications to artificial intelligence. Factor graphs were expounded in [8.10]. Normal graphs were introduced by Forney [8.6].

Our presentation of the material in this chapter is essentially based on [8.1, 8.5, 8.10–8.12]. Our analysis of the sum–product algorithm on cycle-free graphs is derived from [8.7]

8.6   Problems

1. Consider the function $f(x_1, \ldots, x_6)$, with each x_i taking on $|\mathcal{X}|$ values, and its factorization

$$f(x_1, x_2, x_3, x_4, x_5, x_6) = g_1(x_1)g_2(x_1, x_2)g_3(x_1, x_3, x_4)g_4(x_4, x_5, x_6)$$
$$(8.31)$$

We are interested in the marginal

$$f_4(x_4) \triangleq \sum_{\sim x_4} f(x_1, x_2, x_3, x_4, x_5, x_6) \tag{8.32}$$

Compare the complexity of the direct calculation of (8.32) with the complexity of its calculation based on factorization (8.31).

2. Using the sum-product algorithm, compute the message μ_0 shown in Figure 8.34. Assume x_0, x_1, and x_2 to be binary variables taking values in $\{0, 1\}$, and $g(x_0, x_1, x_2) = x_0 \cdot x_1 \cdot x_2$.

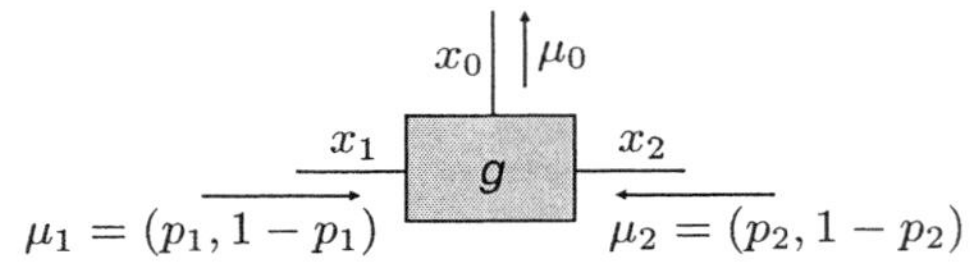

Figure 8.34: *A segment of a factor graph.*

3. Consider the binary code whose Tanner graph is shown in Figure 8.35.

   (a) Draw the normal graph of the code.

   (b) Find the parity-check matrix **H** of the code.

   (c) List all code words.

   (d) Find the minimum Hamming distance.

   (e) Decode a code word using the max–sum algorithm with input data

$$(5, 0) \quad (4, 0) \quad (3, 0) \quad (2, 0) \quad (0, 1)$$

4. (*Soft channel equalization.*) Consider the transmission of n independent binary symbols $\mathbf{x} = (x_1, \ldots, x_n)$ on a common channel, and the observation of a noisy vector $\mathbf{y}$ whose components are known functions of all symbols (for example, linear combinations with known coefficients). The channel is described by the function $p(\mathbf{y} \mid \mathbf{x})$. Draw a factor graph for the estimation of the APPs $p(x_i, \mathbf{y})$, and sketch the corresponding sum–product algorithm. How do the factor graph and the SPA change if $\mathbf{x}$ is a word of the block code $\mathcal{C}$?

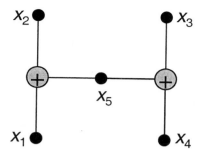

Figure 8.35: *Tanner graph for Problem 3.*

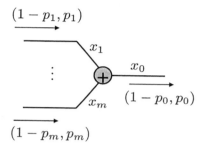

Figure 8.36: *Sum–product algorithm applied to a check node. The variables are binary, and the messages represent probabilities.*

5. Consider the application of the sum–product algorithm to the graph fragment with a check node as shown in Figure 8.36. The variables are binary, and the messages represent probabilities. Prove that

$$p_0 = \frac{1}{2} + \frac{1}{2}\prod_{i=1}^{m}(1 - 2p_i)$$

6. This problem shows how the max–sum algorithm can be viewed as an approximation of the sum–product algorithm. Define the function

$$\max{}^*(x, y) \triangleq \ln(e^x + e^y)$$

   (a) Prove that

$$\max{}^*(x, y) = \max(x, y) + \ln\left(1 + e^{-|x-y|}\right)$$

so that

$$\max{}^*(x, y) \approx \max(x, y)$$

(b) Find upper and lower bounds to the error involved in the approximation above.

(c) Extend (a) to more than two variables.

7. Consider the $(5, 2, 3)$ code $\mathcal{C}$ whose parity-check matrix is

$$\mathbf{H} = \begin{bmatrix} 1 & 0 & 1 & 0 & 1 \\ 1 & 1 & 1 & 0 & 0 \\ 1 & 0 & 0 & 1 & 0 \end{bmatrix}$$

The Tanner graph of this code has cycles. Prove that they can be removed by considering a suitable code equivalent to $\mathcal{C}$.

References

[8.1] S. M. Aji and R. J. McEliece, "The generalized distributive law," *IEEE Trans. Inform. Theory*, Vol. 46, No. 2, pp. 325–343, March 2000.

[8.2] G. Battail, M. C. Decouvelaere, and Oh. Godlewski, "Replication decoding," *IEEE Trans. Inform. Theory*, Vol. 25, No. 3, pp. 332–345, May 1979.

[8.3] D. Divsalar, H. Jin, and R. J. McEliece, "Coding theorems for 'turbo-like' codes," *Proc. 1998 Allerton Conf. Communications, Control and Computers*, Allerton, IL, pp. 201–210, September 1998.

[8.4] T. Etzion, A. Trachtenberg, and A. Vardy, "Which codes have cycle-free Tanner graphs?," *IEEE Trans. Inform. Theory*, Vol. 45, No. 6, pp. 2173–2181, September 1999.

[8.5] G. D. Forney, Jr., "On iterative decoding and the two-way algorithm," *Proc. First International Symp. on Turbo Codes*, Brest, France, pp. 12–25, September 3–5, 1997.

[8.6] G. D. Forney, Jr., "Codes on graphs: Normal realizations," *IEEE Trans. Inform. Theory*, Vol. 47, No. 2, pp. 520–548, February 2001.

[8.7] G. D. Forney, Jr., *Lecture Notes for Course 6.451*, Massachusetts Institute of Technology, Spring 2002.

[8.8] B. J. Frey, *Graphical Models for Machine Learning and Digital Communication*. Cambridge, MA: MIT Press, 1998.

[8.9] R. G. Gallager, *Low-Density Parity-Check Codes*. Cambridge, MA: MIT Press, 1963.

[8.10] F. R. Kschischang, B. J. Frey, and H.-A. Loeliger, "Factor graphs and the sum-product algorithm," *IEEE Trans. Inform. Theory*, Vol. 47, No. 2, pp. 498–519, February 2001.

[8.11] H.-A. Loeliger, "Least squares and Kalman filtering on Forney graphs," in R. E. Blahut and R. Koetter (Eds.), *Codes, Graphs, and Systems*. New York: Kluwer, pp. 113–135, 2002.

[8.12] H.-A. Loeliger, "An introduction to factor graphs," *IEEE Signal Processing Magazine*, Vol. 21, No. 1, pp. 28–41, January 2004.

[8.13] J. Pearl, *Probabilistic Reasoning in Intelligent Systems*. San Mateo, CA: Morgan Kaufmann, 1988.

[8.14] T. J. Richardson and R. L. Urbanke, "The capacity of low-density parity-check codes under message-passing decoding," *IEEE Trans. Inform. Theory*, Vol. 47, No. 2, pp. 599–618, February 2001.

[8.15] R. M. Tanner, "A recursive approach to low complexity codes," *IEEE Trans. Inform. Theory*, Vol. IT-27, pp. 533–547, September 1981.

[8.16] N. Wiberg, "Codes and iterative decoding on general graphs," Ph.D. Dissertation, Linköping University. Linköping, Sweden, 1996.

[8.17] N. Wiberg, H.-A. Loeliger, and R. Kötter, "Codes and iterative decoding on general graphs," *Eur. Trans. Telecomm.*, Vol. 6, pp. 513–525, September/October 1995.

you did your strong nine furlong mile in slick and slapstick record time

LDPC and turbo codes

Classes of codes defined on graphs exist that can approach Shannon's capacity bound quite closely, and with a reasonable decoding complexity. All these codes are obtained by connecting simple component codes through an interleaver. Decoding consists of iterative decodings of these simple codes. In this chapter we describe in some detail turbo codes and low-density parity-check codes, with special attention to their performance and their decoding algorithms. Their distance properties are also given some attention.

9.1 Low-density parity-check codes

A low-density parity-check (LDPC) code is a long linear binary block code whose parity-check matrix $\mathbf{H}$ has a low density of 1s. Specifically, $\mathbf{H}$ is sparse, and contains a small fixed number w_c of 1s in each column and a small fixed number w_r of 1s in each row. If the block length is n, we say that $\mathbf{H}$ characterizes an $\langle n, w_c, w_r \rangle$ LDPC code. These codes may be referred to as *regular* LDPC codes to distinguish them from *irregular* codes, whose values of w_c and w_r are not constant. The matrix $\mathbf{H}$ of the latter has *approximately* w_r 1s in each row and w_c 1s in each column.

The normal graph of a (regular) LDPC code is shown in Figure 9.1. With this representation, we have that an LDPC code is a binary linear code such that every coded symbol participates in exactly w_c parity-check equations, while each one of the m sum-check equations involves exactly w_r bits. For consistency, we have $n w_c = m w_r$.

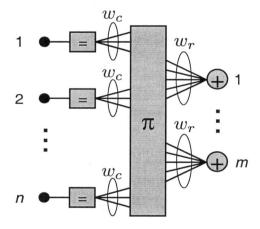

Figure 9.1: *Normal graph of a regular $\langle n, w_c, w_r \rangle$ LDPC code.*

It follows from the definition of an LDPC code that $\mathbf{H}$ has $n w_c / w_r$ rows: in fact, the total number of 1s in $\mathbf{H}$ is $n w_c$; dividing by w_r, we obtain the number of rows. Since $\mathbf{H}$ is in general an $m \times n$ matrix, if $\mathbf{H}$ has full rank the code rate is

$$\frac{n - m}{n} = 1 - \frac{m}{n} = 1 - \frac{w_c}{w_r} \tag{9.1}$$

The above equality yields the constraint $w_c \leq w_r$. Notice that the actual rate ρ of the code might be higher than $m/n = w_c/w_r$ because the parity-check equations summarized by $\mathbf{H}$ might not be all independent. We call $\rho^* \triangleq 1 - w_c/w_r$ the *design rate* of the code.

Example 9.1

The parity-check matrix of a $\langle 20, 3, 4 \rangle$ LDPC code with $\rho^* = 1/4$ is shown below.

$$\mathbf{H} = \begin{bmatrix} 1 & 1 & 1 & 1 & 0 & 0 & 0 & 0 & 0 & 0 & 0 & 0 & 0 & 0 & 0 & 0 & 0 & 0 & 0 & 0 \\ 0 & 0 & 0 & 0 & 1 & 1 & 1 & 1 & 0 & 0 & 0 & 0 & 0 & 0 & 0 & 0 & 0 & 0 & 0 & 0 \\ 0 & 0 & 0 & 0 & 0 & 0 & 0 & 0 & 1 & 1 & 1 & 1 & 0 & 0 & 0 & 0 & 0 & 0 & 0 & 0 \\ 0 & 0 & 0 & 0 & 0 & 0 & 0 & 0 & 0 & 0 & 0 & 0 & 1 & 1 & 1 & 1 & 0 & 0 & 0 & 0 \\ 0 & 0 & 0 & 0 & 0 & 0 & 0 & 0 & 0 & 0 & 0 & 0 & 0 & 0 & 0 & 0 & 1 & 1 & 1 & 1 \\ 1 & 0 & 0 & 0 & 1 & 0 & 0 & 0 & 1 & 0 & 0 & 0 & 1 & 0 & 0 & 0 & 0 & 0 & 0 & 0 \\ 0 & 1 & 0 & 0 & 0 & 1 & 0 & 0 & 0 & 1 & 0 & 0 & 0 & 0 & 0 & 0 & 1 & 0 & 0 & 0 \\ 0 & 0 & 1 & 0 & 0 & 0 & 1 & 0 & 0 & 0 & 0 & 0 & 0 & 1 & 0 & 0 & 0 & 1 & 0 & 0 \\ 0 & 0 & 0 & 1 & 0 & 0 & 0 & 0 & 0 & 0 & 1 & 0 & 0 & 0 & 1 & 0 & 0 & 0 & 1 & 0 \\ 0 & 0 & 0 & 0 & 0 & 0 & 0 & 1 & 0 & 0 & 0 & 1 & 0 & 0 & 0 & 1 & 0 & 0 & 0 & 1 \\ 1 & 0 & 0 & 0 & 0 & 1 & 0 & 0 & 0 & 0 & 0 & 1 & 0 & 0 & 0 & 0 & 0 & 1 & 0 & 0 \\ 0 & 1 & 0 & 0 & 0 & 0 & 1 & 0 & 0 & 0 & 1 & 0 & 0 & 0 & 0 & 1 & 0 & 0 & 0 & 0 \\ 0 & 0 & 1 & 0 & 0 & 0 & 0 & 1 & 0 & 0 & 0 & 0 & 1 & 0 & 0 & 0 & 0 & 0 & 1 & 0 \\ 0 & 0 & 0 & 1 & 0 & 0 & 0 & 0 & 1 & 0 & 0 & 0 & 0 & 1 & 0 & 0 & 1 & 0 & 0 & 0 \\ 0 & 0 & 0 & 0 & 1 & 0 & 0 & 0 & 0 & 1 & 0 & 0 & 0 & 0 & 1 & 0 & 0 & 0 & 0 & 1 \end{bmatrix}$$

In this example we observe that $\mathbf{H}$ can be viewed as composed of three submatrices, each of which contains a single "1" in each column. The second and third submatrices are obtained from the first submatrix by permuting the column order. $\square$

9.1.1 Desirable properties

While the ultimate quality of an LDPC code is defined in terms of its rate, coding gain, and complexity, some simple considerations may guide the selection of a candidate code. First, for good convergence properties of the iterative decoding algorithm, the Tanner graph of the code should have a large girth. In particular, short cycles must be avoided. (Observe that the shortest possible cycle in a bipartite graph has length 4, as shown in Figure 9.2 along with the structure of the parity-check matrix that generates it.) Next, regularity of the code eases implementation. Finally, for small error probability at high $\mathcal{E}_b/N_0$ on the AWGN channel, the minimum Hamming distance of the code must be large. This is especially interesting,

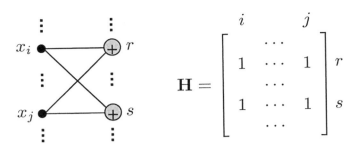

Figure 9.2: *Four-cycle in a Tanner graph, and corresponding parity-check matrix.*

because LDPC codes are known to achieve a large value of $d_{H,min}$. Roughly speaking, if $w_c > 2$ this minimum distance grows linearly with the block length n, and hence a large random LDPC code will exhibit a large $d_{H,min}$ with high probability. More specifically, it has been proved [9.12, 9.18] that, for a large enough block length n, an LDPC code exists with rate $\rho \geq 1 - 1/\lambda$, and minimum distance $d_{H,min} \geq \delta n$, for any $\delta < 0.5$ that satisfies the inequality

$$-\delta \log \delta - (1 - \delta) \log(1 - \delta) < 1/\lambda$$

9.1.2   Constructing LDPC codes

Several techniques for the design of parity-check matrices of LDPC codes have been proposed and analyzed. They can be classified under two main rubrics: *random* and *algebraic* constructions. Here we provide an example of each.

Random constructions

These are based on the generation of a parity-check matrix randomly filled with 0s and 1s, and such that the LDPC properties are satisfied. In particular, after one selects the parameters n, ρ^*, and w_c, for regular codes the row and column weights of $\mathbf{H}$ must be exactly w_r and w_c, respectively, with w_r and w_c small compared to the number of columns and rows. Additional constraints may be included: for example, the number of 1s in common between any two columns (or two rows) should not exceed one (this constraint prevents four-cycles).

In general, randomly constructed codes are good if n is large enough, but their performance may not be satisfactory for intermediate values of n [9.11, 9.16]. Also, usually they are not structured enough to allow simple encoding.

A method for the random construction of $\mathbf{H}$ was developed by Gallager in [9.12]. The transpose of the matrix $\mathbf{H}$ of a regular $\langle n, w_c, w_r \rangle$ has the form

$$\mathbf{H}' = \begin{bmatrix} \mathbf{H}'_1 & \mathbf{H}'_2 & \cdots & \mathbf{H}'_{w_c} \end{bmatrix} \tag{9.2}$$

where $\mathbf{H}_1$ has n columns and n/w_r rows, contains a single 1 in each column, and contains 1s its ith row from column $(i-1)w_r + 1$ to column iw_r. Matrices $\mathbf{H}_2$ to $\mathbf{H}_{w_c}$ are obtained by randomly permuting (with equal probabilities) the columns of $\mathbf{H}_1$. The matrix $\mathbf{H}$ of Example 9.1 is generated in this way, although there the permutations are not random.

Another algorithm for the generation of the parity-check matrix of an $\langle n, w_c, w_r \rangle$ LDPC code works as follows:

Step 1 Set $i = 1$.

Step 2 Generate a random binary vector with length nw_c/w_r and weight w_c. This is the ith column of $\mathbf{H}$.

Step 3 If the weight of each row of $\mathbf{H}$ at this point is $\leq w_r$, and the scalar product of each pair of columns is ≤ 1 (four-cycle constraint), then set $i = i + 1$. Otherwise, go to Step 2.

Step 4 If $i = n$, then stop. Otherwise, go to Step 2.

Since there is no guarantee that there are exactly w_r 1s in each row of $\mathbf{H}$, this algorithm may generate an irregular code. If a regular code is sought, suitable modifications to the procedure should be made.

Algebraic constructions

Algebraic LDPC codes may lend themselves to easier decoding than random codes. In addition, for intermediate n, the error probability of well-designed algebraic codes may be lower [9.1, 9.20].

A simple algebraic construction works as follows [9.10, 9.13]. Choose $p > (w_c - 1)(w_r - 1)$, and consider the $p \times p$ matrix obtained from the identity matrix $\mathbf{I}_p$ by cyclically shifting its rows by one position to the right:

$$\mathbf{J} \triangleq \begin{bmatrix} 0 & 1 & 0 & 0 & \cdots & 0 \\ 0 & 0 & 1 & 0 & \cdots & 0 \\ 0 & 0 & 0 & 1 & \cdots & 0 \\ 0 & 0 & 0 & 0 & \cdots & 1 \\ 1 & 0 & 0 & 0 & \cdots & 0 \end{bmatrix}$$

The ℓth power of $\mathbf{J}$ is obtained from $\mathbf{I}_p$ by cyclically shifting its rows by $\ell \bmod p$ positions to the right. After defining $\mathbf{J}^0 \triangleq \mathbf{I}_p$, construct the matrix

$$
\mathbf{H} = \begin{bmatrix}
\mathbf{J}^0 & \mathbf{J}^0 & \mathbf{J}^0 & \cdots & \mathbf{J}^0 \\
\mathbf{J}^0 & \mathbf{J}^1 & \mathbf{J}^2 & \cdots & \mathbf{J}^{w_r-1} \\
\mathbf{J}^0 & \mathbf{J}^2 & \mathbf{J}^4 & \cdots & \mathbf{J}^{2(w_r-1)} \\
& & \cdots & & \\
\mathbf{J}^0 & \mathbf{J}^{(w_c-1)} & \mathbf{J}^{2(w_c-1)} & \cdots & \mathbf{J}^{(w_c-1)(w_r-1)}
\end{bmatrix}
$$

This matrix has $w_c p$ rows and $w_r p$ columns. The number of 1s in each row and column is exactly w_r and w_c, respectively. Hence, this construction generates a $\langle w_r p, w_c, w_r \rangle$ LDPC code. It can be proved that no four-cycles are present.

Combining random and algebraic constructions

A technique that combines random and algebraic construction is proposed in [9.20]. Start with the $m \times n$ parity-check matrix $\mathbf{H}(0)$ of a good "core" LDPC code. Next, substitute for each 1 in $\mathbf{H}(0)$ a $p_1 \times p_1$ permutation matrix chosen randomly. We obtain the new $mp_1 \times np_1$ parity-check matrix $\mathbf{H}(1)$. Since the probability of repeating the same permutation matrix in the construction of $\mathbf{H}(1)$ is $1/p_1!$, it is suggested to choose $p_1 \geq 5$. The construction is repeated by substituting for each 1 in $\mathbf{H}(1)$ a $p_2 \times p_2$ random permutation matrix, which yields the $mp_1 p_2 \times np_1 p_2$ parity-check matrix $\mathbf{H}(2)$. This procedure can be repeated. In [9.20], it is shown that this construction preserves the girth and the minimum Hamming distance of the core code.

9.1.3 Decoding an LDPC code

Decoding can be performed using the sum–product or the max–sum algorithm, as indicated in the previous chapter. Here, however, since the Tanner graph of the code has cycles, the algorithm is not exact, nor does it necessarily converge in a finite number of steps. An iterative algorithm can be devised that computes alternatively the messages associated with both directions of each branch, and stops according to a preassigned criterion. A possible stopping rule is the following: set $\hat{x}_i = 1$ if $p(x_i = 1 \mid \mathbf{y}) > p(x_i = 0 \mid \mathbf{y})$, and $\hat{x}_i = 0$ otherwise. If the vector $\hat{\mathbf{x}} \triangleq (\hat{x}_1, \dots, \hat{x}_n)$ is a code word (i.e., all parity checks are satisfied) then stop. Otherwise, keep on iterating to some maximum number of iterations, and then stop and declare a failure.

Figure 9.3 represents, in a schematic form, the two basic message-passing steps when an iterative version of the sum–product algorithm is used for decoding an

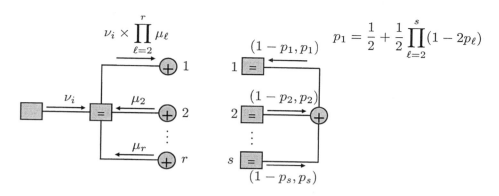

Figure 9.3: *Decoding an LDPC code: message-passing from a symbol node to a check node, and vice versa.*

LDPC code. We assume here that the messages are normalized so as to represent probabilities, and use a result from Problem 4 of Chapter 8. The algorithm starts with the intrinsic probabilities $\nu_i \triangleq p(y_i|x_i)$, and with uniform messages coming out of check nodes: $\mu_\ell = (0.5, 0.5)$ for all ℓ. Application of the SPA first computes the messages passing from symbol nodes to check nodes, and then from check nodes to symbol (repetition) nodes. These two steps represent a single iteration of the SPA.

Figure 9.4 shows the performance of two LDPC codes.

A simple suboptimum decoding algorithm: bit flipping

An LDPC code can be suboptimally decoded by a simple iterative technique called the *bit-flipping algorithm*. First, the symbols are individually "hard decoded" by transforming the channel observations into 1s and 0s so that the received vector **y** is transformed into the binary vector **b**. Consider the syndrome **Hb′**, whose components are the results of the sums computed in the right part of the graph. Each component of **b** affects w_c components of the syndrome. Thus, if only one bit is in error, then w_c syndrome components will equal 1. The bit-flipping algorithm is based on this observation and works as follows. Each iteration step includes the computation of all check sums, as well as the computation of the number of unsatisfied parity checks involving each of the n bits of **b**. Next, the bits of **b** are flipped when they are involved in the largest number of unsatisfied parity checks. The steps are repeated until all checks are satisfied, or a predetermined number of iterations is reached.

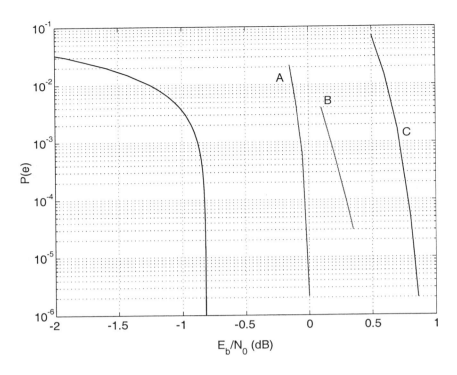

Figure 9.4: *Performance of rate-1/4 codes. Code B [9.7] is an irregular LDPC code with $n = 16,000$. Code C [9.18] is a regular LDPC code with $n = 40,000$. For reference's sake, Code A is a turbo code with $n = 16,384$ (see Figure 9.17 for further details). The leftmost curve is the Shannon limit for $\rho = 1/4$ and unconstrained AWGN channel, as derived in Problem 8 of Section 3 (see also Figure 1.5).*

Example 9.2

For illustration purposes, consider the rate-1/3 code (not exactly an LDPC code, since n is not large enough to yield a sparse $\mathbf{H}$) with parity-check matrix

$$\mathbf{H} = \begin{bmatrix} 1 & 1 & 1 & 0 & 0 & 0 \\ 1 & 0 & 0 & 1 & 1 & 0 \\ 0 & 1 & 0 & 1 & 0 & 1 \\ 0 & 0 & 1 & 0 & 1 & 1 \end{bmatrix}$$

corresponding to the Tanner graph of Figure 9.5. Let the observed vector be

$$(.1, .3, -1.2, .02, .5, .9)$$

The binary 6-tuple obtained by hard decoding is (001000). This is not a code word. The first iteration shows that the parity checks that fail are 1 and 4—a finding that is interpreted as indicating the presence of an error among the symbols whose nodes are connected to adders 1 and 4. Now, symbol 4 corresponds to no failed check, symbols 1, 2, 5, and 6 correspond to one failed check, and symbol 3 to two failed checks. We flip the third bit, thus obtaining the code word (000000), which is accepted, as all parity checks are satisfied. □

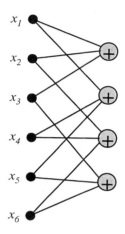

Figure 9.5: *Tanner graph of an LDPC code.*

9.2 Turbo codes

The general scheme of a turbo code based on *parallel concatenation* of two convolutional codes was shown in Figure 8.14. There, $\mathcal{C}_1$ and $\mathcal{C}_2$ are binary terminated convolutional codes or block codes, realized in systematic form. Let the generator matrices of $\mathcal{C}_1$ and $\mathcal{C}_2$ be $\mathbf{G}_1 = [\mathbf{I} \quad \mathbf{P}_1]$ and $\mathbf{G}_2 = [\mathbf{I} \quad \mathbf{P}_2]$, respectively. If the vector to be encoded is $\mathbf{u}$, the first encoder outputs $[\mathbf{u} \quad \mathbf{c}_1]$, with $\mathbf{c}_1 \triangleq \mathbf{u}\mathbf{P}_1$. The interleaver π applies a fixed permutation to the components of $\mathbf{u}$ and sends $\pi\mathbf{u}$ to the second encoder, which generates $[\pi\mathbf{u} \quad \mathbf{c}_2]$, with $\mathbf{c}_2 \triangleq (\pi\mathbf{u})\mathbf{P}_2$.

If $\mathcal{C}_1$ and $\mathcal{C}_2$ have rates ρ_1 and ρ_2, respectively, the turbo-code rate is given by

$$\rho = \frac{\rho_1\rho_2}{\rho_1 + \rho_2 - \rho_1\rho_2} \qquad (9.3)$$

To prove this, neglect the effect of the trellis termination, and observe that if k bits enter the encoder of Figure 8.14, then $\mathbf{u}$ contains k bits, $\mathbf{c}_1$ contains $k/\rho_1 - k$ bits,

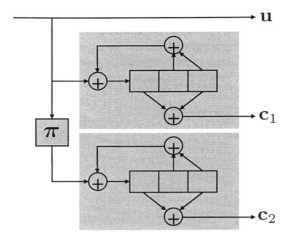

Figure 9.6: *Encoder of a parallel-concatenated turbo code with recursive component encoders, and $\rho = 1/3$.*

and c_2 contains $k/\rho_2 - k$ bits. The ratio between k and the total number of encoded bits yields (9.3). Note that if $\rho_1 = \rho_2$ we simply have

$$\rho = \frac{\rho_1}{2 - \rho_1} \tag{9.4}$$

The most popular turbo-code design has $\rho_1 = \rho_2 = 1/2$ (typically obtained with $\mathcal{C}_1 = \mathcal{C}_2$), and hence $\rho = 1/3$ [9.3, 9.4]. If the even bits of c_1 and the odd bits of c_2 are punctured, then $\rho_1 = \rho_2 = 2/3$, and $\rho = 1/2$.

The most common form of convolutional encoder used in general is nonsystematic and polynomial (as, for example, the rate-$1/2$ encoder of Figure 6.3). Such an encoder cannot be used as a constituent of a turbo code, which requires systematic encoders. Nonrecursive (i.e., feedback-free) encoders should also be ruled out because the resulting turbo code would exhibit poor distance properties. A turbo encoder including two systematic recursive codes is shown in Figure 9.6.

Serially concatenated turbo codes

A serially concatenated turbo code is obtained by cascading two convolutional encoders as shown in Figure 9.7. $\mathcal{C}_o$ is called the *outer* code and $\mathcal{C}_i$ the *inner* code. Their rates are ρ_o and ρ_i, respectively. In practice, the outer code may be chosen as nonrecursive and nonsystematic or recursive and systematic; however, $\mathcal{C}_i$ should be recursive and systematic for better performance. The rate ρ of the concatenated

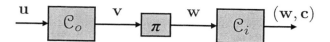

Figure 9.7: *General scheme of a serially concatenated turbo code.*

code is simply given by the product of the two rates:

$$\rho = \rho_o \rho_i \qquad (9.5)$$

For example, the rate $\rho = 1/2$ can be obtained by choosing two component codes with rates $\rho_o = 2/3$ and $\rho_i = 3/4$. Notice that this choice involves constituent codes with higher rates and complexity than for a rate-$1/2$ turbo code with parallel concatenation.

9.2.1 Turbo algorithm

Although, in principle, turbo codes can be optimally decoded by drawing their trellises and using the Viterbi algorithm, the complexity of the resulting decoder would be generally prohibitive. Using an iterative version of the sum–product algorithm (the *turbo algorithm*) provides instead extremely good performance with moderate complexity. This algorithm is conceptually similar to the message-passing algorithm described for LDPC codes, consisting of iterative exchanges of messages from symbol nodes to check nodes and vice versa (see Figure 9.3). With turbo codes, the more complex structure of their factor graph (which includes convolutional codes in lieu of symbol nodes and check nodes: see Figure 8.15) calls for a more complex algorithm. In fact, it requires the separate decoding of the component codes: each decoder operates on the received data, forms an estimate of the transmitted message, and exchanges information with the other decoder. After a number of iterations, this estimate is finally accepted. The algorithm is run for a fixed number of iterations or can be stopped as soon as a code word is obtained (see *supra*, Section 9.1.3).

Figure 9.8 summarizes the general principle, whereby two decoders (one for $\mathcal{C}_1$ and one for $\mathcal{C}_2$) exchange messages back and forth: this decoding mechanism is reminiscent of the working of a turbo-charged engine, which suggested the name *turbo* for the algorithm. Although relatively little is known about its theoretical convergence properties (which will be examined *infra*, in Section 9.2.4), its behavior with graphs having cycles is surprisingly good.

To describe the turbo algorithm, we first examine the behavior of the two decoders of Figure 9.8, and, in particular, the messages they exchange under the SPA.

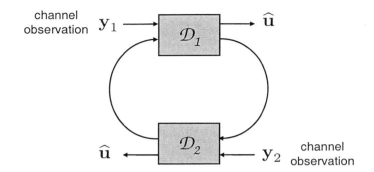

Figure 9.8: *General scheme of turbo decoding algorithm. Here* $\mathbf{y}_1$ *and* $\mathbf{y}_2$ *are channel observations generated by two independent encodings of the same block* $\mathbf{u}$.

Consider a linear binary block code $\mathcal{C}$ with length n and with k information symbols (if a convolutional code is used, let its termination generate a block code with the above parameters). Here we compute explicitly the a posteriori probabilities of the code symbols, examining separately systematic and nonsystematic codes.

SISO decoder: systematic codes

If the code is systematic, the first k entries of each word $\mathbf{x}$ coincide with the information symbols $\mathbf{u}$. We write $\mathbf{x} = (u_1, \dots, u_k, x_{k+1}, \dots, x_n)$, and we have

$$p(\mathbf{x}) = [\mathbf{x} \in \mathcal{C}] \prod_{i=1}^{k} p(u_i)$$

Hence, under our usual assumption of a stationary memoryless channel,

$$p(\mathbf{x} \mid \mathbf{y}) \propto [\mathbf{x} \in \mathcal{C}] \prod_{i=1}^{k} p(y_i \mid u_i) p(u_i) \prod_{j=k+1}^{n} p(y_j \mid x_j) \qquad (9.6)$$

To compute the APPs of the information symbols u_i, $i = 1, \dots, k$, (and hence to soft-decode $\mathcal{C}$) we combine, according to (9.6), the *a priori information* $p(u_1)$, ..., $p(u_k)$ on the source symbols and the *channel information* $p(\mathbf{y} \mid \mathbf{x})$ into one intrinsic message (Figure 9.9).

To describe the message-passing turbo algorithm, it is convenient to introduce a *soft-input, soft-output* (SISO) decoder, as shown in Figure 9.10. This is a system that, based on (9.6), has two sets of inputs: (a) the n conditional probabilities whose

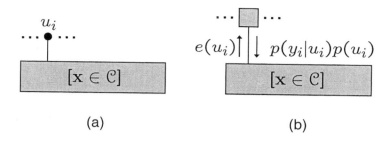

Figure 9.9: *(a) Factor graph for the systematic code* $\mathcal{C}$*. (b) Messages exchanged by the sum–product algorithm applied to the computation of the APPs* $p(u_i|\mathbf{y})$.

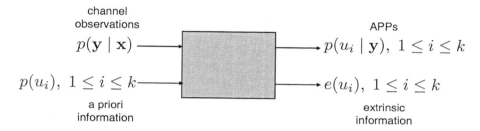

Figure 9.10: *Soft-input, soft-output decoder for systematic codes.*

product forms $p(\mathbf{y} \mid \mathbf{x})$, and (b) the k a priori probabilities $p(u_j)$. It outputs (c) the k APPs $p(u_i \mid \mathbf{y})$ and (d) the k extrinsic messages (the *extrinsic information*). This block may be implemented using the BCJR algorithm, or, if this is computationally too intensive, an approximate version of it.

SISO decoder: nonsystematic codes

In this case, with the assumption of a stationary memoryless channel, the APP $p(\mathbf{x} \mid \mathbf{y})$ takes the form

$$p(\mathbf{x} \mid \mathbf{y}) \propto [\mathbf{x} \in \mathcal{C}] \prod_{i=1}^{n} p(y_i \mid x_i)p(x_i) \qquad (9.7)$$

This equation implies the assumption that the symbols x_i are all independent so that $p(\mathbf{x})$ can be factored as the product of individual probabilities $p(x_i)$. A priori, this assumption does not seem to make sense: however, we shall see in the following that, in turbo decoding algorithms, the role of these probabilities will be taken by the extrinsic messages $e(x_i)$. Since one of the effects of a long interleaver is to

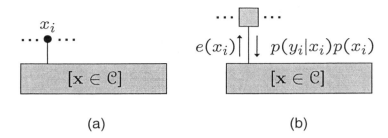

(a)   (b)

Figure 9.11: *(a) Factor graph for the nonsystematic code* $\mathcal{C}$. *(b) Messages exchanged by the sum–product algorithm applied to the computation of the APPs* $p(x_i \mid \mathbf{y})$.

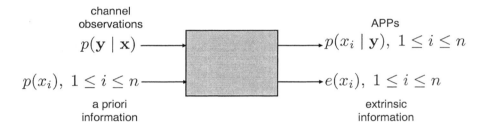

Figure 9.12: *Soft-input, soft-output decoder for nonsystematic codes.*

make the random variables $e(x_i)$ independent (at least approximately), the above assumption becomes realistic for long enough blocks. The corresponding factor graph is shown in Figure 9.11, while Figure 9.12 illustrates the SISO decoder. This system has two sets of inputs: (a) the n conditional probabilities $p(y_j \mid x_j)$, and (b) the n a priori probabilities $p(x_j)$. Its outputs are: (c) the n APPs $p(x_i \mid \mathbf{y})$ and (d) the n extrinsic messages $e(x_i)$. (Notice that the a priori probabilities are unknown here.)

Turbo algorithm for parallel concatenation

Having defined SISO decoders, we can now specialize the general iteration scheme of Figure 9.8. If codes $\mathcal{C}_1$ and $\mathcal{C}_2$ are connected together, they may exchange extrinsic information, as suggested in Figure 9.13. The complete scheme of Figure 9.14 shows how two SISO decoders combine into the turbo algorithm. The algorithms starts by soft-decoding $\mathcal{C}_1$, which is done by the SISO decoder $\mathcal{D}_1$. At this step, the a priori probabilities of each bit are initialized to $1/2$. The output APPs are not

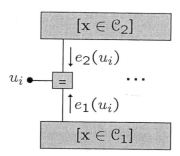

Figure 9.13: *Exchange of extrinsic information between two codes.*

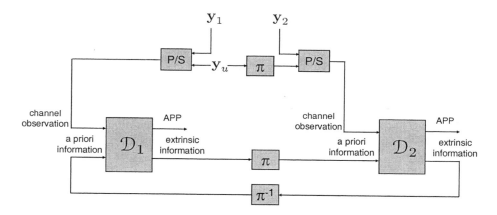

Figure 9.14: *General scheme of an iterative turbo decoder for parallel concatenation. P/S denotes a parallel-to-series converter; $\mathcal{D}_1$ and $\mathcal{D}_2$ are soft-input, soft-output decoders for code $\mathcal{C}_1$ and code $\mathcal{C}_2$, respectively; π denotes the same interleaver used in the encoder; and π^{-1} denotes its inverse.*

used, while the extrinsic messages, suitably normalized to form probabilities, are used, after permutation, *as a priori probabilities* in $\mathcal{D}_2$, the SISO decoder for $\mathcal{C}_2$. The extrinsic messages at the output of $\mathcal{D}_2$ are permuted and used as a priori probabilities for $\mathcal{D}_1$. These operations are repeated until a suitable stopping criterion is met. At this point the output APPs are used to hard-decode the information bits. Notice that in the iterations the channel information gathered from the observation of $\mathbf{y}_u$, $\mathbf{y}_1$, and $\mathbf{y}_2$ does not change: only the *a priori information* inputs to the decoders vary.

By this algorithm, the operation of the individual SISO decoders is relatively easy, because $\mathcal{C}_1$ and $\mathcal{C}_2$ are weak codes. As such, neither $\mathcal{C}_1$ nor $\mathcal{C}_2$ can individu-

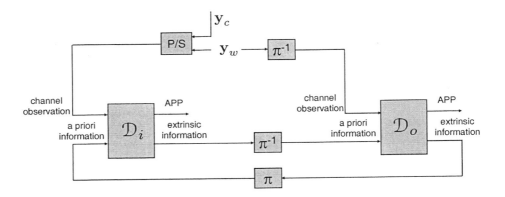

Figure 9.15: *General scheme of an iterative turbo decoder for serial concatenation. P/S denotes a parallel-to-series converter; $\mathcal{D}_i$ and $\mathcal{D}_o$ are soft-input, soft-output decoders for inner code $\mathcal{C}_i$ and outer code $\mathcal{C}_o$, respectively; π denotes the same interleaver used in the encoder; and π^{-1} denotes its inverse.*

ally achieve a high performance. It is their combination that makes for a powerful code and at the same time allows the decoding task to be split into simpler operations.

Turbo algorithm for serial concatenation

We assume here that the inner code is systematic, while the outer code is nonsystematic. Recalling Figure 9.7, let $\mathbf{u}$, $\mathbf{v}$ denote input and output of $\mathcal{C}_o$, respectively; $\mathbf{w} \triangleq \pi\mathbf{v}$ the permuted version of $\mathbf{v}$; and $(\mathbf{w}, \mathbf{c})$ the output of $\mathcal{C}_i$. Finally, let $\mathbf{y} = (\mathbf{y}_w, \mathbf{y}_c)$ denote the observed vector. The block diagram of a turbo decoder for serially concatenated codes is shown in Figure 9.15. The operation of this decoder is similar to that of Figure 9.14; however, the two SISO decoders are different here: $\mathcal{D}_i$ has the structure illustrated in Figure 9.10, while $\mathcal{D}_o$ corresponds to Figure 9.12.

9.2.2   Convergence properties of the turbo algorithm

Figure 9.16 shows qualitatively a typical behavior of the bit error rate of an iteratively decoded turbo code. Three regions can be identified on this chart.   In the low-SNR region, the BER decreases very slowly as the iteration order and the SNR increase. For intermediate values of SNR, the BER decreases rapidly as the SNR increases and improves with increasing the number of iterations. This *water-*

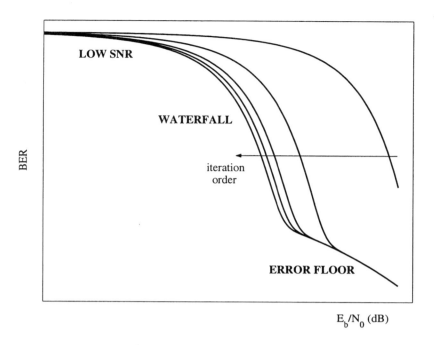

Figure 9.16: *Qualitative aspect of the BER curves vs. $\mathcal{E}_b/N_0$ and the number of iterations for turbo codes.*

fall region is where turbo codes are most useful, as their coding gain approaches the theoretical limit. Finally, for large SNR, an *error floor* effect takes place: the performance is dictated by the minimum Hamming distance of the code, the BER-curve slope changes, and the coding gain decreases.[1]

Figure 9.17 shows the performance of three turbo codes in the waterfall region. Their error probabilities are compared with the Shannon limits for the unconstrained AWGN channel, as derived in Problem 8 of Section 3 (see also Figure 1.5).

---

[1]It has been argued [9.15] that the presence of this error floor makes turbo codes not suitable for applications requiring extremely low BERs. Their poor minimum distance, and their natural lack of error-detection capability, due to the fact that in turbo decoding only information bits are decoded (but see [9.26] for an automatic repeat-request scheme based on punctured turbo codes), make these codes perform badly in terms of block error probability. Poor block error performance also makes these codes not suitable for certain applications. Another relevant factor that may guide in the choice of a coding scheme is the decoding delay one should allow: in fact, turbo codes, as well as LDPC codes, suffer from a substantial decoding delay, and hence their application might be more appropriate for data transmission than for real-time speech.

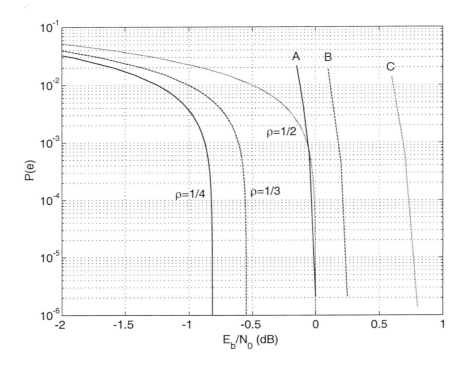

Figure 9.17: *Performance of three turbo codes with block length* $16{,}384$, *obtained by parallel concatenation of two convolutional codes. Code A has* $\rho = 1/4$, $16 + 16$ *states, and is decoded with* 13 *iterations. Code B has* $\rho = 1/3$, $16 + 16$ *states, and is decoded with* 11 *iterations. Code C has* $\rho = 1/2$, $2 + 32$ *states, and is decoded with* 18 *iterations.*

9.2.3 Distance properties of turbo codes

As just observed, for intermediate SNRs the good performance of turbo codes does not depend on their minimum-distance properties: it is rather affected by their small error coefficient (small number of nearest neighbors) in their low-weight words. For high SNRs, on the other hand, the error probability curve of turbo codes exhibits a "floor" caused by a relatively modest minimum distance.[2]

Let n denote the block length of the component codes (and hence the interleaver length), and B the number of parallel codes ($B = 2$ in all preceding discussions,

---

[2]An error floor may also occur with LDPC codes, caused by *near-code words*, i.e., n-tuples with low Hamming weight whose syndrome has also a low weight. See [9.19].

but we can think of a more general turbo coding scheme). Moreover, let the component codes be recursive. Then it can be shown [9.14] that the minimum Hamming distance grows like $n^{1-2/B}$. More precisely, given a code $\mathcal{C}$, let $\mathcal{C}(d)$ denote the set of its nonzero words with weight $1, \ldots, d$. If we choose at random a parallel concatenated code $\mathcal{C}$ using B equal recursive convolutional codes and the block length is n, then as $n \to \infty$ we have, for every $\epsilon > 0$,

$$\mathbb{P}\left[\left|\mathcal{C}\left(n^{1-2/B-\epsilon}\right)\right| = 0\right] \to 1 \quad \text{and} \quad \mathbb{P}\left[\left|\mathcal{C}\left(n^{1-2/B+\epsilon}\right)\right| = 0\right] \to 0 \quad (9.8)$$

Notice how this result implies that a turbo code with only two parallel branches has a minimum distance that does not grow as any power of n, whereas if three branches are allowed, then the growth is $n^{1/3}$.[3]

For serially concatenated codes, the minimum-distance behavior is quite different. Let us pick at random a code from an ensemble of serially concatenated turbo codes. Moreover, let d_o denote the free Hamming distance of the outer code. Then as $n \to \infty$ we have, for every $\epsilon > 0$,

$$\mathbb{P}\left[\left|\mathcal{C}\left(n^{1-2/d_o-\epsilon}\right)\right| = 0\right] \to 1 \quad \text{and} \quad \mathbb{P}\left[\left|\mathcal{C}\left(n^{1-2/d_o+\epsilon}\right)\right| = 0\right] \to 0$$
$$(9.9)$$

We see that if the outer code has a large d_o we can achieve a growth rate close to linear with n.

9.2.4   EXIT charts

Since the turbo algorithm operates by exchanging extrinsic messages between two SISO decoders, its convergence may be studied by examining how these evolve with iterations. A convenient graphical description of this process is provided by EXIT charts [9.28], which yield quite accurate, albeit not exact, results, especially in the waterfall region of the error-probability curve of turbo codes. An EXIT chart is a graph that illustrates the input–output relationship of a SISO decoder by showing the transformations induced on a single parameter associated with input and output extrinsic probabilities. The upside of EXIT-chart analyses is that only simulation of the behavior of the individual decoders is needed, instead of computer-intensive error counting with the full decoding procedure.

Let us focus on the binary alphabet $\mathcal{X} = \{\pm 1\}$ and assume an AWGN channel so that the observed signal is

$$y = x + z$$

---

[3]For $B = 2$, an upper bound to the minimum distance of a turbo code for *all possible* interleavers is derived in [9.5].

with $z \sim \mathcal{N}(0, \sigma_z^2)$. Since

$$p(y \mid x) = \frac{1}{\sqrt{2\pi}\sigma_z} e^{-(y-x)^2/2\sigma_z^2}$$

the *log-likelihood ratio* (LLR)

$$\Lambda(y) \triangleq \ln \frac{p(y|x = +1)}{p(y|x = -1)}$$

takes value

$$\Lambda(y) = \frac{2}{\sigma_z^2}(x + z) \tag{9.10}$$

and hence, given x, Λ is conditionally Gaussian: we write

$$\Lambda(y) \mid x \sim \mathcal{N}\left(\frac{2}{\sigma_z^2}x, \frac{4}{\sigma_z^2}\right) \tag{9.11}$$

In summary, we may say that $\Lambda(y) \mid x \sim \mathcal{N}(\mu, \sigma^2)$ and that it satisfies the *consistency condition* [9.25]

$$\mu = x\sigma^2/2 \tag{9.12}$$

The above allows us to write

$$p(\Lambda \mid x) = \frac{1}{\sqrt{2\pi}\sigma} e^{-(\Lambda - x\sigma^2/2)^2/2\sigma^2} \tag{9.13}$$

EXIT-chart techniques are based on the empirical evidence that extrinsic messages, when expressed in the form of log-likelihood ratios, approach a Gaussian distribution satisfying the consistency condition (9.12). In addition, for large block lengths (and hence large interleavers) the messages exchanged remain approximately uncorrelated from the respective channel observations over many iterations [9.28]. Under the Gaussian assumption, the extrinsic messages are characterized by a single parameter, which is commonly and conveniently chosen to be the mutual information exchanged between the LLR and the random variable x (see [9.29] for experiments that justify this choice):

$$I(x; \Lambda) = \frac{1}{2} \sum_{x \in \{\pm 1\}} \int p(\Lambda|x) \log \frac{p(\Lambda|x)}{p(\Lambda)} \, d\Lambda \tag{9.14}$$

with $p(\Lambda) = 0.5[p(\Lambda|x = -1) + p(\Lambda|x = +1)]$ under the assumption that x takes on equally likely values. In particular, if Λ is conditionally Gaussian, and the

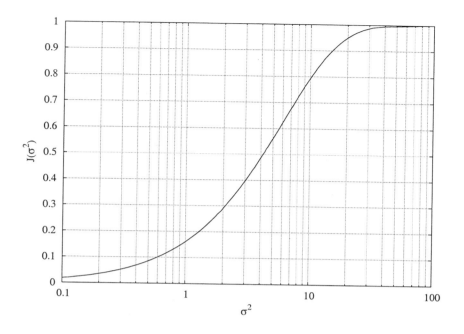

Figure 9.18: *Plot of the function $J(\sigma^2)$ defined in (9.15).*

consistency condition (9.12) is satisfied, then $I(x; \Lambda)$ does not depend on the value of x, and we have explicitly $I(x; \Lambda) = J(\sigma^2)$, where

$$
J(\sigma^2) \triangleq 1 - \int_{-\infty}^{\infty} \frac{1}{\sqrt{2\pi}\sigma} e^{-[(w - x\sigma^2/2)^2/2\sigma^2]} \log(1 + e^{-xz}) \, dw
$$
$$
= 1 - \mathbb{E}\left[\log\left(1 - e^{-x\Lambda}\right)\right] \tag{9.15}
$$

where $\mathbb{E}$ is taken with respect to the pdf $\mathcal{N}\left(x\sigma^2/2, \sigma^2\right)$. The function $J(\sigma^2)$ (plotted in Figure 9.18) is monotonically increasing, and takes values from 0 (for $\sigma \to 0$) to 1 (for $\sigma \to \infty$). If the assumption of conditional Gaussianity on Λ is not made, a convenient approximation of $I(x; \Lambda)$, based on the observation of N samples of the random variable Λ, is based on (9.15):

$$
I(x; \Lambda) \approx 1 - \frac{1}{N} \sum_{i=1}^{N} \log\left(1 + \exp(-x_i \Lambda_i)\right) \tag{9.16}
$$

Recall now that we have four different messages at the input and output of a SISO decoder: a priori, channel observation, soft-decision, and extrinsic. We denote these messages by μ^a, μ^o, μ^d, and μ^e, respectively, and by I^a, I^o, I^d, and I^e,

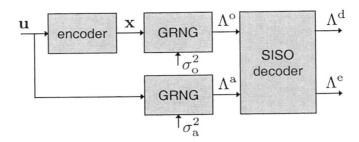

Figure 9.19: *Computing the transfer function T. GRNG is a Gaussian random noise generator.*

respectively, the mutual informations exchanged between their LLRs and x. We describe the behavior of a SISO processor used in iterative decoding by giving its *extrinsic information transfer* (EXIT) function

$$I^{\mathrm{e}} = T(I^{\mathrm{a}}, I^{\mathrm{o}}) \qquad (9.17)$$

Figure 9.19 schematizes the Monte Carlo derivation of the EXIT chart for a given code. Choose first the values of I^{a} and I^{o}. The random vector **u** of uncoded ± 1 symbols is encoded to generate the vector **x**. A Gaussian random noise generator outputs, for each component x of **x**, the LLR Λ^{o} such that

$$\Lambda^{\mathrm{o}}|x \sim \mathcal{N}\left(x\frac{\sigma_{\mathrm{o}}^2}{2}, \sigma_{\mathrm{o}}^2\right)$$

where $\sigma_{\mathrm{o}}^2 = J^{-1}(I^{\mathrm{o}})$. Similarly, another Gaussian random noise generator outputs, for each component u of **u**, the LLR Λ^{a} such that

$$\Lambda^{\mathrm{a}}|u \sim \mathcal{N}\left(u\frac{\sigma_{\mathrm{a}}^2}{2}, \sigma_{\mathrm{a}}^2\right)$$

where $\sigma_{\mathrm{a}}^2 = J^{-1}(I^{\mathrm{a}})$. These two LLRs correspond to messages entering the SISO decoder. This outputs the LLRs Λ^{d} and Λ^{e}. Only the latter is retained, and N values of it are used to approximate I^{e} through (9.16), so no Gaussian assumption is imposed on Λ^{e}.

Once the transfer functions of both decoders have been obtained, they are drawn on a single chart. Since the output of a decoder is the input of the other one, the second transfer functions is drawn after swapping the axes, as shown in the example of Figure 9.20 (here the two decoders are equal). The behavior of the iterative decoding algorithm is described by a trajectory, i.e., a sequence of moves, along

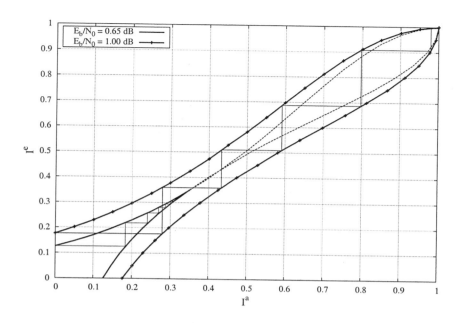

Figure 9.20: *EXIT chart for a rate-1/2 convolutional code and two values of* $\mathcal{E}_b/N_0$.

horizontal and vertical steps, through the pair of transfer functions. Iterations start with no a priori knowledge, so $I^a = 0$. Due to the channel observations, the corresponding value of I^e at the output of the first SISO decoder increases with $\mathcal{E}_b/N_0$. The resulting extrinsic message is fed to the second decoder, which corresponds to moving along a horizontal line joining the two transfer functions. We thus obtain the value of I^e at the output of the second decoder. The corresponding message is fed back to the first decoder, whose output yields the value of I^e obtained by joining the two curves with a vertical line, and so on.

Figure 9.20 shows two examples of convergence behavior. For $\mathcal{E}_b/N_0 = 0.65$ dB, the two curves intersect, the trajectory is blocked, and we experience no convergence to large values of mutual information (which correspond to small error probabilities). For $\mathcal{E}_b/N_0 = 1$ dB, instead, we have convergence.

Estimates of the error probability of a coded system can be superimposed on EXIT charts to offer some extra insight into the performance of the iterative decoder. If the LLR Λ^d is assumed to be conditionally Gaussian, with mean $\mu_d =$

$x\sigma_{\rm d}^2/2$ and variance $\sigma_{\rm d}^2$, the bit error rate (BER) can be approximated in the form

$$P_b(e) = \mathbb{P}(\Lambda^{\rm d} > 0 \mid x = -1) \approx Q\left(\frac{|\mu_{\rm d}|}{\sigma_{\rm d}}\right) = Q\left(\frac{\sigma_{\rm d}}{2}\right) \qquad (9.18)$$

Since $\Lambda^{\rm d} = \Lambda^{\rm o} + \Lambda^{\rm a} + \Lambda^{\rm e}$, the assumption of independent LLRs leads to

$$\sigma_{\rm d}^2 = \sigma_{\rm o}^2 + \sigma_{\rm a}^2 + \sigma_{\rm e}^2$$

which in turn yields

$$P_b(e) \approx Q\left(\frac{\sqrt{J^{-1}(I^{\rm o}) + J^{-1}(I^{\rm a}) + J^{-1}(I^{\rm e})}}{2}\right) \qquad (9.19)$$

Notice that, due to (9.11), we have

$$\sigma_{\rm o}^2 = \frac{4}{\sigma_z^2} = 4\,{\rm SNR} = 8\rho\frac{\mathcal{E}_b}{N_0}$$

where ρ is the rate of the concatenated code. Figure 9.21 superimposes the EXIT chart corresponding to $\mathcal{E}_b/N_0 = 1$ dB onto a set of constant-BER curves. A comparison of this figure with Figure 9.22, obtained by Monte Carlo simulation, shows a good match between the "true" BERs and those predicted by EXIT charts. In addition, observing the evolution of the constant-BER curves, one can observe how traversing the bottleneck region between the two curves corresponds to a slow convergence of the BER. Once the bottleneck is passed, faster convergence of BER is achieved.

Accuracy of EXIT-chart convergence analysis

In the upper portion of the EXIT chart, extrinsic messages become increasingly correlated, and the true evolution of $I^{\rm e}$ deviates from the behavior predicted by the chart. As correlations depend also on the interleaver size, it is expected that EXIT analyses become more accurate as this size increases.

9.3 Bibliographical notes

Low-density parity-check codes were introduced by Gallager in his doctoral thesis [9.12], and rediscovered in the mid-1990s [9.18]. Reference [9.21] reviews techniques for constructing LDPC codes whose graphs have large girths. LDPC

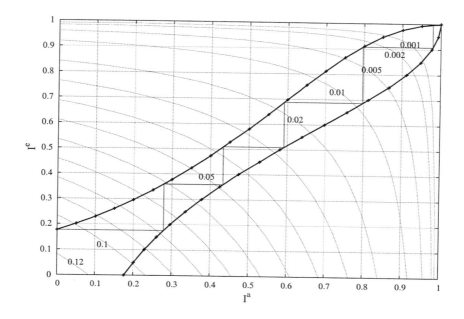

Figure 9.21: *EXIT chart as in Figure 9.20, for* $\mathcal{E}_b/N_0 = 1$ *dB, superimposed to constant-BER curves.*

decoding algorithms are analyzed in [9.12, 9.18]. LDPC codes over nonbinary alphabets are examined in [9.8]. Turbo codes, and their iterative decoding algorithm, were first presented to the scientific community in [9.4]. The iterative (turbo) decoding algorithm was shown in [9.17] to be an instance of J. Pearl's *belief propagation* in graphs [9.22]. Our presentation of SISO decoders follows [9.23].

The capacity-approaching codes described in this chapter are now finding their way into a number of practical applications, ranging from UMTS to wireless local-area networks, deep-space communication, and digital video broadcasting. A list of practical implementations of LDPC codes can be found in [9.24].

Richardson and Urbanke [9.6] have introduced the study of the evolution of the probability distribution of the exchanged messages as a tool to study the convergence behavior of turbo algorithms. EXIT charts, which characterize these distributions using a single parameter, were advocated in [9.28]. Application of EXIT charts to LDPC codes, a topic not considered here, is described in [9.2].

Computation of bounds to the error probability of turbo codes can be found in [9.9, 9.27].

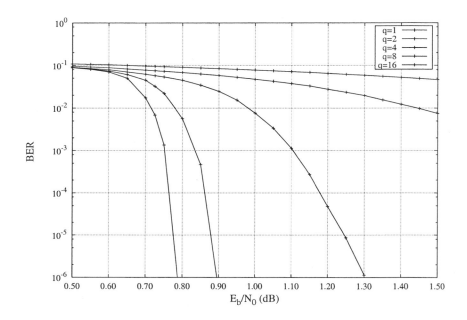

Figure 9.22: *Convergence of a turbo code based on two equal convolutional codes as in Figure 9.20 with block length* 10^5 *(simulation results).*

9.4 Problems

1. Once the matrix **H** of an LDPC code is selected, show how the generator matrix **G** can be obtained. Consider separately the cases of **H** having or not having full rank. Is **G** a sparse matrix?

2. Derive EXIT charts for some simple convolutional codes assuming $I^a = 0$. Interpret the shape of the functions.

3. Extend the EXIT-chart analysis to the frequency-flat, slow independent Rayleigh fading channel.

References

[9.1] B. Ammar, B. Honary, Y. Kou, J. Xu, and S. Lin, "Construction of low-density parity-check codes based on balanced incomplete block designs," *IEEE Trans. Inform. Theory*, Vol. 50, No. 6, pp. 1257–1268, June 2004.

[9.2] M. Ardakani and F. R. Kschischang, "Designing irregular LPDC codes using EXIT charts based on message error rate," *Proc. 2002 IEEE Int. Symp. Inform. Theory*, Lausanne, Switzerland, p. 454, June 30–July 5, 2002.

[9.3] C. Berrou and A. Glavieux, "Near optimum error correcting coding and decoding: Turbo codes," *IEEE Trans. Commun.*, Vol. 44, No. 10, pp. 1261–1271, October 1996.

[9.4] C. Berrou, A. Glavieux, and P. Thitimajshima, "Near Shannon limit error-correcting coding: Turbo codes," *Proc. IEEE 1993 Int. Conf. Commun. (ICC93)*, Geneva, Switzerland, pp. 1064–1070, May 1993.

[9.5] M. Breiling and J. B. Huber, "Combinatorial analysis of the minimum distance of turbo codes," *IEEE Trans. Inform. Theory*, Vol. 47, No. 7, pp. 2737–2750, November 2001.

[9.6] S.-Y. Chung, T. J. Richardson, and R. L. Urbanke, "Analysis of sum–product decoding of low-density parity-check codes using a Gaussian approximation," *IEEE Trans. Inform. Theory*, Vol. 47, No. 2, pp. 657–670, February 2001.

[9.7] M. C. Davey, *Error-correction using Low-Density Parity-Check Codes*. PhD Dissertation, University of Cambridge, December 1999.

[9.8] M. C. Davey and D. J. C. MacKay, "Low density parity check codes over GF(q)," *IEEE Comm. Lett.*, Vol. 2, No. 6, pp. 165–167, June 1998.

[9.9] D. Divsalar and E. Biglieri, "Upper bounds to error probabilities of coded systems beyond the cutoff rate," *IEEE Trans. Commun.*, Vol. 51, No. 12, pp. 2011–2018, December 2003.

[9.10] J. L. Fan, "Array codes as low-density parity-check codes," *Proc. 2nd Int. Symp. on Turbo Codes & Related Topics*, Brest, France, pp. 543–546, September 4–7, 2000.

[9.11] M. P. C. Fossorier, "Quasi-cyclic low-density parity-check codes from circulant permutation matrices," *IEEE Trans. Inform. Theory*, Vol. 50, No. 8, pp. 1788–1793, August 2004.

[9.12] R. G. Gallager, *Low-Density Parity-Check Codes*. Cambridge, MA: MIT Press, 1963.

[9.13] B. Honary et al., "On construction of low density parity check codes," *2nd Int. Workshop on Signal Processing for Wireless Communications (SPWC 2004)*, London, U.K., June 2–4, 2004.

[9.14] N. Kahale and R. Urbanke, "On the minimum distance of parallel and serially concatenated codes," *Proc. IEEE ISIT 1998*, Cambridge, MA, p. 31, August 16–21, 1998.

[9.15] Y. Kou, S. Lin, and M. P. C. Fossorier, "Low-density parity-check codes based on finite geometries: A rediscovery and new results," *IEEE Trans. Inform. Theory*, Vol. 47, No. 7, pp. 2711–2736, November 2001.

[9.16] R. Lucas, M. P. C. Fossorier, Y. Kou, and S. Lin, "Iterative decoding of one-step majority logic decodable codes based on belief propagation," *IEEE Trans. Commun.*, Vol. 48, No. 6, pp. 931–937, June 2000.

[9.17] R. J. McEliece, D. J. C. MacKay, and J.-F. Cheng, "Turbo decoding as an instance of Pearl's 'belief propagation' algorithm," *IEEE Journal Select. Areas Commun.*, Vol. 16, No. 2, pp. 140–152, February 1998.

[9.18] D. J. C. MacKay, "Good error correcting codes based on very sparse matrices," *IEEE Trans. Inform. Theory*, Vol. 45, No. 2, pp. 399–431, March 1999. (See also Errata, *ibidem*, Vol. 47, No. 5, p. 2101, July 2001.)

[9.19] D. J. C. MacKay and M. S. Postol, "Weaknesses of Margulis and Ramanujan-Margulis low-density parity-check codes," *Electron. Notes Theoret. Comput. Sci.*, Vol. 74, pp. 1–8, 2003.

[9.20] N. Miladinovic and M. Fossorier, "Systematic recursive construction of LDPC codes," *IEEE Commun. Letters*, Vol. 8, No. 5, pp. 302–304, May 2004.

[9.21] J. M. F. Moura, J. Lu, and H. Zhang, "Structured low-density parity-check codes," *IEEE Signal Process. Mag.*, Vol. 21, No. 1, pp. 42–55, January 2004.

[9.22] J. Pearl, *Probabilistic Reasoning in Intelligent Systems: Networks of Plausible Inference*. San Francisco, CA: Morgan Kaufmann, 1988.

[9.23] O. Pothier, *Codes Composites Construits à Partir de Graphes et Leur Décodage Itératif*. Ph.D. Dissertation, École Nationale Supérieure des Télécommunications, Paris, France, January 2000.

[9.24] T. Richardson and R. Urbanke, "The renaissance of Gallager's low-density parity-check codes," *IEEE Commun. Magazine*, Vol. 41, No. 8, pp. 126–131, August 2003.

[9.25] T. Richardson, A. Shokrollahi, and R. Urbanke, "Design of provably good low-density parity-check codes," *Proc. 2000 IEEE Int. Symp. Inform. Theory (ISIT 2000)*, Sorrento, Italy, p. 199, June 25–30, 2000.

[9.26] D. N. Rowitch and L. B. Milstein, "On the performance of hybrid FEC/ARQ systems using rate compatible punctured turbo (RCPT) codes," *IEEE Trans. Commun.*, Vol. 48, No. 6, pp. 948–959, June 2000.

[9.27] I. Sason and S. Shamai (Shitz), "On improved bounds on the decoding error probability of block codes over interleaved fading channels, with applications to turbo-like codes," *IEEE Trans. Inform. Theory*, Vol. 47, No. 6, pp. 2275–2299, September 2001.

[9.28] S. ten Brink, "Convergence behavior of iteratively decoded parallel concatenated codes," *IEEE Trans. Commun.*, Vol. 49, No. 10, pp. 1727–1737, October 2001.

[9.29] M. Tüchler, S. ten Brink, and J. Hagenauer, "Measures for tracing convergence of iterative decoding algorithms," *Proc. 4th IEEE/ITG Conf. on Source and Channel Coding*, Berlin, Germany, pp. 53–60, January 2002.

10

What signifieth whole that but, be all the prowess of ten,

Multiple antennas

This chapter focuses on the main theoretical aspects of transmission systems where more than one antenna is used at both ends of a radio link and hence where one channel use involves sending signals simultaneously through several propagation paths. The use of multiple transmit and receive antennas allows one to reach capacities that cannot be obtained with any other technique using present-day technology, and is expected to make it possible to increase the data rates in wireless networks by orders of magnitude. After describing system and channel models, we compute the capacities that can be achieved, and we show how "space–time" codes can be designed and how suboptimum architectures can be employed to simplify the receiver. Finally, we examine the basic trade-off between the data-rate gain, made possible by simultaneous transmission by several antennas, and the diversity gain, achieved by sending the same signal through several independently faded paths.

10.1   Preliminaries

We consider a radio system with t antennas simultaneously transmitting one signal each, and with r antennas receiving these signals (Figure 10.1).   Assuming

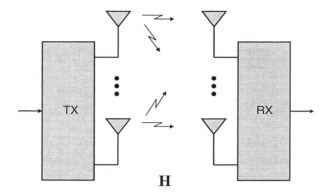

Figure 10.1: *Transmission and reception with multiple antennas.   The channel gains are described by the $r \times t$ matrix* **H**.

two-dimensional elementary constellations throughout, the channel input–output relationship is

$$\mathbf{y} = \mathbf{H}\mathbf{x} + \mathbf{z} \tag{10.1}$$

where $\mathbf{x} \in \mathbb{C}^t$, $\mathbf{y} \in \mathbb{C}^r$, $\mathbf{H} \in \mathbb{C}^{r \times t}$ (i.e., **H** is an $r \times t$ complex, possibly random, matrix, whose entries h_{ij} describe the gains of each transmission path from a transmit to a receive antenna), and **z** is a circularly symmetric, complex Gaussian noise vector. The component x_i, $i = 1, \ldots, t$, of vector **x** is the elementary signal transmitted from antenna i; the component y_j, $j = 1, \ldots, r$, of vector **y** is the signal received by antenna j. We also assume that the complex noise components affecting the different receivers are independent with variance N_0, i.e.,

$$\mathbb{E}[\mathbf{z}\mathbf{z}^\dagger] = N_0 \mathbf{I}_r \tag{10.2}$$

where $\mathbf{I}_r$ is the $r \times r$ identity matrix, and the signal energy is constrained by $\mathbb{E}[\mathbf{x}^\dagger\mathbf{x}] = t\mathcal{E}$, where $\mathcal{E}$ denotes the average energy per elementary signal. The additional assumption that $\mathbb{E}[|h_{ij}|^2] = 1$ for all i, j,[1] yields the average signal-to-

---

[1]The assumption of equal second-order moments for the channel coefficients facilitates the analysis but is somewhat restrictive, as it does not allow consideration of antennas differing in their radiation patterns.

noise ratio (SNR) at the receiver (see (3.28))

$$\zeta = t\frac{\mathcal{E}_b}{N_0}\frac{R_b}{W} \tag{10.3}$$

with R_b the bit rate and W the Shannon bandwidth of the elementary signals. Since we assume here $N = 2$, we have $R_b/W = \log M$, with M the size of the elementary constellation, and hence

$$\zeta = t\frac{\mathcal{E}}{N_0} \tag{10.4}$$

Then, rather than assuming a power or energy constraint, we may refer to an SNR constraint, i.e.,

$$\mathbb{E}[\mathbf{x}^\dagger\mathbf{x}] \le \zeta N_0 \tag{10.5}$$

For later reference, we define

$$m \triangleq \min\{t, r\} \qquad n \triangleq \max\{t, r\} \tag{10.6}$$

Explicitly, we have, from (10.1),

$$y_j = \sum_{i=1}^{t} h_{ji}x_i + z_j, \qquad j = 1, \dots, r \tag{10.7}$$

which shows how every component of the received signal includes a linear combination of the signals emitted by each antenna. We say that $\mathbf{y}$ is affected by *spatial interference*, generated by the signals transmitted from the various antennas. This interference has to be removed, or controlled in some way, in order to separate the single transmitted signals. We shall see in the following how this can be done: for the moment we may just observe that the tools for the analysis of multiple-antenna transmission have much in common with those used in the study of other disciplines centering on interference control, such as digital equalization of linear dispersive channels (where the received signals are affected by *intersymbol* interference: see, e.g., [10.4]) or multiuser detection (where the received signals are affected by *multiple-access* interference: see, e.g., [10.61]). Notice, however, the peculiar feature of multiple-antenna systems, which allow for coördination among transmitted signals: this can be exploited to simplify the receiver's operation.

10.1.1 Rate gain and diversity gain

The upsides of using multiple antennas can be summarized by defining two types of gain. As we shall see in the following, in the presence of fading, a multiplicity

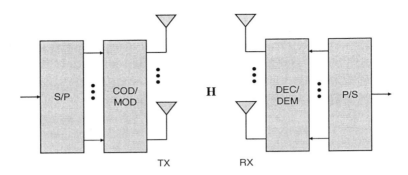

Figure 10.2: *Spatial multiplexing and diversity obtained by transmission and reception with multiple antennas.*

of transmit antennas creates a set of parallel channels, that can be used to potentially increase the data rate up to a factor of $\min\{t, r\}$ (with respect to single-antenna transmission) and hence generate a *rate gain*. This corresponds to the *spatial multiplexing* illustrated by Figure 10.2. Here the serial-to-parallel converter S/P distributes the stream of data across the transmit antennas; after reception, the original stream is reconstituted by the parallel-to-serial converter P/S[2] The other gain is due to the combination of received signals that are independently faded replicas of a single transmitted signal, which allows a more reliable reception. We call *diversity gain* the number of independent paths traversed by each signal, which has a maximum value rt. We hasten to observe here that these two gains are not independent, but there is a fundamental trade-off between the two: and actually it can be said that the problem of designing a multiple-antenna system is based on this trade-off. As an example, Figure 10.3 illustrates the diversity–rate trade-off for a multiple-input multiple-output (MIMO) system with $t = 2$ transmit and $r = 2$ receive antennas. Figure 10.3(a) assumes the channels are orthogonal so that the rate is maximum (twice as large as the single-channel rate), but there is no diversity gain, since each symbol traverses only one path. Figure 10.3(b) assumes that the transmitter replicates the same signal over the two channels so that there is no

---

[2]Here we limit ourselves to considering only transmissions with the same rate on all antennas. However, different (and possibly adaptive) modulation rates can also be envisaged.

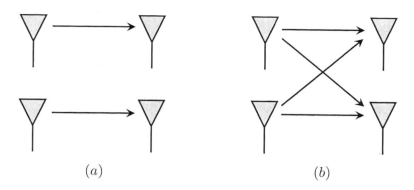

Figure 10.3: *Diversity–rate trade-off with $r = t = 2$. (a) Transmission of two signals over two orthogonal channels. (b) Transmission of one signal over four paths.*

rate gain, but the diversity is now four, since the signal traverses four independent paths. We shall discuss this point in more depth in Section 10.14.

The problems we address in this chapter are the following:

1. What is the limiting performance (channel capacity) of this multiple-antenna system?

2. What is its error probability?

3. How can we design "space–time" codes matched to the channel structure?

4. How can we design architectures allowing simple decoding of space–time codes, and what is their performance?

10.2 Channel models

Here we focus on two models simple enough to allow tractable analysis.

For fast, frequency-nonselective channels, we have, if the index n denotes discrete time,

$$\mathbf{y}_n = \mathbf{H}_n \mathbf{x}_n + \mathbf{z}_n \tag{10.8}$$

with $\mathbf{H}_n$, $-\infty < n < \infty$, an ergodic random process. This channel is consequently ergodic.

For slow, frequency-nonselective channels, the model becomes

$$\mathbf{y}_n = \mathbf{H}\mathbf{x}_n + \mathbf{z}_n \qquad (10.9)$$

and each code word, however long, experiences only one channel state. This fading model is nonergodic.

10.2.1 Narrowband multiple-antenna channel models

Assume that the $r \times t$ channel matrix $\mathbf{H}$ remains constant during the transmission of an entire code word. Analysis of this channel requires the joint pdf of the rt entries of $\mathbf{H}$. A number of relatively simple models for this pdf have been proposed in the technical literature, based on experimental results and analyses. Among these we consider the following:

Rich scattering The entries of $\mathbf{H}$ are independent, circularly symmetric complex zero-mean Gaussian random variables.

Completely correlated The entries of $\mathbf{H}$ are correlated, circularly symmetric complex zero-mean Gaussian random variables. To specify this model, the correlation coefficients of all pairs of elements are required.

Separately correlated The entries of $\mathbf{H}$ are correlated, circularly symmetric complex zero-mean Gaussian random variables, with the correlation between two entries of $\mathbf{H}$ separated in two factors accounting for the receive and transmit correlation:

$$\mathbb{E}[(\mathbf{H})_{i,j}(\mathbf{H})^*_{i',j'}] = (\mathbf{R})_{i,i'}(\mathbf{T})_{j,j'} \qquad (10.10)$$

for two Hermitian, nonnegative definite matrices $\mathbf{R}$ ($r \times r$) and $\mathbf{T}$ ($t \times t$). This model is justified by the fact that only the objects surrounding the receiver and the transmitter cause the local antenna-elements correlation, while they have no impact on the correlation at the other end of the link. In other words, this model does not account for correlation between transmit and receive antennas: for this to be actually negligible, the distance between transmitter and receiver must be large. The channel matrix can be expressed in the form

$$\mathbf{H} = \mathbf{R}^{1/2}\mathbf{H}_u\mathbf{T}^{1/2} \qquad (10.11)$$

where $\mathbf{H}_u$ is a matrix of uncorrelated, circularly symmetric complex zero-mean Gaussian random variables with unit variance, and $(\cdot)^{1/2}$ denotes matrix square root.[3] For a fair comparison of different correlation cases, one

---

[3]The square root of matrix $\mathbf{A} \geq \mathbf{0}$ whose singular-value decomposition (SVD: see Appendix B) is $\mathbf{A} = \mathbf{U}\mathbf{D}\mathbf{V}^\dagger$ is defined as $\mathbf{A}^{1/2} \triangleq \mathbf{U}\mathbf{D}^{1/2}\mathbf{V}^\dagger$.

should assume that the total average received power is constant, i.e.,

$$
\begin{aligned}
\mathbb{E}[\mathrm{Tr}\,(\mathbf{T}\mathbf{H}_u\mathbf{R}\mathbf{H}_u^\dagger)] &= \sum_{i,j,k,\ell} \mathbb{E}[(\mathbf{T})_{ij}(\mathbf{H}_u)_{jk}(\mathbf{R})_{k\ell}(\mathbf{H}_u)_{i\ell}^*] \\
&= \sum_{i,k}(\mathbf{T})_{ii}(\mathbf{R})_{kk} \\
&= \mathrm{Tr}\,(\mathbf{T})\mathrm{Tr}\,(\mathbf{R}) \\
&= tr \quad\quad\quad\quad\quad\quad\quad\quad (10.12)
\end{aligned}
$$

Since $\mathbf{H}$ is not affected if $\mathbf{T}$ is scaled by a factor $\alpha \neq 0$ and $\mathbf{R}$ by a factor α^{-1}, one can assume, without loss of generality, that

$$
\mathrm{Tr}\,(\mathbf{T}) = t \quad\quad \mathrm{Tr}\,(\mathbf{R}) = r \quad\quad\quad\quad (10.13)
$$

Uncorrelated keyhole The rank of $\mathbf{H}$ may be smaller than $\min\{t, r\}$. A special case occurs when $\mathbf{H}$ has rank one (a *keyhole* channel). Assume $\mathbf{H} = \mathbf{h}_r\mathbf{h}_t^\dagger$, with the entries of the vectors $\mathbf{h}_r$ and $\mathbf{h}_t$ being independent, circularly symmetric complex zero-mean Gaussian random variables. This model applies in the presence of walls where the propagating signal passes through a small aperture, such as a keyhole. In this way, the incident electric field is a linear combination of the electric fields arriving from the transmit antennas and irradiates through the hole after scalar multiplication by the scattering cross-section of the keyhole. As a result, the channel matrix can be written as the product of a column vector by a row vector. Similar phenomena arise in indoor propagation through hallways or tunnels.

Rice channel The channel models listed above are zero-mean. However, for certain applications, the channel matrix $\mathbf{H}$ should be modeled as having entries whose means are nonzero.

10.2.2 Channel state information

As we discussed in Chapter 4, a crucial factor in determining the performance of transmission over a channel affected by fading is the availability, at the transmitting or at the receiving terminal, of *channel-state information* (CSI), that is, the value taken on by the fading gains in a transmission path. In a fixed wireless environment, the fading gains can be expected to vary slowly, so their estimate can be obtained by the receiver with a reasonable accuracy, even in a system with a large number of antennas, and possibly fed back to the transmitter. In some cases, we may assume that a partial knowledge of the CSI is available. One way of obtaining this estimate is by periodically sending pilot signals on the same channel used for data (these pilot signals are used in wireless systems also for acquisition, synchronization, etc.). We shall address this issue in Section 10.7.

10.3   Channel capacity

In this section we evaluate the capacity of the MIMO transmission system described by (10.1). Several models for the matrix $\mathbf{H}$ can be considered:

   (a) $\mathbf{H}$ is deterministic.

   (b) $\mathbf{H}$ is a random matrix, and each channel use (viz., each transmission of one symbol from each of the t transmit antennas) corresponds to an *independent* realization of $\mathbf{H}$ (ergodic channel).

   (c) $\mathbf{H}$ is a random matrix, but once it is chosen it remains fixed for the whole transmission (nonergodic channel).

   When $\mathbf{H}$ is random (cases (b) and (c) above) we assume here that its entries are iid and $\sim \mathcal{N}_c(0, 1)$, i.e., Gaussian with zero-mean, independent real and imaginary parts, each with variance $1/2$. Equivalently, each entry of $\mathbf{H}$ has uniform phase and Rayleigh magnitude. This choice models Rayleigh fading with enough separation between antennas such that the fades for each TX/RX antenna pair are independent. We also assume, unless otherwise stated, that the CSI (that is, the realization of $\mathbf{H}$) is known at the receiver, while only the probability distribution of $\mathbf{H}$ is perfectly known at the transmitter (the latter assumption is necessary for capacity computations, since the transmitter must choose an optimum code for that specific channel).

10.3.1   Deterministic channel

Assume first that the nonrandom value of $\mathbf{H}$ is known at both transmitter and receiver. We derive the channel capacity by maximizing the average mutual information $I(\mathbf{x}; \mathbf{y})$ between input and output of the channel over the choice of the distribution of $\mathbf{x}$. Singular-value decomposition of the matrix $\mathbf{H}$ yields (Section B.6.4 of Appendix B)

$$\mathbf{H} = \mathbf{U}\mathbf{D}\mathbf{V}^{\dagger} \tag{10.14}$$

where $\mathbf{U} \in \mathbb{C}^{r \times r}$ and $\mathbf{V} \in \mathbb{C}^{t \times t}$ are unitary, and $\mathbf{D} \in \mathbb{R}^{r \times t}$ is diagonal. We can write

$$\mathbf{y} = \mathbf{U}\mathbf{D}\mathbf{V}^{\dagger}\mathbf{x} + \mathbf{z} \tag{10.15}$$

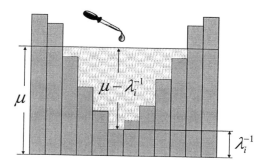

Figure 10.4: *Illustration of water-filling. The height of each patch is λ_i^{-1}. The region is flooded to a level μ by using the total amount of water ζ.*

Premultiplication of (10.15) by $\mathbf{U}^\dagger$ shows that the original channel is equivalent to the channel described by the input–output relationship

$$\tilde{\mathbf{y}} = \mathbf{D}\tilde{\mathbf{x}} + \tilde{\mathbf{z}} \tag{10.16}$$

where $\tilde{\mathbf{y}} \triangleq \mathbf{U}^\dagger\mathbf{y}$, $\tilde{\mathbf{x}} \triangleq \mathbf{V}^\dagger\mathbf{x}$ (so that $\mathbb{E}[\tilde{\mathbf{x}}^\dagger\tilde{\mathbf{x}}] = \mathbb{E}[\mathbf{x}^\dagger\mathbf{x}]$), and $\tilde{\mathbf{z}} \triangleq \mathbf{U}^\dagger\mathbf{z} \sim \mathcal{N}_c(0, N_0\mathbf{I}_r)$. Now, the rank of $\mathbf{H}$ is at most $m \triangleq \min\{t, r\}$, and hence, at most, m of its singular values are nonzero. Denote these by $\sqrt{\lambda_i}$, $i = 1, \ldots, m$, and rewrite (10.16) componentwise in the form

$$\tilde{y}_i = \begin{cases} \sqrt{\lambda_i}\tilde{x}_i + \tilde{z}_i, & i = 1, \ldots, m \\ 0, & i = m+1, \ldots, r \end{cases} \tag{10.17}$$

which shows how this channel is equivalent to a set of m parallel independent channels, each corresponding to a nonzero singular value of $\mathbf{H}$. In addition, we see that, for $i > m$, $\tilde{y}_i$ is *independent of the transmitted signal*, and $\tilde{x}_i$ plays no role.

Maximization of the mutual information requires independent $\tilde{x}_i$, $i = 1, \ldots, m$, each with independent Gaussian, zero-mean real and imaginary parts. Their SNRs should be chosen, as indicated in Section A.5 of Appendix A, via *water-filling* (Figure 10.4):

$$\zeta_i \triangleq \frac{1}{N_0}\mathbb{E}\,|\tilde{\mathbf{x}}|^2 = \left(\mu - \lambda_i^{-1}\right)_+ \tag{10.18}$$

where $(\cdot)_+ \triangleq \max(0, \cdot)$. With μ chosen so as to meet the SNR constraint, we see that the SNR, as parametrized by μ, is

$$\zeta(\mu) = \sum_i \left(\mu - \lambda_i^{-1}\right)_+ \tag{10.19}$$

and the capacity takes on the value (in bits per dimension pair)

$$C = \sum_i \log(1 + \zeta_i \lambda_i) \tag{10.20}$$

Observation 10.3.1   Since the nonzero eigenvalues of $\mathbf{H}^\dagger\mathbf{H}$ are the same as those of $\mathbf{H}\mathbf{H}^\dagger$, the capacities of the channels corresponding to $\mathbf{H}$ and to $\mathbf{H}^\dagger$ are the same. A sort of "reciprocity" holds in this case.

Example 10.1

Take $t = r = m$, and $\mathbf{H} = \mathbf{I}_m$. Due to the structure of $\mathbf{H}$, there is no spatial interference here, and transmission occurs over m parallel additive white Gaussian noise (AWGN) channels, each with SNR ζ/m and hence with capacity $\log(1+\zeta/m)$ bit/dimension pair. Thus,

$$C = m\log(1 + \zeta/m) \tag{10.21}$$

We see here that we have a rate gain, since the capacity is proportional to the number of transmit antennas. Notice also that, as $m \to \infty$, the capacity tends to the limiting value $C = \zeta \log e$. $\square$

Example 10.2

Consider as $\mathbf{H}$ the all-1 matrix, a limiting case of spatial interference. Its SVD is

$$\mathbf{H} = \begin{bmatrix} \sqrt{1/r} \\ \sqrt{1/r} \\ \vdots \\ \sqrt{1/r} \end{bmatrix} (\sqrt{rt})\,[\sqrt{1/t}\cdots\sqrt{1/t}] \tag{10.22}$$

Here we have $m = 1$, $\sqrt{\lambda_1} = \sqrt{rt}$, and hence $\lambda_1 = rt$. Thus, for $\zeta > 0$,

$$\zeta = \left(\mu - \frac{1}{rt}\right)_+ = \mu - \frac{1}{rt} \tag{10.23}$$

and hence the capacity is

$$C = \log\left[\left(\zeta + \frac{1}{rt}\right)rt\right] = \log(1 + rt\,\zeta) \tag{10.24}$$

The signals achieving this capacity can be described as follows. Vector $\tilde{\mathbf{x}}$ has only one component, and

$$\mathbf{V} = \sqrt{\frac{1}{t}} \begin{bmatrix} 1 \\ 1 \\ \vdots \\ 1 \end{bmatrix} \tag{10.25}$$

Thus, the components of $\mathbf{x} = \mathbf{V}\tilde{\mathbf{x}}$ are all equal, i.e., the transmit antennas all send the same signal. Each transmit antenna sends an energy $\mathcal{E}$. Because of the structure of $\mathbf{H}$, the signals *add coherently* at the receiver, so at each receiver we have the voltage $t\sqrt{\mathcal{E}}$ and hence the energy $t^2\mathcal{E}$. Since each receiver sees the same signal, and the noises are uncorrelated, the overall SNR is $rt^2\mathcal{E}/N_0 = rt\zeta$, as shown by the capacity formula (10.24). In this case we see no rate gain, but a diversity gain is obtained through proper combination of the received signals. $\qquad\square$

10.3.2 Independent Rayleigh fading channel

We assume here that $\mathbf{H}$ is independent of both $\mathbf{x}$ and $\mathbf{z}$, with entries $\sim \mathcal{N}_c(0, 1)$, and that for each channel use an independent realization of $\mathbf{H}$ is drawn so that the channel is ergodic. If the receiver has perfect CSI, the mutual information between the channel input (the vector $\mathbf{x}$) and its output (the pair $\mathbf{y}, \mathbf{H}$), is (Appendix A)

$$I(\mathbf{x}; \mathbf{y}, \mathbf{H}) = I(\mathbf{x}; \mathbf{H}) + I(\mathbf{x}; \mathbf{y} \mid \mathbf{H}) \tag{10.26}$$

Since $\mathbf{H}$ and $\mathbf{x}$ are independent, then $I(\mathbf{x}; \mathbf{H}) = 0$, and hence

$$I(\mathbf{x}; \mathbf{y}, \mathbf{H}) = I(\mathbf{x}; \mathbf{y} \mid \mathbf{H}) = \mathbb{E}_{\widetilde{\mathbf{H}}}[I(\mathbf{x}; \mathbf{y} \mid \mathbf{H} = \widetilde{\mathbf{H}})] \tag{10.27}$$

where $\widetilde{\mathbf{H}}$ denotes a realization of the random matrix $\mathbf{H}$. The maximum of the mutual information $I(\mathbf{x}; \mathbf{y}, \mathbf{H})$, taken with respect to the distribution of $\mathbf{x}$, yields the channel capacity C. From the results of Appendix A we know that the capacity, achieved by a transmitted signal $\mathbf{x} \sim \mathcal{N}_c(0, (\zeta/t)\mathbf{I}_t)$, is equal to

$$C = \mathbb{E}\left[\log \det \left(\mathbf{I}_r + \frac{\zeta}{t}\mathbf{H}\mathbf{H}^\dagger\right)\right] \tag{10.28}$$

The exact computation of (10.28) will be examined soon. For the moment, note that, if r is fixed and $t \to \infty$, the strong law of large numbers yields

$$\frac{1}{t}\mathbf{H}\mathbf{H}^\dagger \to \mathbf{I}_r \qquad \text{a.s.} \tag{10.29}$$

Thus, as $t \to \infty$, the capacity tends to

$$
\begin{aligned}
\log \det \left(\mathbf{I}_r + \zeta \mathbf{I}_r \right) &= \log(1 + \zeta)^r \\
&= r \log(1 + \zeta)
\end{aligned}
\tag{10.30}
$$

and hence increases *linearly* with r, thus exhibiting a rate gain (compare this result with (10.24), where C increases with r only *logarithmically*).

One may be tempted to interpret the above result by qualifying fading as *beneficial* to MIMO transmission, since independent path gains generate r independent spatial channels. Actually, high capacity is generated by a multiplicity of nonzero singular values in $\mathbf{H}$, which is typically achieved if $\mathbf{H}$ is a random matrix, but not if it is deterministic.

Exact computation of C

Exact calculation of (10.28) yields

$$
C = \log(e) \frac{m!}{(n-1)!} \sum_{\ell=0}^{m-1} \sum_{\mu=0}^{m} \sum_{p=0}^{\ell+\mu+n-m} \frac{(-1)^{\ell+\mu}(\ell+\mu+n-m)!}{\ell!\mu!}
\tag{10.31}
$$

$$
\cdot e^{t/\zeta} E_{p+1}(t/\zeta) \left[\binom{n-1}{m-1-\ell}\binom{n}{m-1-\mu} - \binom{n-1}{m-2-\ell}\binom{n}{m-\mu} \right]
$$

where

$$
E_n(x) \triangleq \int_1^\infty e^{-xy} y^{-n}\, dy
$$

is the exponential integral function of order n.

Proof

Observe first that, since the matrices $\mathbf{H}\mathbf{H}^\dagger$ and $\mathbf{H}^\dagger\mathbf{H}$ share the same set of eigenvalues, we have

$$
\det \left(\mathbf{I}_r + \frac{\zeta}{t}\mathbf{H}\mathbf{H}^\dagger \right) = \det \left(\mathbf{I}_t + \frac{\zeta}{t}\mathbf{H}^\dagger\mathbf{H} \right)
$$

Next, define the $m \times m$ matrix

$$
\mathbf{W} \triangleq \begin{cases} \mathbf{H}\mathbf{H}^\dagger, & r < t \\ \mathbf{H}^\dagger\mathbf{H}, & t \leq r \end{cases}
\tag{10.32}
$$

where again $m \triangleq \min\{t, r\}$. This is a nonnegative definite random matrix and thus has real, nonnegative random eigenvalues. The joint pdf of the ordered eigenvalues of $\mathbf{W}$ is known (see Appendix C, Equation (C.28)). The expectation to be computed can be expressed in terms of one of the unordered eigenvalues of $\mathbf{W}$ (say, λ_1) as follows:

$$
\begin{aligned}
C = \mathbb{E} \log \prod_{i=1}^{m}\left(1+\frac{\zeta}{t}\lambda_i\right) &= \mathbb{E} \sum_{i=1}^{m} \log\left(1+\frac{\zeta}{t}\lambda_i\right) \\
&= \sum_{i=1}^{m} \mathbb{E} \log\left(1+\frac{\zeta}{t}\lambda_i\right) \\
&= m\,\mathbb{E} \log(1+\frac{\zeta}{t}\lambda_1)
\end{aligned}
\tag{10.33}
$$

To compute the marginal pdf of λ_1, use

$$
p(\lambda_1) = \int \cdots \int p(\lambda_1, \lambda_2, \cdots, \lambda_m)\, d\lambda_2 \cdots d\lambda_m
\tag{10.34}
$$

To perform this computation, we resort to an orthogonalization of the power sequence $1, \lambda, \lambda^2, \ldots, \lambda^{m-1}$ in the Hilbert space of real functions defined in $(0, \infty)$ with inner product

$$
(f, g) \triangleq \int_0^{\infty} f(\lambda)g(\lambda)\lambda^{n-m}e^{-\lambda}\, d\lambda
\tag{10.35}
$$

Explicitly, we express the pdf (10.34) in terms of the polynomials

$$
\phi_{k+1}(\lambda) \triangleq \sqrt{\frac{k!}{(k+n-m)!}} L_k^{n-m}(\lambda)
\tag{10.36}
$$

Here, $L_k^{\alpha}(\lambda)$ is an associated Laguerre polynomial [10.54], defined as

$$
\begin{aligned}
L_k^{\alpha}(\lambda) &= \frac{1}{k!} e^{\lambda} \lambda^{-\alpha} \frac{d^k}{d\lambda^k}(e^{-\lambda}\lambda^{k+\alpha}) \\
&= \sum_{\ell=0}^{k} (-1)^{\ell} \binom{k+\alpha}{k-\ell} \frac{\lambda^{\ell}}{\ell!}
\end{aligned}
\tag{10.37}
$$

The polynomials $\phi_i(\lambda)$ satisfy the orthonormality relation

$$
\int_0^{\infty} \phi_i(\lambda)\phi_j(\lambda)\lambda^{n-m}e^{-\lambda}\, d\lambda = \delta_{ij}
\tag{10.38}
$$

In order to calculate (10.34), we first observe that the term $\prod_{j=i+1}^{m}(\lambda_i - \lambda_j)$ appearing in (C.28) can be expressed as the determinant of the Vandermonde matrix

$$
\mathbf{D}(\lambda_1, \lambda_2, \ldots, \lambda_m) \triangleq \begin{bmatrix} 1 & \cdots & 1 \\ \lambda_1 & \cdots & \lambda_m \\ \vdots & & \vdots \\ \lambda_1^{m-1} & \cdots & \lambda_m^{m-1} \end{bmatrix}
\tag{10.39}
$$

so we can write

$$p(\lambda_1, \lambda_2, \ldots, \lambda_m) = \frac{1}{\Gamma_m(m)\Gamma_m(n)} \det\left[\mathbf{D}(\lambda_1, \lambda_2, \ldots, \lambda_m)\right]^2 \prod_{i=1}^{m} \lambda_i^{n-m} e^{-\lambda_i}$$

$$(10.40)$$

with $\Gamma_m(a) \triangleq \prod_{i=0}^{m-1} \Gamma(a-i)$. Next, with row operations we transform matrix $\mathbf{D}(\lambda_1, \lambda_2, \ldots, \lambda_m)$ into

$$\widetilde{\mathbf{D}}(\lambda_1, \lambda_2, \ldots, \lambda_m) \triangleq \begin{bmatrix} \phi_1(\lambda_1) & \cdots & \phi_1(\lambda_m) \\ \vdots & & \vdots \\ \phi_m(\lambda_1) & \cdots & \phi_m(\lambda_m) \end{bmatrix} \qquad (10.41)$$

so the determinant of $\mathbf{D}$ equals (apart from multiplicative constants generated by the row operations) the determinant of $\widetilde{\mathbf{D}}$, that is,

$$\det \widetilde{\mathbf{D}}(\lambda_1, \ldots, \lambda_m) = \sum_{\alpha} (-1)^{\pi(\alpha)} \prod_{i=1}^{m} \phi_{\alpha_i}(\lambda_i) \qquad (10.42)$$

where the summation is over all permutations of $\{1, \ldots, m\}$, and

$$\pi(\alpha) = \begin{cases} 0, & \alpha \text{ is an even permutation} \\ 1, & \text{otherwise} \end{cases}$$

Thus, with $c(m, n)$ a normalization constant, we have

$$p(\lambda_1, \ldots, \lambda_m) = c(m, n) \sum_{\alpha, \beta} (-1)^{\pi(\alpha)+\pi(\beta)} \prod_i \phi_{\alpha_i}(\lambda_i)\phi_{\beta_i}(\lambda_i)\lambda_i^{n-m} e^{-\lambda_i}$$

$$(10.43)$$

and, integrating over $\lambda_2, \ldots, \lambda_m$, we obtain

$$\begin{aligned} p(\lambda_1) &= c(m, n) \sum_{\alpha, \beta} (-1)^{\pi(\alpha)+\pi(\beta)} \phi_{\alpha_1}(\lambda_1)\phi_{\beta_1}(\lambda_1)\lambda_1^{n-m} e^{-\lambda_1} \prod_{i=2}^{m} \delta_{\alpha_i \beta_i} \\ &= c(m, n)(m-1)! \sum_{i=1}^{m} \phi_i^2(\lambda_1)\lambda_1^{n-m} e^{-\lambda_1} \\ &= \frac{1}{m} \sum_{i=1}^{m} \phi_i^2(\lambda_1)\lambda_1^{n-m} e^{-\lambda_1} \end{aligned} \qquad (10.44)$$

where the second equality follows from the fact that, if $\alpha_i = \beta_i$ for $i \geq 2$, then also $\alpha_1 = \beta_1$ (since both α and β are permutations of the same set) and thus $\alpha = \beta$. The last equality follows from the fact that $\phi_i^2(\lambda_1)\lambda_1^{n-m} e^{-\lambda_1}$ integrates to unity, which entails $c(m, n) = 1/m!$. In conclusion, the capacity can be given the form

$$C = \int_0^\infty \log\left(1 + \frac{\zeta}{t}\lambda\right) \sum_{k=0}^{m-1} \frac{k!}{(k+n-m)!} [L_k^{n-m}(\lambda)]^2 \lambda^{n-m} e^{-\lambda} \, d\lambda \qquad (10.45)$$

where $m \triangleq \min\{t, r\}$, $n \triangleq \max\{t, r\}$.

Consider now the Christoffel–Darboux identity, valid for orthonormal polynomials [10.54]:

$$\sum_{k=1}^{m} \phi_k(x)\phi_k(y) = \frac{A_m}{A_{m+1}} \frac{\phi_{m+1}(x)\phi_m(y) - \phi_m(x)\phi_{m+1}(y)}{x - y} \qquad (10.46)$$

where A_k denotes the coefficient of x^{k-1} in $\phi_k(x)$. Taking the limit as $y \to x$, the above identity yields

$$\sum_{k=1}^{m} \phi_k(x)^2 = \frac{A_m}{A_{m+1}} [\phi_m(x)\phi'_{m+1}(x) - \phi_{m+1}(x)\phi'_m(x)] \qquad (10.47)$$

When specialized to associated Laguerre polynomials, (10.36) and (10.37) yield

$$A_k = \frac{(-1)^{k-1}}{\sqrt{(k-1)!(k-1+\alpha)!}}$$

so

$$\begin{aligned}
\sum_{k=1}^{m} \phi_k^2(x) &= \frac{m!}{(m+\alpha)!} [L_m^\alpha(x)(L_{m-1}^\alpha(x))' - L_{m-1}^\alpha(x)(L_m^\alpha(x))'] \\
&= \frac{m!}{(m+\alpha)!} [L_{m-1}^\alpha(x)L_{m-1}^{\alpha+1}(x) - L_{m-2}^{\alpha+1}(x)L_m^\alpha(x)] \quad (10.48)
\end{aligned}$$

where we used the relation $[L_k^\alpha(x)]' = L_{k-1}^{\alpha+1}(x)$ [10.54]. Then, we can rewrite Equation (10.45) as

$$\begin{aligned}
C &= \frac{m!}{(n-1)!} \int_0^\infty [L_{m-1}^{n-m}(\lambda)L_{m-1}^{n-m+1}(\lambda) - L_{m-2}^{n-m+1}(\lambda)L_m^{n-m}(\lambda)] \\
&\quad \log\left(1 + \frac{\varsigma}{t}\lambda\right) \lambda^{n-m} e^{-\lambda} d\lambda
\end{aligned} \qquad (10.49)$$

and further expand it using (10.37) as follows:

$$\begin{aligned}
C &= \frac{m!}{(n-1)!} \sum_{\ell=0}^{m-1} \sum_{\mu=0}^{m} \frac{(-1)^{\ell+\mu}}{\ell!\mu!} \int_0^\infty \log\left(1 + \frac{\varsigma}{t}\lambda\right) \lambda^{\ell+\mu+n-m} e^{-\lambda} d\lambda \\
&\quad \left[\binom{n-1}{m-1-\ell}\binom{n}{m-1-\mu} - \binom{n-1}{m-2-\ell}\binom{n}{m-\mu}\right] \quad (10.50)
\end{aligned}$$

Finally, using the equality (from [10.50])

$$\int_0^\infty \ln(1 + \varsigma\lambda)\lambda^\mu e^{-\lambda} d\lambda = \mu! \, e^{1/\varsigma} \sum_{p=0}^{\mu} E_{p+1}(1/\varsigma) \qquad (10.51)$$

we obtain (10.31). $\qquad\qquad\qquad\qquad\qquad\qquad\qquad\qquad\qquad\qquad\qquad\square$

Some capacity values for $\zeta = 20$ dB are plotted in Figure 10.5 and 10.6. Special cases, as well as asymptotic approximations to the values of C, are examined in the examples that follow.

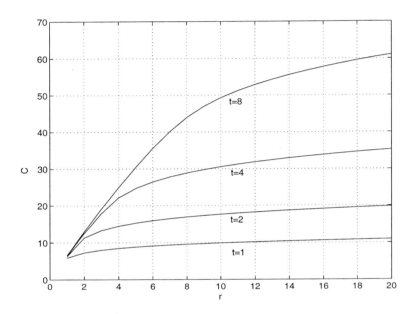

Figure 10.5: *Capacity (in bit/s/Hz, or bit/dimension pair) of the ergodic Rayleigh MIMO channel with $\zeta = 20$ dB.*

Example 10.3 ($r \gg t$)

Consider first $t = 1$, so that $m = 1$ and $n = r$. Application of (10.31) yields

$$C = \log(e) \sum_{k=1}^{r} e^{1/\zeta} E_k(1/\zeta) \, . \tag{10.52}$$

This is plotted in Figure 10.7. An asymptotic expression of C, valid as $r \to \infty$, can be obtained as follows. Using in (10.52) the approximation, valid for large k,

$$e^x E_k(x) \sim \frac{1}{x + k} \tag{10.53}$$

we obtain

$$C \sim \log(e) \sum_{k=1}^{r} \frac{1}{1/\zeta + k} \approx \log(e) \int_0^r \frac{1}{1/\zeta + x} \, dx = \log(1 + r\zeta) \tag{10.54}$$

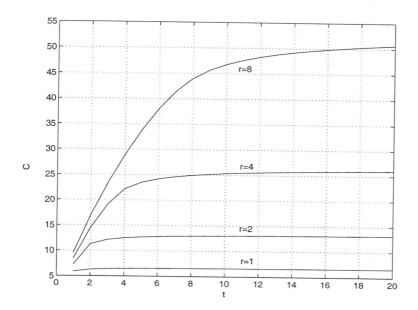

Figure 10.6: *Capacity (in bit/s/Hz, or bit/dimension pair) of the ergodic Rayleigh MIMO channel with $\zeta = 20$ dB.*

This approximation to the capacity is also plotted in Figure 10.7. We see here that if $t = 1$ the capacity increases only logarithmically as the number of receive antennas is increased—hardly an efficient way of enhancing capacity. In addition, this capacity is much smaller than that of a system with $t = r$. Increasing r generates an increase of SNR from ζ to $r\zeta$, but no rate gain.

For finite $t > 1$ (and $r \to \infty$), we set $\mathbf{W} = \mathbf{H}^\dagger\mathbf{H} \to r\mathbf{I}_t$ a.s. Hence, the following asymptotic expression holds:

$$C = \log\det\left(\mathbf{I}_t + (\zeta/t)\mathbf{W}\right) \sim t\log(1 + (\zeta/t)r) \tag{10.55}$$

$\square$

Example 10.4 ($t \gg r$)

Consider first $r = 1$, so that $m = 1$ and $n = t$. Application of (10.31) yields

$$C = \log(e)\sum_{k=1}^{t} e^{t/\zeta}E_k(t/\zeta)$$

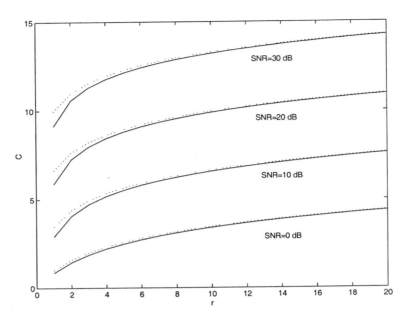

Figure 10.7: *Capacity (in bit/s/Hz, or bit/dimension pair) of the ergodic Rayleigh MIMO channel with $t = 1$ (continuous line). The asymptotic approximation $C \sim \log(1 + \zeta r)$ is also shown (dotted line).*

This is plotted in Figure 10.8. Proceeding as in Example 10.2, an asymptotic expression of C as $t \to \infty$ can be obtained, yielding $C \sim \log(1 + \zeta)$. This approximation to the capacity is also plotted in Figure 10.8. Since $\log(1 + \zeta)$ is the capacity of the AWGN channel, we can see that letting $t \to \infty$ closes the gap between the AWGN channel and the independent Rayleigh fading channel. Since this gap is not wide (see Figure 4.6), having a large t when $r = 1$ corresponds to an inefficient expenditure of spatial resources (recall that we are assuming no CSI at the transmitter).

For finite $r > 1$ (and $t \to \infty$), we have the result (10.29). $\qquad\square$

Example 10.5 ($r = t$)

With $r = t$ we have $m = n = r$, so application of (10.31) yields

$$
C = r \log(e) \sum_{\ell=0}^{r-1} \sum_{\mu=0}^{r} \sum_{p=0}^{\ell+\mu} \frac{(-1)^{\ell+\mu}(\ell+\mu)!}{\ell!\mu!} e^{t/\zeta} E_{p+1}(t/\zeta)
$$
$$
\left[\binom{r-1}{\ell}\binom{r}{\mu+1} - \binom{r-1}{\ell+1}\binom{r}{\mu} \right]
$$

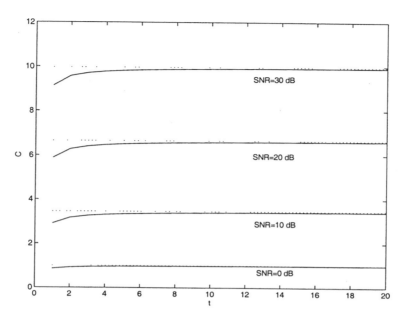

Figure 10.8: *Capacity (in bit/s/Hz, or bit/dimension pair) of the ergodic Rayleigh MIMO channel with $r = 1$ (continuous line). The asymptotic approximation $C \sim \log(1 + \zeta)$ is also shown (dotted line).*

The capacity is plotted in Figure 10.9. □

The results of Figure 10.9 show that capacity increases almost linearly with $m \triangleq \min\{t, r\}$. This fact can be analyzed in a general setting by showing that, when t and r both grow to infinity, the capacity per antenna tends to a constant. To prove this, observe that (10.33) becomes

$$\frac{C}{m} \sim \mathbb{E}\left[\log\left(1 + \zeta\frac{m}{t}\nu\right)\right] \tag{10.56}$$

where $\nu \triangleq \lambda_1/m$ is now a random variable whose pdf is known (see Theorem C.3.2, Appendix C): as $m \to \infty$ and n/m approaches a limit $\tau \geq 1$,

$$p(\nu) = \frac{1}{2\pi\nu}\sqrt{(\nu_+ - \nu)(\nu - \nu_-)} \tag{10.57}$$

with

$$\nu_\pm \triangleq (1 \pm \sqrt{\tau})^2$$

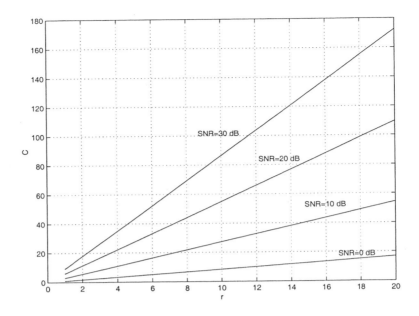

Figure 10.9: *Capacity with independent Rayleigh fading and $t = r$ antennas.*

for $\nu_- \leq \nu \leq \nu_+$. With the aid of some algebra, expectation in (10.56) can be computed in closed form [10.43, 10.62], yielding

$$\frac{C}{m} \sim (\log(w_+\zeta) + (1 - \alpha)\log(1 - w_-) - (w_-\alpha)\log e) \cdot \max\{1, 1/\alpha\} \quad (10.58)$$

where

$$w_\pm \triangleq (w \pm \sqrt{w^2 - 4/\alpha})/2 \quad (10.59)$$

and

$$w \triangleq 1 + \frac{1}{\alpha} + \frac{1}{\zeta} \quad (10.60)$$

This asymptotic result can be used to approximate the value of C for finite r, t, by setting $\alpha = t/r$. This approximation provides values very close to the true capacity even for small r and t, as shown in Figures 10.10 and 10.11. The figures show the asymptotic value of C/m (for $t, r \to \infty$ with $t/r \to \alpha$) versus α and the nonasymptotic values of C/m corresponding to $r = 2$ and 4, respectively.

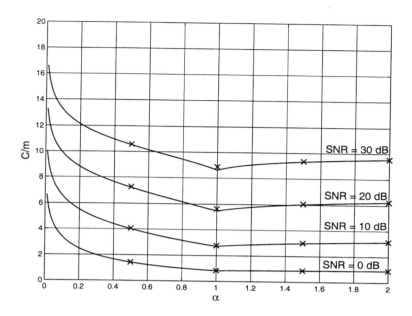

Figure 10.10: *Asymptotic ergodic capacity per antenna (C/m) with independent Rayleigh fading as $t, r \rightarrow \infty$ and $t/r \rightarrow \alpha$ (solid curves). The exact ergodic capacity per antenna for $r = 2$ is also shown for comparison ($\times$).*

Observation 10.3.2 We can observe, from (10.58) and a modicum of algebra, that, for large SNR, i.e., for $\zeta \rightarrow \infty$, the ergodic capacity is asymptotically equal to $m \log \zeta$: comparing this result with the asymptotic capacity of the single-input, single-output channel $C \sim \log \zeta$ (see Section 4.2.2), we see that use of multiple antennas increases the capacity by a factor m. That is, multiple antennas generate m independent parallel channels and hence a rate gain m. This explains why m is sometimes called the *number of degrees of freedom* generated by the MIMO system.

Observation 10.3.3 For the validity of (10.58), it is not necessary to assume that the entries of $\mathbf{H}$ are Gaussian, as needed for the preceding nonasymptotic results: a sufficient condition is that $\mathbf{H}$ have iid entries with unit variance (Appendix C, Theorem C.3.2).

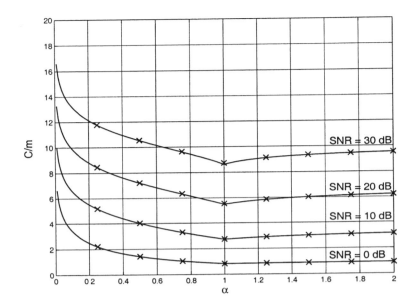

Figure 10.11: *Same as Figure 10.10, but $r = 4$.*

Observation 10.3.4   The reciprocity valid for deterministic channels (Observation 10.3.1) does not hold in this case. If $C(r, t, \zeta)$ denotes the capacity of a channel with t transmit and r receive antennas, and SNR ζ, we have

$$C(a, b, \hat{\zeta}b) = C(b, a, \hat{\zeta}a) \tag{10.61}$$

Thus, for example, $C(r, 1, \hat{\zeta}) = C(1, r, r\hat{\zeta})$, which shows that with transmit rather than receive diversity we need r times as much transmit power to achieve the same capacity.

Observation 10.3.5   Choose $t = r = 1$ as the baseline; this yields *one* more bit per dimension pair for every 3 dB of SNR increase. In fact, for large ζ,

$$C = \log(1 + \zeta) \sim \log \zeta \tag{10.62}$$

and hence, if $\zeta \to 2\zeta$ we have

$$\log(2\zeta) = 1 + \log \zeta \tag{10.63}$$

For multiple antennas with $t = r$, (10.56) shows that for every 3 dB of SNR increase we have t more bits per dimension pair.

10.4 Correlated fading channels

The separately correlated MIMO channel model was introduced in Section 10.2.1. The entries of the channel matrix are correlated, circularly symmetric, complex zero-mean Gaussian random variables, and the channel matrix can be written as

$$\mathbf{H} = \mathbf{R}^{1/2}\mathbf{H}_u\mathbf{T}^{1/2} \tag{10.64}$$

where $\mathbf{H}_u$ is a matrix of independent, circularly symmetric complex zero-mean Gaussian random variables with unit variance. Calculation of the ergodic capacity in this case is an open problem, its solution being known only for some special cases. A lower bound C^* to capacity can be obtained under the constraint that the signals across the transmit antennas are independent (if the correlations among the entries of $\mathbf{H}$ are unknown at the transmitter, then independent signal are the sensible choice, and hence C^* yields the actual capacity). We have

$$C^* = \mathbb{E}\left[\log\det\left(\mathbf{I}_r + \frac{\zeta}{t}\mathbf{H}_u\mathbf{T}\mathbf{H}_u^\dagger\mathbf{R}\right)\right] \tag{10.65}$$

In the special case $t = r$, for high SNR the following asymptotic approximation can be derived [10.50]:

$$C^* \sim m\log(\zeta/t) + \log(m!) + \log\det(\mathbf{TR}) \tag{10.66}$$

This result can be interpreted by saying that, when $t = r = m$, the asymptotic loss in C^* due to correlation is $\log\det(\mathbf{TR})/m$ bit/s/Hz. To prove that correlation actually causes a loss, let $t_i, i = 1, \ldots, m$, denote the positive eigenvalues of $\mathbf{T}$, and recall the trace constraint (10.13). We obtain

$$\det(\mathbf{T})^{1/m} = \prod_i t_i^{1/m} \leq \frac{1}{m}\sum_i t_i = 1$$

Since a similar result applies to $\mathbf{R}$, we obtain

$$-\log\det(\mathbf{TR})/m \geq 0$$

with equality if and only if $\mathbf{T} = \mathbf{R} = \mathbf{I}_m$. This confirms that, under the "fair comparison" conditions dictated by (10.13), the asymptotic power loss due to separate correlation is always nonnegative and is zero only in the uncorrelated case. This proves the following asymptotic (in the SNR) statements:

- (Separate) correlation degrades system performance.

- The linear growth of capacity with respect to the minimum number of transmit/receive antennas is preserved.

The above can be extended to the case $t \neq r$ [10.50].

Example 10.6

Consider the case of a constant separately correlated $m \times m$ MIMO fading channel with correlation matrices

$$\mathbf{T} = \begin{pmatrix} 1 & \varepsilon_T & \cdots & \varepsilon_T \\ \varepsilon_T & 1 & \cdots & \varepsilon_T \\ \vdots & \vdots & \ddots & \vdots \\ \varepsilon_T & \varepsilon_T & \cdots & 1 \end{pmatrix} \tag{10.67}$$

and

$$\mathbf{R} = \begin{pmatrix} 1 & \varepsilon_R & \cdots & \varepsilon_R \\ \varepsilon_R & 1 & \cdots & \varepsilon_R \\ \vdots & \vdots & \ddots & \vdots \\ \varepsilon_R & \varepsilon_R & \cdots & 1 \end{pmatrix} \tag{10.68}$$

Some algebra leads to the asymptotic approximation

$$C \sim m \log(\zeta/t) + \log(m!) + (m-1)\log(1-\varepsilon_T)$$
$$+ \log(1 - \varepsilon_T + m\varepsilon_T) + (m-1)\log(1-\varepsilon_R) + \log(1 - \varepsilon_T + m\varepsilon_R)$$

When m is large, the asymptotic capacity loss is about

$$\log[(1 - \varepsilon_T)(1 - \varepsilon_R)]$$

□

10.5   A critique to asymptotic analyses

The previous results derived under the assumption $r \to \infty$ should be taken *cum grano salis*. Our assumption that the entries of the channel-gain matrix $\mathbf{H}$ are independent random variables becomes increasingly questionable as r increases. In fact, for this assumption to be justified, the antennas should be separated by some multiple of the wavelength, which cannot be obtained when a large number of antennas is packed in a finite volume. Thus, as r increases, the effects of correlation invalidate the assumption of independent channel gains. In addition, if the variance of the entries of $\mathbf{H}$ does not depend on r, increasing r leads to an increased total received power, which becomes physically unacceptable beyond a certain value. It follows that capacity calculations for large r and a finite volume become quite involved. A simple, yet instructive, analysis is possible if the effects of varying correlation are disregarded and a MIMO system is assumed whereby not only the total transmit power remains constant as t increases but also the average received power remains constant when r increases [10.16]. This is obtained by rescaling $\mathbf{H}$ by a factor $r^{-1/2}$ so that the capacity (10.28)–(10.33) becomes

$$C = \mathbb{E}\left[\log\det\left(\mathbf{I}_r + \frac{\zeta}{rt}\mathbf{H}\mathbf{H}^\dagger\right)\right] = \sum_{i=1}^{m}\mathbb{E}\log\left(1 + \frac{\zeta}{rt}\lambda_i\right) \qquad (10.69)$$

One simple heuristic way of dealing with this situation consists of rewriting C in the form

$$C = \mathbb{E}\left[\log\det\left(\mathbf{I}_t + \frac{\zeta}{rt}\mathbf{H}^\dagger\mathbf{H}\right)\right]$$

and observing that, due to the strong law of large numbers, $(1/r)\mathbf{H}^\dagger\mathbf{H} \to \mathbf{I}_t$ almost surely. Thus,

$$C \to t\log(1 + \zeta/t) \qquad (10.70)$$

that is, the channel is transformed into a set of t independent parallel channels, each with capacity $\log(1 + \zeta/t)$. As t also grows to infinity, from (10.70) we obtain

$$C \to \zeta\log e \qquad (10.71)$$

a conclusion in contrast with our previous result that capacity increases linearly with the number of antennas.

10.6 Nonergodic Rayleigh fading channel

When $\mathbf{H}$ is chosen randomly at the beginning of the transmission, and held fixed for all channel uses, average capacity has no meaning, as the channel is nonergodic. In this case the quantity to be evaluated is, rather than capacity, outage probability, that is, the probability that the transmission rate ρ exceeds the mutual information of the channel. The instantaneous mutual information is the random variable

$$C(\mathbf{H}) = \log\det\left(\mathbf{I}_r + \frac{\zeta}{t}\mathbf{H}\mathbf{H}^\dagger\right) \qquad (10.72)$$

and the outage probability is defined as

$$P_{\text{out}}(\rho) \triangleq \mathbb{P}(C(\mathbf{H}) < \rho) \qquad (10.73)$$

The maximum rate that can be supported by the channel with a given outage probability is referred to as *outage capacity*.

The evaluation of (10.73) should be done by Monte Carlo simulation. However, one can profitably use an asymptotic result which states that, as t and r grow to infinity, the instantaneous mutual information $C(\mathbf{H})$ tends to a Gaussian random variable. Thus, by computing its asymptotic mean μ_C and variance σ_C^2, one can characterize its asymptotic behavior. The value of this asymptotic result is strongly enhanced by the fact that $C(\mathbf{H})$ is very well approximated by a Gaussian random variable even for small numbers of antennas. Thus, the outage probability for any pair t, r is given by

$$P_{\text{out}}(\rho) \approx Q\left(\frac{\mu_C - \rho}{\sigma_C}\right) \tag{10.74}$$

where

$$\mu_C \triangleq -t\Big\{(1+\beta)\log w + q_0 r_0 \log e + \log r_0 + \beta \log(q_0/\beta)\Big\}$$
$$\sigma_C^2 \triangleq -\log e \cdot \log(1 - q_0^2 r_0^2/\beta)$$

expressed in bit/dimension pair and (bit/dimension pair)^2, respectively, with $w \triangleq \sqrt{1/\zeta}$, $\beta \triangleq \alpha^{-1}$, and

$$
\begin{cases}
q_0 = \dfrac{\beta - 1 - w^2 + \sqrt{(\beta - 1 - w^2)^2 + 4w^2\beta}}{2w} \\[3mm]
r_0 = \dfrac{1 - \beta - w^2 + \sqrt{(1 - \beta - w^2)^2 + 4w^2}}{2w}
\end{cases}
\tag{10.75}
$$

Figure 10.12, which plots P_{out} versus ζ for $r = t = 4$ and two values of SNR, shows the quality of the Gaussian approximation for $r = t = 4$ and a Rayleigh channel.

Based on these results, we can use (10.74) to approximate closely the outage probabilities as in Figures 10.13 and 10.14. These figures show the rate that can be supported by the channel for a given SNR and a given outage probability, that is, from (10.74):

$$\rho = \mu_C - \sigma_C Q^{-1}(P_{\text{out}}) \tag{10.76}$$

Notice how, as r, t increase, the outage probabilities curves come closer to each other: this fact can be interpreted by saying that, as r and t grow to infinity, the channel tends to an ergodic channel.

Figure 10.15 shows the outage capacity (at $P_{\text{out}} = 0.01$) of a nonergodic Rayleigh fading MIMO channel.

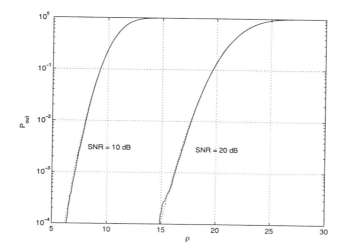

Figure 10.12: *Outage probability for* $r = t = 4$ *and a nonergodic Rayleigh channel vs.* ρ, *the transmission rate in bits per dimension pair. The continuous line shows the results obtained by Monte Carlo simulation, while the dashed line shows the normal approximation.*

10.6.1   Block-fading channel

Here we take the approach of choosing a block-fading channel model, introduced in Section 4.3 and shown in Figure 10.16. Here the channel is characterized by the F matrices $\mathbf{H}_k$, $k = 1, \ldots, F$, each describing the fading gains in a block. The channel input–output equation is

$$\mathbf{y}_k[n] = \mathbf{H}_k \mathbf{x}_k[n] + \mathbf{z}_k[n] \tag{10.77}$$

for $k = 1, \ldots, F$ (block index) and $n = 1, \ldots, N$ (symbol index along a block), $\mathbf{y}_k, \mathbf{z}_k \in \mathbb{C}^r$, and $\mathbf{x}_k \in \mathbb{C}^t$. Moreover, the additive noise $\mathbf{z}_k[n]$ is a vector of circularly symmetric complex Gaussian RVs with zero mean and variance N_0: hence,

$$\mathbb{E}[\mathbf{z}_k[n]\mathbf{z}_k^\dagger[n]] = N_0 \mathbf{I}_r$$

It is convenient to use the SVD

$$\mathbf{H}_k = \mathbf{U}_k \mathbf{D}_k \mathbf{V}_k^\dagger \tag{10.78}$$

where $\mathbf{D}_k$ is an $r \times t$ real matrix whose main-diagonal entries are the ordered singular values $\sqrt{\lambda_{k,1}} \geq \cdots \geq \sqrt{\lambda_{k,m}}$, with $\lambda_{k,i}$ the ith largest eigenvalue of the

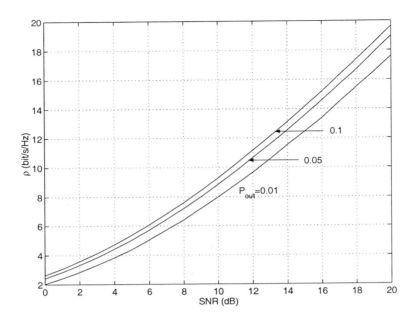

Figure 10.13: *Transmission rate that can be supported with $r = t = 4$ and a given outage probability by a nonergodic Rayleigh channel. The results are based on the Gaussian approximation.*

Hermitian matrix $\mathbf{H}_k\mathbf{H}_k^\dagger$, and $m \triangleq \min\{r, t\}$. Since $\mathbf{U}_k$ and $\mathbf{V}_k$ are unitary, by premultiplying $\mathbf{y}_k[n]$ by $\mathbf{U}_k^\dagger$ the input–output relation (10.77) can be rewritten in the form

$$\tilde{\mathbf{y}}_k[n] = \mathbf{D}_k\tilde{\mathbf{x}}_k[n] + \tilde{\mathbf{z}}_k[n] \qquad (10.79)$$

where $\tilde{\mathbf{y}}_k[n] \triangleq \mathbf{U}_k^\dagger\mathbf{y}_k[n]$, $\tilde{\mathbf{x}}_k[n] \triangleq \mathbf{V}_k^\dagger\mathbf{x}_k[n]$, $\tilde{\mathbf{z}}_k[n] \triangleq \mathbf{U}_k^\dagger\mathbf{z}_k[n]$, and $\tilde{\mathbf{z}}_k[n] \sim \mathcal{N}_c(\mathbf{0}, N_0\mathbf{I}_r)$ since

$$\mathbb{E}[\tilde{\mathbf{z}}_k[n]\tilde{\mathbf{z}}_k[n]^\dagger] = \mathbf{U}^\dagger\mathbb{E}[\tilde{\mathbf{z}}_k[n]\tilde{\mathbf{z}}_k[n]^\dagger]\mathbf{U} = N_0\mathbf{I}_r$$

No delay constraints. When the random matrix process $\{\mathbf{H}_k\}_{k=1}^F$ is iid, as $F \to \infty$ the channel is ergodic, and the average capacity is the relevant quantity. When the entries of the channel matrices are uncorrelated, and perfect CSI is available to

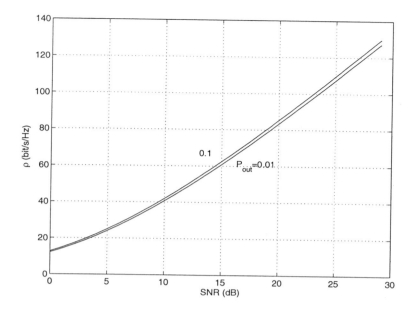

Figure 10.14: *Transmission rate that can be supported with $r = t = 16$ and a give outage probability by a nonergodic Rayleigh channel. The results are based on the Gaussian approximation.*

the receiver only, this is given by

$$C = \mathbb{E}\left[\sum_{i=1}^{m} \log\left(1 + \frac{\zeta}{t}\lambda_i\right)\right] \tag{10.80}$$

If perfect CSI is available to transmitter and receiver,

$$C = \mathbb{E}\left[\sum_{i=1}^{m} (\log(\mu\lambda_i))_+\right] \tag{10.81}$$

where μ is the solution of the water-filling equation

$$\mathbb{E}\left[\sum_{i=1}^{m}(\mu - 1/\lambda_i)_+\right] = \zeta \tag{10.82}$$

For all block lengths $N = 1, 2, \ldots$, the capacities (10.80) and (10.81) are achieved by code sequences with length FNt with $F \to \infty$. Capacity (10.80) is achieved

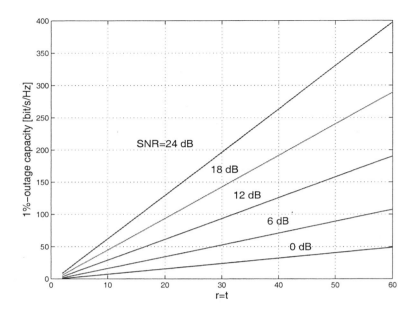

Figure 10.15: *Outage capacity (at $P_{\mathrm{out}} = 0.01$) with independent Rayleigh fading and $r = t$ antennas.*

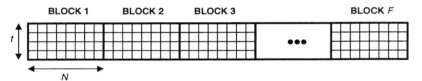

Figure 10.16: *One code word in an F-block fading channel.*

by random codes whose symbols are iid complex $\sim \mathcal{N}_c(0, \zeta/t)$. Thus, all antennas transmit the same average energy per symbol. Capacity (10.81) can be achieved by generating a random code with iid components $\sim \mathcal{N}_c(0, 1)$ and having each code word split into F blocks of N vectors $\tilde{\mathbf{x}}_k[n]$ with t components each. For block k, the optimal linear transformation

$$\mathbf{W}_k = \mathbf{V}_k \operatorname{diag}\left(\sqrt{\zeta_{k,1}}, \ldots, \sqrt{\zeta_{k,m}}, \underbrace{0, \ldots, 0}_{t-m}\right) \tag{10.83}$$

is computed, where $\zeta_{k,i} \triangleq (\mu - 1/\lambda_{k,i})_+$. The vectors $\mathbf{x}_k[n] = \mathbf{W}_k \tilde{\mathbf{x}}_k[n]$ are transmitted from the t antennas. This optimal scheme can be viewed as the concatenation of an optimal encoder for the unfaded AWGN channel, followed by a linear transformation ("beamforming") described by the weighting matrix $\mathbf{W}_k$ varying from block to block [10.6].

Delay constraints. Consider now a delay constraint that forces F to take on a finite value. Define $\mathbf{\Lambda} \triangleq \{\lambda_{k,i}\}_{k=1,i=1}^{F,m}$, $\mathbf{\Gamma} \triangleq \{\zeta_{k,i}\}_{k=1,i=1}^{F,m}$, the instantaneous mutual information

$$I(\mathbf{\Lambda}, \mathbf{\Gamma}) \triangleq \frac{1}{F} \sum_{k=1}^{F} \sum_{i=1}^{m} \log(1 + \lambda_{k,i}\zeta_{k,i}) \tag{10.84}$$

and the *instantaneous* SNR per block

$$\zeta_F \triangleq \frac{1}{F} \sum_{k=1}^{F} \sum_{i=1}^{m} \zeta_{k,i} \tag{10.85}$$

Assuming that the receiver has perfect knowledge of the CSI (and hence of $\mathbf{\Lambda}$), we can define a power allocation rule depending on $\mathbf{\Lambda}$ such that $\zeta_{k,i}$ and ζ_F are functions of $\mathbf{\Lambda}$. We may consider two power constraints:

$$\zeta_F(\mathbf{\Lambda}) \leq \zeta \qquad \text{(short-term)} \tag{10.86}$$
$$\mathbb{E}[\zeta_F(\mathbf{\Lambda})] \leq \zeta \qquad \text{(long-term)} \tag{10.87}$$

The optimum power allocation rules minimizing the outage probability

$$P_{\text{out}}(\rho \triangleq \mathbb{P}\left(I(\mathbf{\Lambda}, \mathbf{\Gamma}) < \rho\right) \tag{10.88}$$

under constraints (10.86) and (10.87) are derived in [10.6] and summarized in the following.

1. With the short-term power constraint, we have

$$\mathbf{\Gamma}(\mathbf{\Lambda}) = \begin{cases} \mathbf{\Gamma}^{\text{st}}(\mathbf{\Gamma}, \zeta) & \text{if } \mathbf{\Lambda} \in \mathcal{R}_{\text{on}}(\rho, \zeta) \\ \mathbf{G}(\mathbf{\Gamma}) & \text{if } \mathbf{\Lambda} \in \mathcal{R}_{\text{off}}(\rho, \zeta) \end{cases} \tag{10.89}$$

where

(a) The (k, i)-th SNR is given by

$$\zeta_{k,i}^{\text{st}} = (\mu^{\text{st}}(\mathbf{\Lambda}, \zeta) - 1/\lambda_{k,i})_+ \tag{10.90}$$

where

$$\mu^{\mathrm{st}}(\mathbf{\Lambda}, \zeta) = \frac{F}{|\mathcal{F}(\zeta)|}\zeta + \frac{1}{|\mathcal{F}(\zeta)|}\sum_{(k,i)\in\mathcal{F}(\zeta)}\frac{1}{\lambda_{k,i}} \tag{10.91}$$

and $\mathcal{F}(\zeta)$ is the unique set of indexes (k, i) such that $1/\lambda_{k,i} \leq \mu^{\mathrm{st}}(\mathbf{\Lambda}, \zeta)$ for all $(k, i) \in \mathcal{F}(\zeta)$ and $1/\lambda_{k,i} > \mu^{\mathrm{st}}(\mathbf{\Lambda}, \zeta)$ for all $(k, i) \notin \mathcal{F}(\zeta)$.

(b) The set

$$\mathcal{R}_{\mathrm{on}}(\rho, \zeta) \triangleq \left\{\mathbf{\Lambda} : I(\mathbf{\Lambda}, \mathbf{\Gamma}^{\mathrm{st}}(\mathbf{\Lambda}, \zeta)) \geq \rho\right\} \tag{10.92}$$

is called the *power-on region*.

(c) The set

$$\mathcal{R}_{\mathrm{off}}(\rho, \zeta) \triangleq \left\{\mathbf{\Lambda} : I(\mathbf{\Lambda}, \mathbf{\Gamma}^{\mathrm{st}}(\mathbf{\Lambda}, \zeta)) < \rho\right\} \tag{10.93}$$

is called the *outage*, or *power-off, region*.

(d) $\mathbf{G}(\mathbf{\Lambda})$ is an arbitrary power allocation function satisfying the short-term constraint, i.e., $\zeta_F(\mathbf{G}) \leq \zeta$.

2. With the long-term power constraint, we have

$$\mathbf{\Gamma}(\mathbf{\Lambda}) = \begin{cases} \mathbf{\Gamma}^{\mathrm{lt}}(\mathbf{\Lambda}, \rho) & \text{if } \mathbf{\Lambda} \in \mathcal{R}_{\mathrm{on}}^*(\rho, \zeta^*) \\ 0 & \text{if } \mathbf{\Lambda} \in \mathcal{R}_{\mathrm{off}}^*(\rho, \zeta^*) \end{cases} \tag{10.94}$$

where

(a) The (k, i)-th SNR is given by

$$\zeta_{k,i}^{\mathrm{lt}} = (\mu^{\mathrm{lt}}(\mathbf{\Lambda}, \rho) - 1/\lambda_{k,i})_+ \tag{10.95}$$

where

$$\mu^{\mathrm{lt}}(\mathbf{\Lambda}, \rho) = \left(\frac{2^{F\rho}}{\prod_{(k,i)\in\mathcal{F}^*(\rho)}\lambda_{k,i}}\right)^{1/|\mathcal{F}^*(\rho)|} \tag{10.96}$$

and $\mathcal{F}^*(\rho)$ is the unique set of indexes (k, i) such that $1/\lambda_{k,i} \leq \mu^{\mathrm{lt}}(\mathbf{\Lambda}, \rho)$ for all $(k, i) \in \mathcal{F}^*(\rho)$ and $1/\lambda_{k,i} > \mu^{\mathrm{lt}}(\mathbf{\Lambda}, \rho)$ for all $(k, i) \notin \mathcal{F}^*(\rho)$.

(b) The set

$$\mathcal{R}_{\mathrm{on}}^*(\rho, \zeta^*) \triangleq \left\{\mathbf{\Lambda} : \zeta_F(\mathbf{\Gamma}^{\mathrm{lt}}(\mathbf{\Lambda}, \rho)) \leq \zeta^*\right\} \tag{10.97}$$

is called the *power-on region*.

(c) The set

$$\mathcal{R}_{\text{off}}^*(\rho, \zeta^*) \triangleq \left\{ \mathbf{\Lambda} : \zeta_F(\mathbf{\Gamma}^{\text{lt}}(\mathbf{\Lambda}, \rho)) > \zeta^* \right\} \tag{10.98}$$

is called the *outage*, or *power-off, region*.

(d) The *threshold* $\zeta^* > 0$ is set in order to satisfy the long-term contraint (10.87) with equality, i.e., it is the solution of

$$\mathbb{E}\{\zeta_F(\mathbf{\Gamma}^{\text{lt}}(\mathbf{\Lambda}, \rho)) [\mathbf{\Lambda} \in \mathcal{R}_{\text{on}}(\rho, \zeta^*)]\} = \zeta$$

where $[\mathcal{A}] \triangleq 1$ if $\mathcal{A}$ is true, and 0 otherwise.

In other words, the outage probability is minimized under a long-term power constraint by setting a threshold ζ^*. If the instantaneous SNR per block necessary to avoid an outage exceeds ζ^*, then transmission is turned off and an outage is declared. If it is below ζ^*, transmission is turned on and power is allocated to the blocks according to a rule that depends on the fading statistics only through the threshold value ζ^* (see [10.6]).

Figure 10.17 illustrates the concept of an outage region for a single transmit and receive antenna system ($t = r = 1$) with $F = 2$, $\rho = 1$ bit/dimension pair, and $\zeta = 1$ dB. The outage region is the inner region corresponding to smaller values of the channel matrix eigenvalues ($\sqrt{\lambda_{k,1}}$, $k = 1, 2$) reflecting the occurrence of a deep fade.

It is interesting to note that the short-term and long-term outage regions $\mathcal{R}_{\text{off}}(\rho, \zeta)$ and $\mathcal{R}_{\text{off}}^*(\rho, \zeta)$ exhibit the same functional dependence on ρ and ζ in spite of their very different definitions of (10.93) and (10.98) [10.6]. This is again illustrated by Figure 10.17. The figure also shows that though the outage regions $\mathcal{R}_{\text{off}}(1, 10^{0.1})$ and $\mathcal{R}_{\text{off}}^*(1, 10^{0.1})$ coincide, the boundaries of constant-$|\mathcal{F}|$ regions differ in the two cases (short-term and long-term) [10.6].

Another important concept related to outage probability is given in the following definition [10.6, 10.59]:

Definition 10.6.1 *The* zero-outage capacity, *sometimes also referred to as* delay-limited capacity, *is the maximum rate for which the minimum outage probability is zero under a given power constraint.*

It was shown in [10.6] that, under a long-term power constraint, the zero-outage capacity of a block-fading channel is positive if the channel is *regular*. A regular channel is defined as follows.

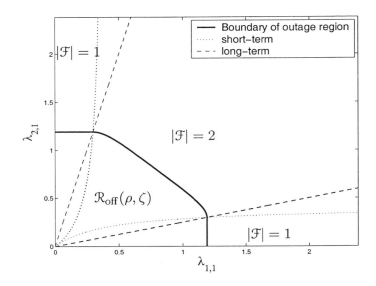

Figure 10.17: *Outage region* $\mathcal{R}_{\text{off}}(\rho, \zeta)$ *of a single transmit and receive antenna system* $(t = r = 1)$ *with* $F = 2$, $\rho = 1$ *bit/dimension pair, and* $\zeta = 1$ *dB. The boundaries of constant-*$|\mathcal{F}|$ *regions are also indicated, for the short-term and long-term constraints, as dotted and dashed lines, respectively.*

Definition 10.6.2 *A block-fading channel is said to be* regular *if the fading distribution is continuous and*

$$\mathbb{E}[1/\bar{\lambda}_F] < \infty \tag{10.99}$$

where $\bar{\lambda}_F$ *is the geometric mean of the* $\lambda_{k,i}$:

$$\bar{\lambda}_F \triangleq \prod_{k,i} \lambda_{k,i}^{1/mF} \tag{10.100}$$

where, as usual, $m \triangleq \min\{t, r\}$.

Example 10.7

The Rayleigh fading channel with $F = m = 1$ is not regular, and its zero-outage capacity is null. The Rayleigh block-fading channel is regular if $mF > 1$ (see [10.6] for a proof). For example, if $F > 1$ and $m = 1$, we have

$$\mathbb{E}[1/\bar{\lambda}_F] = (\mathbb{E}[\lambda_1^{-1/F}])^F = [\Gamma(1 - 1/F)]^F < \infty$$

$\square$

10.6.2 Asymptotics

Under a long-term power constraint and with optimal transmit power allocation, the zero-outage capacity of a regular block-fading channel as $\zeta \to \infty$ is given by [10.6]

$$C_{\text{zero-outage}} \sim m \log \left(\frac{\zeta}{m\mathbb{E}\left[1/\bar{\lambda}_F\right]} \right) \qquad (10.101)$$

As $m \to \infty$ and $n \to \alpha > 0$, the limiting value of the normalized zero-outage capacity per degree of freedom C/m coincides with the limiting normalized ergodic capacity [10.6].

10.7 Influence of channel-state information

As we have seen, in a system with t transmit and r receive antennas and an ergodic Rayleigh fading channel modeled by a $t \times r$ matrix with random iid complex Gaussian entries, the average channel capacity with perfect CSI at the receiver is about $m \triangleq \min\{t, r\}$ times larger than that of a single-antenna system for the same transmitted power and bandwidth. The capacity increases by about m bit/s/Hz for every 3-dB increase in SNR. Due to the assumption of perfect CSI available at the receiver, this result can be viewed as a fundamental limit for coherent multiple-antenna systems.

Perfect CSI at the receiver

The most commonly studied situation is that of perfect CSI available at the receiver, which is the assumption under which we developed our study of multiple-antenna systems above.

No channel state information

Fundamental limits of noncoherent communication, i.e., one taking place in an environment where estimates of the fading coefficients are not available, will now be derived. Consider a block-fading channel model. To compute the capacity of this channel, we assume that coding is performed using blocks, each of them consisting of tN elementary symbols being transmitted by t antennas in N time instants. Each block is represented by the $t \times N$ matrix $\mathbf{X}$. We further assume that the $r \times N$ noise matrix $\mathbf{Z}$ has iid $\mathcal{N}_c(0, N_0)$ entries. The received signal is the $r \times N$ matrix

$$\mathbf{Y} = \mathbf{HX} + \mathbf{Z} \tag{10.102}$$

and the entries of $\mathbf{Y}$ have the explicit expression

$$y_{in} = \sum_{j=1}^{t} h_{ij} x_{jn} + z_{in}, \qquad i = 1, \ldots, r, \quad n = 1, \ldots, N \tag{10.103}$$

Given $\mathbf{X}$, these are random variables whose mean value is zero and whose covariance is

$$\mathbb{E}[y_{in} y_{i'n'}^* \mid \mathbf{X}] = \sum_{j=1}^{t} \sum_{j'=1}^{t} \mathbb{E}[h_{ij} h_{i'j'}^*] x_{jn} x_{j'n'}^* + \mathbb{E}[z_{in} z_{i'n'}^*] \tag{10.104}$$

Now, under the assumptions that $\mathbf{H}$ and $\mathbf{Z}$ are temporally and spatially white, that is,

$$\mathbb{E}[h_{ij} h_{i'j'}^*] = \delta_{ii'} \qquad \mathbb{E}[z_{in} z_{i'n'}^*] = \delta_{ii'} \delta_{nn'} \tag{10.105}$$

we have

$$\mathbb{E}[y_{in} y_{i'n'}^* \mid \mathbf{X}] = \delta_{ii'} \left[\sum_{j=1}^{t} x_{jn} x_{jn'}^* + \delta_{nn'} \right] \tag{10.106}$$

The previous equality expresses the fact that the rows of $\mathbf{Y}$ are independent, while the columns have a nonzero correlation. This observation allows us to write down the relation connecting the rows $(\mathbf{Y})_i$ of $\mathbf{Y}$ with those of $\mathbf{H}$, denoted $(\mathbf{H})_i$, and those of $\mathbf{Z}$, denoted $(\mathbf{Z})_i$, so that

$$(\mathbf{Y})_i = (\mathbf{H})_i \mathbf{X} + (\mathbf{Z})_i \qquad i = 1, \ldots, r \tag{10.107}$$

Each row of $\mathbf{Y}$ is a zero-mean Gaussian vector with covariance matrix

$$\mathbb{E}[(\mathbf{Y})_i^\dagger (\mathbf{Y})_i \mid \mathbf{X}] = \mathbf{X}^\dagger \mathbf{X} + \mathbf{I}_N \tag{10.108}$$

and, writing the pdf of matrix $\mathbf{Y}$ as the product of the pdfs of its rows, we obtain

$$p(\mathbf{Y} \mid \mathbf{X}) = \prod_{i=1}^{r} p(\mathbf{Y}_i \mid \mathbf{X})$$

$$= \frac{1}{\pi^r \det{}^r[\mathbf{X}^\dagger\mathbf{X} + \mathbf{I}_N]} \prod_{i=1}^{r} \exp\{-(\mathbf{Y})_i(\mathbf{X}^\dagger\mathbf{X} + \mathbf{I}_N)^{-1}(\mathbf{Y})_i^\dagger\}$$

$$= \frac{1}{\pi^r \det{}^r[\mathbf{X}^\dagger\mathbf{X} + \mathbf{I}_N]} \exp\{-\mathrm{Tr}\,((\mathbf{X}^\dagger\mathbf{X} + \mathbf{I}_N)^{-1}\mathbf{Y}^\dagger\mathbf{Y})\} \quad (10.109)$$

We observe the following:

(a) The pdf of $\mathbf{Y}$ depends on its argument only through the product $\mathbf{Y}^\dagger\mathbf{Y}$, which consequently plays the role of a sufficient statistic. If $N < r$, the $N \times N$ matrix $\mathbf{Y}^\dagger\mathbf{Y}$ provides a representation of the received signals that is more economical than the $r \times N$ matrix $\mathbf{Y}$.

(b) The pdf (10.109) depends on the transmitted signal $\mathbf{X}$ only through the $N \times N$ matrix $\mathbf{X}^\dagger\mathbf{X}$.

Observation (b) above is the basis of the following theorem, which says that there is no increase in capacity if we have $t > N$, and hence there is no point in making the number of transmit antennas greater than N if there is no CSI. In particular, if $N = 1$ (an independent fade occurs at each symbol period), only one transmit antenna is useful. Note how this result contrasts sharply with its counterpart of CSI known at the receiver, where the capacity grows linearly with $\min\{t, r\}$.

Theorem 10.7.1 *If the entries of $\mathbf{H}$ are iid, then the channel capacity for $t > N$ equals the capacity for $t = N$.*

Proof

Suppose that the capacity is achieved for a particular pdf of matrix $\mathbf{X}$ with $t > N$. Recalling (b) above, the capacity is determined by the matrix $\mathbf{X}^\dagger\mathbf{X}$: if we prove that an $\mathbf{X}$ can be found that generates the same matrix with only N transmit antennas, then the theorem is proved. Now, perform the Cholesky factorization (Section B.6.1, Appendix B) $\mathbf{X}^\dagger\mathbf{X} = \mathbf{L}\mathbf{L}^\dagger$, with $\mathbf{L}$ an $N \times N$ lower-triangular matrix. Using N transmit antennas with a signal matrix that has the same pdf as $\mathbf{L}^\dagger$, we obtain the same pdf that achieves capacity. In fact, if $\mathbf{X}$ satisfies that average-power constraint

$$\frac{1}{N}\mathbb{E}\,[\mathrm{Tr}\,\mathbf{X}^\dagger\mathbf{X}] = \zeta N_0 \quad (10.110)$$

so does $\mathbf{L}^\dagger$. □

From [10.37], the signal matrix that achieves capacity can be written in the form:

$$\mathbf{X} = \mathbf{D}\boldsymbol{\Phi} \tag{10.111}$$

where $\boldsymbol{\Phi}$ is a $t \times N$ matrix such that $\boldsymbol{\Phi}\boldsymbol{\Phi}^\dagger = \mathbf{I}_t$. Moreover, $\boldsymbol{\Phi}$ has a pdf that is unchanged when the matrix is multiplied by a deterministic unitary matrix (this is the matrix counterpart of a complex scalar having unit magnitude and uniformly distributed phase). $\mathbf{D}$ is a $t \times t$ real nonnegative diagonal matrix independent of $\boldsymbol{\Phi}$, whose role is to scale $\mathbf{X}$ to meet the power constraint. In general, the optimizing $\mathbf{D}$ is unknown, as is the exact expression of capacity. However, for the high-SNR regime ($\zeta \gg 1$), the following results are available (see [10.37, 10.66], but also [10.34] for the observation that they depend critically on the assumed fading model):

(a) If $N \gg t$ and $t \le \min\{N/2, r\}$, then capacity is attained when $\mathbf{D} = \sqrt{\zeta N N_0/t}\,\mathbf{I}_t$, so $\mathbf{X} = \sqrt{\zeta N N_0/t}\,\boldsymbol{\Phi}$.

(b) For every 3-dB increase of ζ, the capacity increase is $t^*(1 - t^*/N)$, where $t^* \triangleq \min\{t, r, \lfloor N/2 \rfloor\}$.

(c) If $N \ge 2r$, there is no capacity increase by using $r > t$.

An obvious upper bound to capacity can be obtained if we assume that the receiver is provided with perfect knowledge of the realization of $\mathbf{H}$. Hence, the bound to capacity per block of N symbols is

$$C \le N \log \det \left[\mathbf{I}_t + \frac{\zeta}{t}\mathbf{H}^\dagger\mathbf{H} \right] \tag{10.112}$$

We can reasonably expect that the actual capacity tends to the right-hand side of previous inequality, because a certain (small) fraction of the coherence time can be reserved for sending training data to be used by the receiver for its estimate of $\mathbf{H}$.

10.7.1 Imperfect CSI at the receiver: General guidelines

Assume now that the receiver has some knowledge, albeit imperfect, of the CSI. Let the CSI be obtained by transmitting a preamble in the form of a known $t \times N_p$ code matrix $\mathbf{X}_p$ with total energy $\mathrm{Tr}\left(\mathbf{X}_p\mathbf{X}_p^\dagger\right) = tN_p\mathcal{E}_p$, with $\mathcal{E}_p$ the average symbol energy. Since to estimate the $r \times t$ matrix $\mathbf{H}$ we need at least rt measurements, and each symbol time yields r measurements at the receiver, we need $N_p \ge t$. Moreover, the matrix $\mathbf{X}_p$ must have full rank t, since otherwise t linearly independent columns would not be available to yield rt independent measurements. As a consequence, $\mathbf{X}_p\mathbf{X}_p^\dagger$ must be nonsingular. The corresponding received signal is denoted by

$$\mathbf{Y}_p = \mathbf{H}\mathbf{X}_p + \mathbf{Z}_p \qquad (10.113)$$

Among the several receiver structures that can be envisaged, we focus on the following:

(a) The simplest receiver inserts directly the maximum-likelihood (ML) estimate of the channel into the ML metric conditioned on $\mathbf{H}$. The detection problem consists of computing first

$$\widehat{\mathbf{H}} \triangleq \arg\max_{\mathbf{H}} p(\mathbf{Y}_p \mid \mathbf{X}_p, \mathbf{H}) \qquad (10.114)$$

and then

$$\widehat{\mathbf{X}} \triangleq \arg\max_{\mathbf{X}} \tilde{\mu}(\mathbf{X}) \qquad (10.115)$$

where

$$\tilde{\mu}(\mathbf{X}) \triangleq \|\mathbf{Y} - \widehat{\mathbf{H}}\mathbf{X}\|^2 \qquad (10.116)$$

Since (10.116) is commonly referred to as a *mismatched* metric, we call this a *mismatched receiver*.

(b) The receiver estimates the channel matrix $\widehat{\mathbf{H}}$ from $\mathbf{Y}_p$ and $\mathbf{X}_p$ by an ML criterion and uses this result to detect the transmitted signal $\mathbf{X}$. The detection problem consists of computing

$$\widehat{\mathbf{H}} \triangleq \arg\max_{\mathbf{H}} p(\mathbf{Y}_p \mid \mathbf{X}_p, \mathbf{H}) \qquad (10.117)$$

and

$$\widehat{\mathbf{X}} \triangleq \arg\max_{\mathbf{X}} p(\mathbf{Y} \mid \mathbf{X}, \mathbf{H} = \widehat{\mathbf{H}}) \qquad (10.118)$$

where $p(\mathbf{Y} \mid \mathbf{X}, \mathbf{H} = \widehat{\mathbf{H}})$ denotes the probability density function of $\mathbf{Y}$ given $\mathbf{X}$ and $\mathbf{H}$, with $\mathbf{H}$ equal to $\widehat{\mathbf{H}}$.

(c) The receiver detects the transmitted signal $\mathbf{X}$ by jointly processing $\mathbf{Y}$, $\mathbf{Y}_p$, and $\mathbf{X}_p$ without explicit estimation of $\mathbf{H}$. In this case, the detection problem can be written as

$$\begin{aligned} \widehat{\mathbf{X}} &\triangleq \arg\max_{\mathbf{X}} p(\mathbf{Y}, \mathbf{Y}_p \mid \mathbf{X}, \mathbf{X}_p) \\ &= \mathbb{E}_{\mathbf{H}}\Big[p(\mathbf{Y} \mid \mathbf{X}, \mathbf{H})\, p(\mathbf{Y}_p \mid \mathbf{X}_p, \mathbf{H})\Big] \qquad (10.119) \end{aligned}$$

since, conditionally on $\mathbf{H}$, $\mathbf{X}$, and $\mathbf{X}_p$, the received signals $\mathbf{Y}$ and $\mathbf{Y}_p$ are independent.

Approach (a) is the simplest. Approach (b) is more efficient (see [10.14] for the single-input, single-output case) and allows one to study the impairments caused by imperfect knowledge of $\mathbf{H}$ and by the presence of noise in the received pilot signal $\mathbf{Y}_p$. Approach (c) is optimum: disregarding CSI recovery, it focuses on the detection of the transmitted signal $\mathbf{X}$. In the following, we examine the second and third receivers under the simplifying assumption $\mathbf{X}_p\mathbf{X}_p^\dagger = N_p\mathcal{E}_p\mathbf{I}_t$ (for results not depending on this assumption, see [10.55]).

Approach (b): Receiver based on channel estimate

The ML estimate of $\mathbf{H}$ based on the observation of $\mathbf{Y}_p$ is obtained by maximizing $p(\mathbf{Y}_p \mid \mathbf{H}, \mathbf{X}_p)$ or, equivalently, by minimizing $\|\mathbf{Y}_p - \mathbf{H}\mathbf{X}_p\|$ with respect to $\mathbf{H}$, yielding

$$\widehat{\mathbf{H}} = \mathbf{Y}_p\mathbf{X}_p^\dagger(\mathbf{X}_p\mathbf{X}_p^\dagger)^{-1} = \mathbf{H} + \mathbf{E} \tag{10.120}$$

where

$$\mathbf{E} \triangleq \mathbf{Z}_p\mathbf{X}_p^\dagger(\mathbf{X}_p\mathbf{X}_p^\dagger)^{-1} \tag{10.121}$$

is the matrix error on the estimate $\widehat{\mathbf{H}}$. Now, $\mathbf{H}$ and $\mathbf{E}$ are independent, and, denoting by $(\cdot)_i$ the ith row of a matrix $(\cdot)$, we can write

$$\mathbf{E}_i = (\mathbf{Z}_p)_i\mathbf{X}_p^\dagger(\mathbf{X}_p\mathbf{X}_p^\dagger)^{-1} \tag{10.122}$$

Thus, the rows of $\mathbf{E}$ are independent vectors of zero-mean circularly symmetric complex Gaussian random variables with covariance matrix

$$\begin{aligned}
\boldsymbol{\Sigma}_e &\triangleq \mathbb{E}[\mathbf{E}_i^\dagger\mathbf{E}_i] \\
&= \mathbb{E}[(\mathbf{X}_p\mathbf{X}_p^\dagger)^{-1}\mathbf{X}_p(\mathbf{Z}_p)_i^\dagger(\mathbf{Z}_p)_i\mathbf{X}_p^\dagger(\mathbf{X}_p\mathbf{X}_p^\dagger)^{-1}] \\
&= N_0(\mathbf{X}_p\mathbf{X}_p^\dagger)^{-1} \tag{10.123}
\end{aligned}$$

With our assumption on $\mathbf{X}_p$, the entries of $\mathbf{E}$ are independent, circularly symmetric complex Gaussian random variables with mean zero and variance $N_0/(N_p\mathcal{E}_p)$.

We now calculate the ML metric from the a posteriori probability $p(\mathbf{Y} \mid \mathbf{X}, \widehat{\mathbf{H}})$. First, we note that it can be written as

$$p(\mathbf{Y} \mid \mathbf{X}, \widehat{\mathbf{H}}) = \prod_{i=1}^r p(\mathbf{Y}_i \mid \mathbf{X}, \widehat{\mathbf{H}}_i) \tag{10.124}$$

where $\widehat{\mathbf{H}}_i$ and $\mathbf{Y}_i$ denote the ith rows of $\widehat{\mathbf{H}}$ and $\mathbf{Y}$, respectively, since it is plain to see that, conditionally on $\mathbf{X}$, $\mathbf{Y}_i$ depends only on $\mathbf{H}_i$ and $\mathbf{Z}_i$. Thus, we can apply the following theorem [10.11]:

Theorem 10.7.2 *Let $\mathbf{z}_1$ and $\mathbf{z}_2$ be circularly symmetric complex Gaussian random vectors with zero means and full-rank covariance matrices $\boldsymbol{\Sigma}_{ij} \triangleq \mathbb{E}[\mathbf{z}_i \mathbf{z}_j^\dagger]$. Then, conditionally on $\mathbf{z}_2$, the random vector $\mathbf{z}_1$ is circularly symmetric complex Gaussian with mean $\boldsymbol{\Sigma}_{12}\boldsymbol{\Sigma}_{22}^{-1}\mathbf{z}_2$ and covariance matrix $\boldsymbol{\Sigma}_{11} - \boldsymbol{\Sigma}_{12}\boldsymbol{\Sigma}_{22}^{-1}\boldsymbol{\Sigma}_{21}$.*

Letting

$$\mathbf{z}_1 = \mathbf{Y}_i^\dagger = \mathbf{X}^\dagger \mathbf{H}_i^\dagger + \mathbf{Z}_i^\dagger \qquad \text{and} \qquad \mathbf{z}_2 = \widehat{\mathbf{H}}_i^\dagger = \mathbf{H}_i^\dagger + \mathbf{E}_i^\dagger \qquad (10.125)$$

in Theorem 10.7.2, we have

$$
\begin{aligned}
\boldsymbol{\Sigma}_{11} &= N_0 \mathbf{I}_N + \mathbf{X}^\dagger \mathbf{X} \\
\boldsymbol{\Sigma}_{12} &= \mathbf{X}^\dagger \\
\boldsymbol{\Sigma}_{22} &= \mathbf{I}_t + N_0/(N_p \mathcal{E}_p)
\end{aligned}
$$

Then the conditional probability density function of $\mathbf{Y}_i^\dagger$, given $\mathbf{X}$ and $\widehat{\mathbf{H}}_i$, is a circularly symmetric complex Gaussian distribution, with

$$
\begin{aligned}
\text{mean} &= \mu \mathbf{X}^\dagger \widehat{\mathbf{H}}_i^\dagger & (10.126) \\
\text{covariance matrix} &= N_0 \mathbf{I}_N + (1-\xi)\mathbf{X}^\dagger \mathbf{X} & (10.127)
\end{aligned}
$$

where

$$\xi \triangleq \frac{1}{1 + N_0/(N_p \mathcal{E}_p)} \qquad (10.128)$$

As a result, we have

$$p(\mathbf{Y} \mid \mathbf{X}, \widehat{\mathbf{H}}) = \frac{\operatorname{etr}\left(-(\mathbf{Y} - \xi\widehat{\mathbf{H}}\mathbf{X})(N_0\mathbf{I}_N + (1-\xi)\mathbf{X}^\dagger\mathbf{X})^{-1}(\mathbf{Y} - \mu\widehat{\mathbf{H}}\mathbf{X})^\dagger\right)}{\det\left(\pi(N_0\mathbf{I}_N + (1-\xi)\mathbf{X}^\dagger\mathbf{X})\right)^r}$$

$$(10.129)$$

corresponding to the metric

$$
\begin{aligned}
\mu(\mathbf{X}) &= \operatorname{Tr}\left((\mathbf{Y} - \xi\widehat{\mathbf{H}}\mathbf{X})(\mathbf{I}_N + (1-\xi)\mathbf{X}^\dagger\mathbf{X}/N_0)^{-1}(\mathbf{Y} - \xi\widehat{\mathbf{H}}\mathbf{X})^\dagger\right) \\
&\quad + rN_0 \ln\det\left(\mathbf{I}_N + (1-\xi)\mathbf{X}^\dagger\mathbf{X}/N_0\right) & (10.130)
\end{aligned}
$$

Approach (c): Optimum receiver

In this case the receiver detects the transmitted word $\mathbf{X}$ maximizing the probability density function $p(\mathbf{Y}, \mathbf{Y}_p \mid \mathbf{X}, \mathbf{X}_p)$ without any prior estimate of the channel matrix $\mathbf{H}$. We use the following theorem [10.44, Appendix B]:

Theorem 10.7.3 *Given a Hermitian square matrix* $\mathbf{A}$ *such that* $\mathbf{I} + \mathbf{A} > 0$, *a size-compatible complex matrix* $\mathbf{B}$, *and a matrix* $\mathbf{Z}$ *of iid zero-mean circularly symmetric complex Gaussian random variables with unit variance, the following identity holds:*

$$\mathbb{E}[\text{etr}\,(-\mathbf{Z}\mathbf{A}\mathbf{Z}^\dagger - \mathbf{Z}\mathbf{B}^\dagger - \mathbf{B}\mathbf{Z}^\dagger)] = \det\,(\mathbf{I} + \mathbf{A})^{-r}\text{etr}\,[\mathbf{B}(\mathbf{I} + \mathbf{A})^{-1}\mathbf{B}^\dagger] \quad (10.131)$$

where $\text{etr}\,(\,\cdot\,) \triangleq \exp[\text{Tr}\,(\,\cdot\,)]$.

Applying Theorem 10.7.3, we obtain

$$
\begin{aligned}
&p(\mathbf{Y}, \mathbf{Y}_p \mid \mathbf{X}, \mathbf{X}_p) \\
&= \mathbb{E}_{\mathbf{H}}\left[\frac{\exp(-(\|\mathbf{Y} - \mathbf{H}\mathbf{X}\|^2 + \|\mathbf{Y}_p - \mathbf{H}\mathbf{X}_p\|^2)/N_0)}{(\pi N_0)^{(N_p+N)r}}\right] \\
&= (\pi N_0)^{-(N_p+N)r} \\
&\quad \mathbb{E}_{\mathbf{H}}\Big[\text{etr}\,(-(\mathbf{H}^\dagger\mathbf{H}(\mathbf{X}\mathbf{X}^\dagger + N_p\mathcal{E}_p\mathbf{I}_t) - \mathbf{H}(\mathbf{X}\mathbf{Y}^\dagger + \mathbf{X}_p\mathbf{Y}_p^\dagger) \\
&\quad -(\mathbf{Y}\mathbf{X}^\dagger + \mathbf{Y}_p\mathbf{X}_p^\dagger)\mathbf{H}^\dagger + (\mathbf{Y}\mathbf{Y}^\dagger + \mathbf{Y}_p\mathbf{Y}_p^\dagger))/N_0)\Big] \\
&= (\pi N_0)^{-(N_p+N)r}\det\left[\mathbf{I}_t + (\mathbf{X}\mathbf{X}^\dagger + N_p\mathcal{E}_p\mathbf{I}_t)/N_0\right]^{-r} \\
&\quad \text{etr}\,\Big((\mathbf{Y}\mathbf{X}^\dagger + \mathbf{Y}_p\mathbf{X}_p^\dagger)[\mathbf{I}_t + (\mathbf{X}\mathbf{X}^\dagger + N_p\mathcal{E}_p\mathbf{I}_t)/N_0]^{-1} \\
&\quad (\mathbf{X}\mathbf{Y}^\dagger + \mathbf{X}_p\mathbf{Y}_p^\dagger)/N_0^2 - (\mathbf{Y}\mathbf{Y}^\dagger + \mathbf{Y}_p\mathbf{Y}_p^\dagger)/N_0\Big) \quad (10.132)
\end{aligned}
$$

The logarithm of (10.132) yields the corresponding metric to be minimized by the optimum receiver:

$$
\begin{aligned}
\mu(\mathbf{X}) &= r\ln\det\left[\mathbf{I}_t + (\mathbf{X}\mathbf{X}^\dagger + N_p\mathcal{E}_p\mathbf{I}_t)/N_0\right] \quad (10.133) \\
&\quad -\text{Tr}\Big\{(\mathbf{Y}\mathbf{X}^\dagger + \mathbf{Y}_p\mathbf{X}_p^\dagger)[\mathbf{I}_t + (\mathbf{X}\mathbf{X}^\dagger + N_p\mathcal{E}_p\mathbf{I}_t)/N_0]^{-1} \\
&\quad (\mathbf{X}\mathbf{Y}^\dagger + \mathbf{X}_p\mathbf{Y}_p^\dagger)/N_0^2\Big\}
\end{aligned}
$$

From this result we can verify the fact, a priori rather surprising, that the metrics (10.133) and (10.130) are equivalent (see [10.55] for details).

Example 10.8

Figure 10.18 shows the word-error probability versus the fraction of pilot symbols $N_p/(N_p + N)$ at fixed $\mathcal{E}_b/N_0 = 10$ dB. It refers to a $t = 2, r = 4$ MIMO system

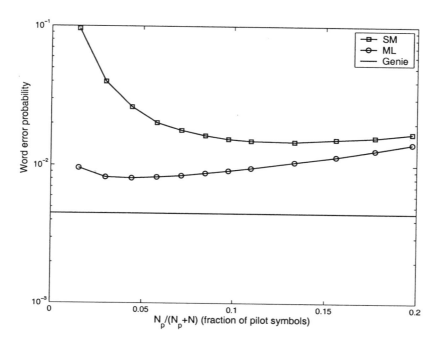

Figure 10.18: *Word error probability of a $t = 2, r = 4$, independent Rayleigh fading MIMO channel with a trellis space–time code versus the fraction of pilot symbols $N_p/(N_p + N)$ at $\mathcal{E}_b/N_0 = 4$ dB. Solid curves with □ show the performance with the suboptimum metric (10.116). Solid curves with ○ show the performance with the ML metric (10.130). The lowest straight line shows the performance of a genie-aided receiver with perfect CSI.*

with a trellis space–time code (see *infra*) and word length $N = 130$. The ML receiver performance is close to that of a "genie-aided" receiver having perfect CSI, and the optimum number of pilot symbols is about 4 for the ML receiver and 16 for the mismatched receiver. □

10.7.2   CSI at transmitter and receiver

It is also possible to envisage a situation in which channel state information is known to the receiver and to the transmitter: the latter can take the appropriate measures to counteract the effect of channel attenuations by suitably modulating its power. To assure causality, the assumption of CSI available at the transmitter is valid if it is applied to a multicarrier transmission scheme in which the available frequency band (over which the fading is selective) is split into a number of subbands, as with OFDM. The subbands are so narrow that fading is frequency flat in each of them, and they are transmitted simultaneously, via orthogonal subcarriers. From a practical point of view, the transmitter can obtain the CSI either from a dedicated feedback channel (some existing systems already implement a fast power-control feedback channel) or by time-division duplex, where the uplink and the downlink time-share the same subchannels and the fading gains can be estimated from the incoming signal.

10.8   Coding for multiple-antenna systems

Given that considerable gains are achievable by a multiantenna system, the challenge is to design coding schemes that perform close to capacity: space–time trellis codes, space–time block codes, and layered space–time codes have been advocated.

A space–time code word with block length N is described by the $t \times N$ matrix $\mathbf{X} \triangleq (\mathbf{x}[1], \ldots, \mathbf{x}[N])$. The code has M words. The row index of $\mathbf{X}$ indicates space, while the column index indicates time: to wit, the ith component of the t-vector $\mathbf{x}[n]$, denoted $x_i[n]$, is a complex number representing the two-dimensional signal transmitted by the ith antenna at discrete time n, $n = 1, \ldots, N$, $i = 1, \ldots, t$. The received signal is the $r \times N$ matrix

$$\mathbf{Y} = \mathbf{HX} + \mathbf{Z} \qquad (10.134)$$

where $\mathbf{Z}$ is matrix of zero-mean circularly symmetric complex Gaussian RVs with variance N_0. Thus, the noise affecting the received signal is spatially and temporally independent, with $\mathbb{E}[\mathbf{ZZ}^\dagger] = NN_0\mathbf{I}_r$, where $\mathbf{I}_r$ denotes the $r \times r$ identity matrix and $(\cdot)^\dagger$ denotes Hermitian transposition. The channel is described by the $r \times t$ matrix $\mathbf{H}$. Here we assume that $\mathbf{H}$ is independent of both $\mathbf{X}$ and $\mathbf{Z}$, it remains constant during the transmission of an entire code word, and its realization (the CSI) is known at the receiver.

10.9   Maximum-likelihood detection

Under the assumptions of known CSI and additive white Gaussian noise, ML decoding corresponds to choosing the code word $\mathbf{X}$ that minimizes the squared Frobenius norm $\|\mathbf{Y} - \mathbf{HX}\|^2$. Explicitly, ML detection and decoding corresponds to the minimization of the quantity

$$\|\mathbf{Y} - \mathbf{H}\mathbf{X}\|^2 = \sum_{i=1}^{r}\sum_{n=1}^{N}\left|y_{in} - \sum_{j=1}^{t} h_{ij}x_{jn}\right|^2 \qquad (10.135)$$

10.9.1 Pairwise error probability

For computations, since calculation of exact error probability is out of the question, we resort to the union bound

$$P(e) \le \frac{1}{M}\sum_{\mathbf{X}}\sum_{\widehat{\mathbf{X}}\neq\mathbf{X}} P(\mathbf{X}\to\widehat{\mathbf{X}}) \qquad (10.136)$$

The pairwise error probability (PEP) $P(\mathbf{X}\to\widehat{\mathbf{X}})$ admits a closed-form expression:

$$
\begin{aligned}
P(\mathbf{X}\to\widehat{\mathbf{X}}) &\triangleq P(\|\mathbf{Y}-\mathbf{H}\widehat{\mathbf{X}}\|^2 < \|\mathbf{Y}-\mathbf{H}\mathbf{X}\|^2)\\
&= P(\|\mathbf{H}\mathbf{\Delta}+\mathbf{Z}\|^2 < \|\mathbf{Z}\|^2)\\
&= P((\mathbf{H}\mathbf{\Delta}+\mathbf{Z},\mathbf{H}\mathbf{\Delta}+\mathbf{Z})-(\mathbf{Z},\mathbf{Z})) < 0)\\
&= P(\|\mathbf{H}\mathbf{\Delta}\|^2 + 2(\mathbf{H}\mathbf{\Delta},\mathbf{Z})) < 0) \qquad (10.137)
\end{aligned}
$$

where $\mathbf{\Delta}\triangleq\mathbf{X}-\widehat{\mathbf{X}}$. The variance of the Gaussian random variable $\nu\triangleq(\mathbf{A},\mathbf{Z})$ can be obtained as follows. Setting $\mathbf{A}=\mathbf{A}_1+j\mathbf{A}_2$ and $\mathbf{Z}=\mathbf{Z}_1+j\mathbf{Z}_2$ (where $\mathbf{A}_1,\mathbf{A}_2,\mathbf{Z}_1,$ and $\mathbf{Z}_2$ are real matrices), we have

$$
\begin{aligned}
\mathbb{E}[\nu^2] &= \mathbb{E}[(\mathrm{Tr}\,(\mathbf{A}_1\mathbf{Z}_1 - \mathbf{A}_2\mathbf{Z}_2))^2]\\
&= \mathbb{E}\left[\left(\sum_i\sum_j(\mathbf{A}_1)_{ij}(\mathbf{Z}_1)_{ji} - (\mathbf{A}_2)_{ij}(\mathbf{Z}_2)_{ji}\right)^2\right]\\
&= \sum_i\sum_j((\mathbf{A}_1)_{ij}^2\mathbb{E}[(\mathbf{Z}_1)_{ji}^2] + (\mathbf{A}_2)_{ij}^2\mathbb{E}[(\mathbf{Z}_2)_{ji}^2])\\
&= \frac{N_0}{2}\|\mathbf{A}\|^2 \qquad (10.138)
\end{aligned}
$$

since $\mathbf{Z}_1$ and $\mathbf{Z}_2$ are independent and have zero mean. Then the pairwise error probability becomes

$$P(\mathbf{X}\to\widehat{\mathbf{X}}) = \mathbb{E}\left[Q\left(\frac{\|\mathbf{H}\mathbf{\Delta}\|}{\sqrt{2N_0}}\right)\right] \qquad (10.139)$$

By writing

$$\|\mathbf{H}\mathbf{\Delta}\|^2 = \mathrm{Tr}\,\left(\mathbf{H}^\dagger\mathbf{H}\mathbf{\Delta}\mathbf{\Delta}^\dagger\right) \qquad (10.140)$$

we see that the exact pairwise error probability, and hence the union bound to $P(e)$, is given by the expected value of a function of the $t \times r$ matrix $\mathbf{H}^\dagger \mathbf{H}$. This matrix can be interpreted as representing the effect of the random spatial interference on error probability: in particular, if $\mathbf{H}^\dagger \mathbf{H} = \mathbf{I}_t$, then (10.139) becomes

$$P(\mathbf{X} \to \widehat{\mathbf{X}}) = Q\left(\frac{\|\boldsymbol{\Delta}\|}{\sqrt{2N_0}}\right) \tag{10.141}$$

This is the PEP we would obtain on a set of t parallel independent AWGN channels, each transmitting a code word consisting of a row of $\mathbf{X}$, with ML detection consisting of minimizing $\|\mathbf{Y} - \mathbf{X}\|^2$.

A useful approximation to the pairwise error probability (10.139) can be computed by substituting exponential functions for Q functions. This is obtained by applying the bound, asymptotically tight for large arguments:

$$Q\left(\frac{\|\mathbf{H}\boldsymbol{\Delta}\|}{\sqrt{2N_0}}\right) \le \exp\left(-\|\mathbf{H}\boldsymbol{\Delta}\|^2/4N_0\right) \tag{10.142}$$

Under the assumption of Rayleigh fading, that is, when $h_{ij} \sim \mathcal{N}_c(0, 1)$, with independent entries in the matrix $\mathbf{H}$, we can compute the exact expectation of the right-hand side of (10.142) using Theorem C.3.1 of Appendix C. We obtain

$$P(\mathbf{X} \to \widehat{\mathbf{X}}) \le \det\left[\mathbf{I}_t + \boldsymbol{\Delta}\boldsymbol{\Delta}^\dagger/4N_0\right]^{-r} \tag{10.143}$$

10.9.2   The rank-and-determinant criterion

Since the determinant of a matrix is equal to the product of its eigenvalues, (10.143) yields

$$P(\mathbf{X} \to \widehat{\mathbf{X}}) \le \prod_{j=1}^{t} (1 + \lambda_j/4N_0)^{-r} \tag{10.144}$$

where λ_j denotes the jth eigenvalue of $\boldsymbol{\Delta}\boldsymbol{\Delta}^\dagger$. We can also write

$$P(\mathbf{X} \to \widehat{\mathbf{X}}) \le \prod_{j \in \mathcal{J}} (\lambda_j/4N_0)^{-r} \tag{10.145}$$

where $\mathcal{J}$ is the index set of the nonzero eigenvalues of $\boldsymbol{\Delta}\boldsymbol{\Delta}^\dagger$. Denoting by ν the number of elements in $\mathcal{J}$, and rearranging the indexes so that $\lambda_1, \dots, \lambda_\nu$ are the nonzero eigenvalues, we have

$$P(\mathbf{X} \to \widehat{\mathbf{X}}) \le \left(\prod_{j=1}^{\nu} \lambda_j\right)^{-r} \gamma^{-r\nu} \tag{10.146}$$

where $\gamma \triangleq 1/4N_0$. From this expression we see that the total diversity order of the coded system is $r\nu_{\min}$, where $\nu_{\min}$ is the minimum rank of $\boldsymbol{\Delta}\boldsymbol{\Delta}^\dagger$ across all possible pairs $\mathbf{X}$, $\widehat{\mathbf{X}}$ ($r\nu_{\min}$ is the *diversity gain*). In addition, the pairwise error probability depends on the power r of the product of eigenvalues of $\boldsymbol{\Delta}\boldsymbol{\Delta}^\dagger$. This does not depend on the SNR (which is proportional to γ), and displaces the error probability curve instead of changing its slope. We call this the *coding gain*. Thus, for high enough SNR we can design a space–time code for which we choose as a criterion the maximization of the coding gain as well as of the diversity gain.

Notice that if $\nu_{\min} = t$, i.e., $\boldsymbol{\Delta}\boldsymbol{\Delta}^\dagger$ is full rank for all code word pairs, we have

$$\prod_{j=1}^{t} \lambda_j = \det\left[\boldsymbol{\Delta}\boldsymbol{\Delta}^\dagger\right] \qquad (10.147)$$

An obvious necessary condition for $\boldsymbol{\Delta}\boldsymbol{\Delta}^\dagger$ to be full rank is that $N \geq t$ (the code block length must be at least equal to the number of transmit antennas).

Observation 10.9.1 Note that, based on the above discussion, the maximum achievable diversity gain is tr. In Section 10.14 we shall discuss how this gain is generally not compatible with the maximum rate gain m.

10.9.3 The Euclidean-distance criterion

Observe that the term in the right-hand side of (10.143) can be written as a negative power of

$$\det\left(\mathbf{I}_t + \gamma\boldsymbol{\Delta}\boldsymbol{\Delta}^\dagger\right) = 1 + \gamma\mathrm{Tr}\left(\boldsymbol{\Delta}\boldsymbol{\Delta}^\dagger\right) + \ldots + \gamma^t\det\left(\boldsymbol{\Delta}\boldsymbol{\Delta}^\dagger\right) \qquad (10.148)$$

We see that if $\gamma \ll 1$ then the left-hand side of (10.148), and hence the PEP, depends essentially on $\mathrm{Tr}\left(\boldsymbol{\Delta}\boldsymbol{\Delta}^\dagger\right)$, which is the squared Euclidean distance between $\mathbf{X}$ and $\widehat{\mathbf{X}}$, while if $\gamma \gg 1$ it depends essentially on $\det\left(\boldsymbol{\Delta}\boldsymbol{\Delta}^\dagger\right)$, that is, on the product of the eigenvalues of $\boldsymbol{\Delta}\boldsymbol{\Delta}^\dagger$. This suggests that, for low SNR, the upper bound (10.148) to error probability depends on the Euclidean distance between code words, as one would expect because the system performance is dictated by additive noise rather than by fading. Conversely, as the SNR increases, the fading effects become more and more relevant, and the rank and determinant of $\boldsymbol{\Delta}\boldsymbol{\Delta}^\dagger$ dictate the behavior of the PEP.

A different perspective can be obtained by allowing the number r of receive antennas to grow to infinity. To do this, we first renormalize the entries of $\mathbf{H}$ so that their variance is now $1/r$ rather than 1: this prevents the total receive power from diverging as $r \to \infty$. We obtain the following new form of (10.143):

$$P(\mathbf{X} \to \widehat{\mathbf{X}}) \leq \det \left[\mathbf{I}_t + \boldsymbol{\Delta}\boldsymbol{\Delta}^\dagger/4rN_0\right]^{-r} \tag{10.149}$$

which yields, in lieu of (10.148):

$$\det\left(\mathbf{I}_t + (\gamma/r)\boldsymbol{\Delta}\boldsymbol{\Delta}^\dagger\right) = 1 + (\gamma/r)\mathrm{Tr}\left(\boldsymbol{\Delta}\boldsymbol{\Delta}^\dagger\right) + \ldots + (\gamma/r)^t\det\left(\boldsymbol{\Delta}\boldsymbol{\Delta}^\dagger\right) \tag{10.150}$$

This shows that as $r \to \infty$ the rank-and-determinant criterion is appropriate for a SNR increasing as fast as r, while the Euclidean-distance criterion is appropriate for finite SNRs.[4] This situation is illustrated in the example of Figure 10.19, which shows the union upper bound on the word-error probability $P(e)$ of the space–time code obtained by splitting evenly the code words of the $(24, 8, 12)$ extended Golay binary code between two transmit antennas (the calculations are based on the techniques described in Appendix D). This space–time code has the minimum rank of $\boldsymbol{\Delta}$ equal to 1, and hence a diversity gain r. Now, it is seen from Figure 10.19 how the slope predicted by (10.143), and exhibited by a linear behavior in the $P(e)$-vs.-$\mathcal{E}_b/N_0$ chart, can be reached only for very small values of error probability (how small generally depends on the code under scrutiny). To justify this behavior, observe from Figure 10.19 that for a given value of r the error-probability curve changes its behavior from a waterfall shape (for small to intermediate SNR) to a linear shape (high SNR). As the number of receive antennas grows, this change of slope occurs for values of $P(e)$ that are smaller and smaller as r increases. Thus, to study the error-probability curve in its waterfall region, it makes sense to examine its asymptotic behavior as $r \to \infty$. The case $r \to \infty$, $t < \infty$ can easily be dealt with by using the strong law of large numbers: this yields $\mathbf{H}^\dagger\mathbf{H} \to \mathbf{I}_t$ a.s., $\mathbf{I}_t$ the $t \times t$ identity matrix. As $r \to \infty$,

$$\|\mathbf{H}\boldsymbol{\Delta}\|^2 \to \|\boldsymbol{\Delta}\|^2 \tag{10.151}$$

and hence

$$P(\mathbf{X} \to \widehat{\mathbf{X}}) \to Q\left(\frac{\|\boldsymbol{\Delta}\|}{\sqrt{2N_0}}\right) \tag{10.152}$$

This result shows that, as the number of receiving antennas grows large, the union bound on the error probability of the space–time code depends only on the Euclidean distances between pairs of code words. This is the result one would get with

---

[4]Other design criteria can also be advocated. See, e.g., [10.27].

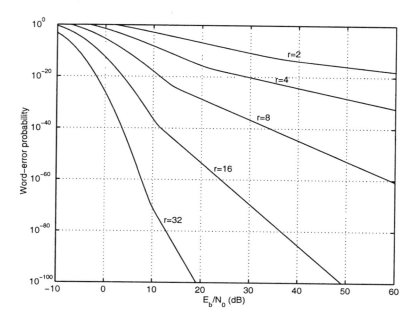

Figure 10.19: *Word-error probability of the binary* $(24, 8, 12)$ *extended Golay code with binary PSK over a channel with* $t = 2$ *transmit antennas and* r *receive antennas with ML decoding.*

a transmission occurring over a nonfading additive white Gaussian noise (AWGN) channel whose transfer matrix $\mathbf{H}$ has orthogonal columns, i.e., is such that $\mathbf{H}^\dagger \mathbf{H}$ is a scalar matrix. In this situation the smallest error probability, at the expense of a larger complexity, can be achieved by using a single code, optimized for the AWGN channel, whose words of length tN are evenly split among the transmit antennas. Within this framework, the number of transmit antennas does not affect the PEP but only the transmission rate, which, expressed in bits per channel use, increases linearly with t.

For another example, observe Figure 10.20. This shows how for intermediate SNRs the Euclidean-distance criterion may yield codes better than the rank-and-determinant criterion. It compares the simulated performances, in terms of frame-error rate, of the four-state, rate-1/2 space–time code of [10.56] and a comparable space–time code obtained by choosing a good binary, four-state, rate-2/4 convolutional code [10.15] and mapping its symbols onto QPSK (the first and second encoded bits are Gray mapped onto the QPSK symbol transmitted by the first antenna, while the third and fourth encoded bits are Gray mapped onto the QPSK

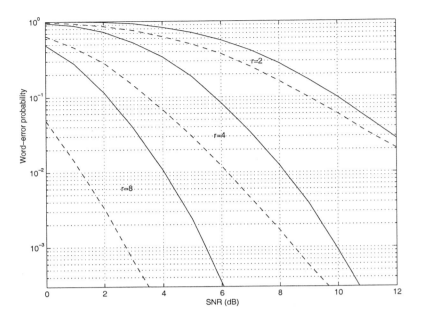

Figure 10.20: *Word-error probability of two space–time codes with four states, rate 1/2, and QPSK. Number of transmit antennas: $t = 2$; number of receive antennas: $r = 2, 4, 8$. Continuous line: code from [10.56]. Dashed line: code obtained from a binary convolutional code good for the AWGN channel [10.15].*

symbol transmitted by the second antenna). The frame length N is 130 symbols for both codes, including one symbol for trellis termination. The decoder has perfect CSI, and uses the Viterbi algorithm. It is seen that, in the error-probability range of these two figures, the "standard" convolutional code generally outperforms the space–time code of [10.56] even for small values of r.

10.10   Some practical coding schemes

10.10.1   Delay diversity

One of the first coding schemes proposed is called *delay diversity*. This is a rate-$1/t$ repetition code, each symbol of which is transmitted from a different antenna after being delayed. For example, with $t = 2$, the transmitted code matrix is

$$\mathbf{X} = \begin{bmatrix} x_1 & x_2 & x_3 & \cdots \\ 0 & x_1 & x_2 & \cdots \end{bmatrix}$$

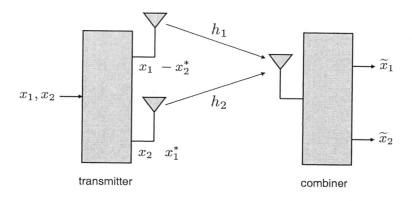

Figure 10.21: *Alamouti code with $t = 2$ and $r = 1$.*

We can see that each symbol traverses rt paths, so diversity rt is achieved. On the other hand, this comes at the cost of having a rate of only one symbol per channel use. Also, observe that delay diversity transforms the frequency-flat channel into an intersymbol-interference (and hence frequency-selective) channel. Optimum detection can be accomplished by using the Viterbi algorithm or spatial-interference-canceling techniques (see *infra*, our discussion of V-BLAST).

10.10.2 Alamouti code

We first describe this code by considering the simple case $t = 2$, $r = 1$, which yields the scheme illustrated in Figure 10.21 [10.1]. The code matrix $\mathbf{X}$ has the form

$$\mathbf{X} = \begin{bmatrix} x_1 & -x_2^* \\ x_2 & x_1^* \end{bmatrix} \tag{10.153}$$

This means that, during the first symbol interval, signal x_1 is transmitted from antenna 1, while signal x_2 is transmitted from antenna 2. During the next symbol period, antenna 1 transmits signal $-x_2^*$, and antenna 2 transmits signal x_1^*. Thus, the signals received in two adjacent time slots are

$$y_1 = h_1 x_1 + h_2 x_2 + z_1$$

and

$$y_2 = -h_1 x_2^* + h_2 x_1^* + z_2$$

where h_1, h_2 denote the path gains from the two transmit antennas to the receive antenna. The combiner of Figure 10.21, which has perfect CSI and hence knows

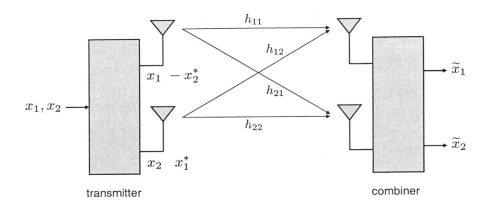

Figure 10.22: *Alamouti code with t = 2 and r = 2.*

the values of the path gains h_1 and h_2, generates the signals

$$\tilde{x}_1 = h_1^* y_1 + h_2 y_2^*$$

and

$$\tilde{x}_2 = h_2^* y_1 - h_1 y_2^*$$

so that

$$
\begin{aligned}
\tilde{x}_1 &= h_1^*(h_1 x_1 + h_2 x_2 + z_1) + h_2(-h_1^* x_2 + h_2^* x_1 + z_2^*) \\
&= (|h_1|^2 + |h_2|^2)x_1 + (h_1^* z_1 + h_2 z_2^*) \qquad (10.154)
\end{aligned}
$$

and similarly

$$\tilde{x}_2 = (|h_1|^2 + |h_2|^2)x_2 + (h_2^* z_1 - h_1 z_2^*) \qquad (10.155)$$

Thus, x_1 is separated from x_2. Provided that each transmit antenna transmits the same power as the single antenna for $t = 1$, this code has the same performance as one with $t = 1$, $r = 2$, and maximal-ratio combining (Section 4.4.1). To prove the last statement, observe that if the signal x_1 is transmitted, the two receive antennas observe $h_1 x_1 + z_1$ and $h_2 x_1 + z_2$, respectively, and after maximal-ratio combining the decision variable is

$$h_1^*(h_1 x_1 + z_1) + h_2^*(h_2 x_1 + z_2) = (|h_1|^2 + |h_2|^2)x_1 + (h_1^* z_1 + h_2 z_2^*) = \tilde{x}_1$$

This code can be generalized to other values of r. For example, with $t = r = 2$ and the same transmission scheme as before (see Figure 10.22), if $y_{11}, y_{12}, y_{21}, y_{22}$, denote the signals received by antenna 1 at time 1, by antenna 1 at time 2, by antenna 2 at time 1, and by antenna 2 at time 2, respectively, we have

$$
\begin{bmatrix} y_{11} & y_{12} \\ y_{21} & y_{22} \end{bmatrix} = \begin{bmatrix} h_{11} & h_{12} \\ h_{21} & h_{22} \end{bmatrix} \begin{bmatrix} x_1 & -x_2^* \\ x_2 & x_1^* \end{bmatrix} + \begin{bmatrix} z_{11} & z_{12} \\ z_{21} & z_{22} \end{bmatrix}
$$

$$
= \begin{bmatrix} h_{11}x_1 + h_{12}x_2 + z_{11} & -h_{11}x_2^* + h_{12}x_1^* + z_{12} \\ h_{21}x_1 + h_{22}x_2 + z_{21} & -h_{21}x_2^* + h_{22}x_1^* + z_{22} \end{bmatrix}
$$

The combiner generates

$$
\tilde{x}_1 = h_{11}^* y_{11} + h_{12} y_{12}^* + h_{21}^* y_{21} + h_{22} y_{22}^*
$$

and

$$
\tilde{x}_2 = h_{12}^* y_{11} - h_{11} y_{12}^* + h_{22}^* y_{21} - h_{21} y_{22}^*
$$

which yields

$$
\tilde{x}_1 = (|h_{11}|^2 + |h_{12}|^2 + |h_{21}|^2 + |h_{22}|^2)x_1 + \text{noise}
$$

and

$$
\tilde{x}_2 = (|h_{11}|^2 + |h_{12}|^2 + |h_{21}|^2 + |h_{22}|^2)x_2 + \text{noise}
$$

As above, it can be easily shown that the performance of this $t = 2$, $r = 2$ code is equivalent to that of a $t = 1$, $r = 4$ code with maximal-ratio combining (again, provided that each transmit antenna transmits the same power as with $t = 1$).

A general code, with $t = 2$ and r unrestricted, can also be exhibited: it has the same performance of a single-transmit-antenna code with $2r$ receive antennas and maximal-ratio combining.

10.10.3 Alamouti code revisited: Orthogonal designs

We can rewrite the transmitted signal in the Alamouti code with $t = 2$ and $r = 1$ in the following equivalent form:

$$
\begin{bmatrix} y_1 \\ y_2^* \end{bmatrix} = \begin{bmatrix} h_1 & h_2 \\ h_2^* & -h_1^* \end{bmatrix} \begin{bmatrix} x_1 \\ x_2 \end{bmatrix} + \begin{bmatrix} z_1 \\ z_2 \end{bmatrix} \tag{10.156}
$$

Now, if we define

$$
\check{\mathbf{H}} \triangleq \begin{bmatrix} h_1 & h_2 \\ h_2^* & -h_1^* \end{bmatrix}
$$

we see that

$$
\check{\mathbf{H}}^\dagger \check{\mathbf{H}} = (|h_1|^2 + |h_2|^2)\mathbf{I}_2 \tag{10.157}
$$

Recalling (10.139)–(10.140), this shows that the error probability for this Alamouti code is the same as without spatial interference, and with a signal-to-noise ratio increased by a factor $(|h_1|^2 + |h_2|^2)$. For this reason the Alamouti code is called an *orthogonal design*. There are also orthogonal designs with $t > 2$. For example, with $t = 3$, $r = 1$, and $N = 4$, we have

$$\mathbf{X} = \begin{bmatrix} x_1 & -x_2^* & -x_3^* & 0 \\ x_2 & x_1^* & 0 & -x_3^* \\ x_3 & 0 & x_1^* & x_2^* \end{bmatrix}$$

so that the equation $\mathbf{Y} = \mathbf{HX} + \check{\mathbf{z}}$ can be rewritten in the equivalent form

$$\begin{bmatrix} y_1 \\ y_2^* \\ y_3^* \\ y_4^* \end{bmatrix} = \check{\mathbf{H}} \begin{bmatrix} x_1 \\ x_2 \\ x_3 \end{bmatrix} + \check{\mathbf{z}} \tag{10.158}$$

where

$$\check{\mathbf{H}} \triangleq \begin{bmatrix} h_1 & h_2 & h_3 \\ h_2^* & -h_1^* & 0 \\ h_3^* & 0 & h_1^* \\ 0 & h_3^* & -h_2^* \end{bmatrix} \tag{10.159}$$

and $\check{\mathbf{z}}$ is a noise 4-vector. In this case we can verify that

$$\check{\mathbf{H}}^\dagger \check{\mathbf{H}} = (|h_1|^2 + |h_2|^2 + |h_3|^2)\mathbf{I}_3$$

Notice that with this code we transmit three signals in four time intervals (that is, $3/4$ signals per channel use), while the original Alamouti codes transmit 1 signal per channel use. It has been proved that orthogonal designs with $t > 2$ cannot transmit more than $3/4$ signals per channel use [10.63].

10.10.4   Linear space–time codes

Alamouti codes and orthogonal designs share the property of having simple decoders due to the linearity of their space–time map from symbols to transmit antennas. Schemes with this property form the class of *linear space–time codes*. These can be used for any number of transmit and receive antennas and may outperform orthogonal designs.

In these codes, the L symbols $x_1, \ldots, x_L$ are transmitted by t antennas in N time intervals. The code matrix $\mathbf{X}$ has the form

$$\mathbf{X} = \sum_{\ell=1}^{L} (\alpha_\ell \mathbf{A}_\ell + j\beta_\ell \mathbf{B}_\ell) \tag{10.160}$$

where α_ℓ and β_ℓ are the real and imaginary part of x_ℓ, respectively, and $\mathbf{A}_\ell, \mathbf{B}_\ell$, $\ell = 1, \ldots, L$, are $t \times N$ complex matrices.

Example 10.9

With Alamouti codes we may write

$$
\begin{aligned}
\mathbf{X} &= \begin{bmatrix} x_1 & -x_2^* \\ x_2 & x_1^* \end{bmatrix} \\
&= \begin{bmatrix} \alpha_1 + j\beta_1 & -\alpha_2 + j\beta_2 \\ \alpha_2 + j\beta_2 & \alpha_1 - j\beta_1 \end{bmatrix} \\
&= \alpha_1 \begin{bmatrix} 1 & 0 \\ 0 & 1 \end{bmatrix} + j\beta_1 \begin{bmatrix} 1 & 0 \\ 0 & -1 \end{bmatrix} + \alpha_2 \begin{bmatrix} 0 & -1 \\ 1 & 0 \end{bmatrix} + j\beta_2 \begin{bmatrix} 0 & 1 \\ 1 & 0 \end{bmatrix}
\end{aligned}
$$

which shows them to be a special case of linear space–time codes. □

Define the column vectors

$$
\check{\mathbf{x}} \triangleq [\alpha_1 \quad \beta_1 \quad \cdots \quad \alpha_L \quad \beta_L]' \qquad \check{\mathbf{z}} \triangleq \mathrm{vec}(\mathbf{Z})
$$

and the $Nr \times 2L$ matrix

$$
\check{\mathbf{H}} \triangleq [\mathrm{vec}(\mathbf{HA}_1) \quad \mathrm{vec}(j\mathbf{HB}_1) \quad \cdots \quad \mathrm{vec}(\mathbf{HA}_L) \quad \mathrm{vec}(j\mathbf{HB}_L)]
$$

Then we can write the received signal in the form

$$
\check{\mathbf{y}} \triangleq \mathrm{vec}(\mathbf{Y}) = \mathrm{vec}(\mathbf{HX} + \mathbf{Z}) = \sum_{\ell=1}^{L} (\alpha_\ell \mathrm{vec}(\mathbf{HA}_\ell) + \beta_\ell \mathrm{vec}(j\mathbf{HB}_\ell)) = \check{\mathbf{H}}\check{\mathbf{x}} + \check{\mathbf{z}}
$$

Notice that, since L signals are transmitted and $\check{\mathbf{y}}$ has Nr components, to be able to recover $\check{\mathbf{x}}$ from $\check{\mathbf{y}}$ we must have $L \leq Nr$.

The observed signal $\check{\mathbf{y}}$ can be decoded as follows. Perform the QR factorization of $\check{\mathbf{H}}$ (Section B.6.2, Appendix B):

$$
\check{\mathbf{H}} = \check{\mathbf{Q}}\check{\mathbf{R}}
$$

where $\check{\mathbf{Q}}$ is unitary and $\check{\mathbf{R}}$ is an upper triangular matrix. Thus, if we make a linear transformation on $\check{\mathbf{y}}$ consisting of its premultiplication by $\check{\mathbf{Q}}^\dagger$, we obtain (disregarding noise for simplicity) a vector $\check{\mathbf{R}}\check{\mathbf{x}}$, whose last entry is proportional to β_L. From this, β_L can be detected. The next-to-last entry is a linear combination of α_L and β_L: thus, since β_L has already been detected, and hence its contribution to spatial interference can be canceled, we may use this entry to detect α_L. The third-from-last entry is a linear combination of β_{L-1}, α_L, and β_L. This can be used to detect β_{L-1}, and so on. This *nulling-and-canceling* idea will be reprised *infra*, with some additional details, in our discussion of zero-forcing V-BLAST. More generally, our treatment of V-BLAST can be applied, *mutatis mutandis*, to linear space–time codes.

10.10.5   Trellis space–time codes

Trellis space–time codes are trellis-coded modulation (TCM) schemes, in which every transition among states, described by a trellis branch, is labeled by t signals, each being associated with one transmit antenna. Trellis space–time codes can achieve higher rates than orthogonal designs, but they suffer from a complexity that grows exponentially with the number of transmit antennas.

Example 10.10

> Examples of space–time codes are shown in Figures 10.23 and 10.24 through their trellises. The code in Figure 10.23 has $t = 2$, has four states, uses a quaternary constellation (whose signals are denoted $0, 1, 2, 3$), and transmits one signal (2 bits) per channel use. Its diversity is $2r$. Label xy means that signal x is transmitted by antenna 1, while signal y is simultaneously transmitted by antenna 2. The code in Figure 10.24 has again $t = 2$, has eight states, uses an octonary constellation (whose signals are denoted $0, 1, \ldots, 7$), and transmits one signal (3 bits) per channel use. Its diversity is $2r$.   □

10.10.6   Space–time codes when CSI is not available

In a rapidly changing mobile environment, or when long training sequences are not allowed, the assumption of perfect CSI at the receiver may not be valid. In the absence of CSI at the receiver, *unitary space–time modulation* has been advocated [10.30, 10.38]. This is a technique that circumvents the use of training symbols. Here the information is carried on the subspace spanned by orthonormal signals that are transmitted. This subspace survives multiplication by the unknown channel-gain matrix **H**. A scheme based on *differential unitary space–time* signals is described in [10.32]. High-rate constellations with excellent performance, obtained via algebraic techniques, are described in [10.29].

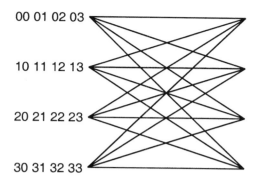

Figure 10.23: *A 4-PSK trellis space–time coding scheme with* $t = 2$ *and diversity* $2r$.

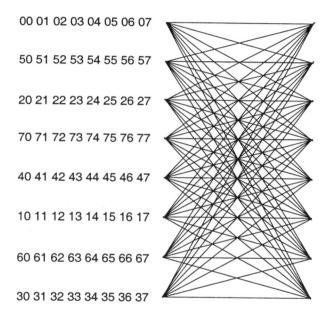

Figure 10.24: *An 8-PSK trellis space–time coding scheme with* $t = 2$ *and diversity* $2r$.

10.11 Suboptimum receiver interfaces

The capacity results described above show that extremely large spectral efficiencies can be achieved on a wireless link if the number of transmit and receive antennas is large. Now, as t and r increase, the complexity of space–time coding with maximum-likelihood detection may become too large. This motivates the design of suboptimal receivers whose complexity is lower than with ML detection and yet that perform close to it. In a receiver we distinguish an *interface*, which is a system accepting as its input the channel observation $\mathbf{Y}$ and generating a "soft estimate" $\widetilde{\mathbf{Y}}$ of the code matrix $\mathbf{X}$, and a decoder, whose input and output are $\widetilde{\mathbf{Y}}$ and the decoded matrix $\widehat{\mathbf{X}}$, respectively.

We describe here some of these interfaces, categorized as *linear* and *nonlinear*.[5]

10.12   Linear interfaces

A linear interface makes a linear transformation $\mathbf{A}$ of the received signal, under the assumption of perfect CSI at the receiver. $\mathbf{A} = \mathbf{A}(\mathbf{H})$ is a $t \times r$ matrix chosen so as to allow a simplification of the metrics used in the Viterbi algorithm employed for decoding. The conditional PEP for this linear interface with the metric $\|\mathbf{AY} - \mathbf{X}\|^2$ is given by

$$
\begin{aligned}
&\mathbb{P}(\mathbf{X} \to \widehat{\mathbf{X}} \mid \mathbf{H}) \\
&= \mathbb{P}(\|\mathbf{AY} - \widehat{\mathbf{X}}\|^2 < \|\mathbf{AY} - \mathbf{X}\|^2 \mid \mathbf{H}) \\
&= \mathbb{P}(\|\mathbf{AHX} - \widehat{\mathbf{X}} + \mathbf{AZ}\|^2 < \|\mathbf{AHX} - \mathbf{X} + \mathbf{AZ}\|^2 \mid \mathbf{H}) \\
&= \mathbb{P}(\|\boldsymbol{\Delta}\|^2 + 2((\mathbf{AH} - \mathbf{I})\mathbf{X}, \boldsymbol{\Delta}) + 2(\mathbf{AZ}, \boldsymbol{\Delta}) < 0 \mid \mathbf{H}) \quad (10.161)
\end{aligned}
$$

By reproducing computations done to derive (10.138), we obtain that $(\mathbf{AZ}, \boldsymbol{\Delta})$ is a zero-mean circularly symmetric complex Gaussian RV with variance $N_0 \|\mathbf{A}^\dagger \boldsymbol{\Delta}\|^2$. Thus, the unconditional PEP becomes

$$
\mathbb{P}(\mathbf{X} \to \widehat{\mathbf{X}}) = \mathbb{E}\left[Q\left(\frac{\|\boldsymbol{\Delta}\|^2 + 2((\mathbf{AH} - \mathbf{I})\mathbf{X}, \boldsymbol{\Delta})}{\sqrt{2N_0 \|\mathbf{A}^\dagger \boldsymbol{\Delta}\|^2}} \right) \right] \quad (10.162)
$$

10.12.1   Zero-forcing interface

A zero-forcing interface consists of choosing $\mathbf{A} = \mathbf{H}^+$, where the superscript $^+$ denotes the Moore–Penrose pseudoinverse of a matrix (Section B.7, Appendix B). For future reference, we note that we have

$$
\mathbf{H}^+ (\mathbf{H}^+)^\dagger = (\mathbf{H}^\dagger \mathbf{H})^{-1} \quad (10.163)
$$

---

[5]Other reduced-complexity receiver interfaces can be envisaged. For example, in [10.39] a scheme is advocated where $r' < r$ antennas are used, by selecting the r' best received signals. As long as $r' \geq t$, the capacity achieved by this system is close to that of a full-complexity system.

If we assume $r \geq t$, then $\mathbf{H}^\dagger\mathbf{H}$ is invertible with probability 1, and we have

$$\mathbf{H}^+ = (\mathbf{H}^\dagger\mathbf{H})^{-1}\mathbf{H}^\dagger \qquad (10.164)$$

so

$$\mathbf{H}^+\mathbf{Y} = \mathbf{X} + \mathbf{H}^+\mathbf{Z} \qquad (10.165)$$

which shows that the spatial interference in completely removed from the received signal, thus justifying the name *zero forcing* associated with this interface. The metric used here is then $\|\mathbf{H}^+\mathbf{Y} - \mathbf{X}\|^2$.

From (10.161), the conditional PEP becomes

$$P(\mathbf{X} \to \widehat{\mathbf{X}} \mid \mathbf{H}) = Q\left(\frac{\|\boldsymbol{\Delta}\|^2}{2\sigma}\right) \qquad (10.166)$$

where, due to (10.162),

$$\begin{aligned}
\sigma^2 &\triangleq \mathbb{V}[(\boldsymbol{\Delta}, \mathbf{H}^+\mathbf{Z})] \\
&= \frac{N_0}{2}\mathrm{Tr}\,[\boldsymbol{\Delta}^\dagger\mathbf{H}^+(\mathbf{H}^+)^\dagger\boldsymbol{\Delta}] \\
&= \frac{N_0}{2}\mathrm{Tr}\,[\boldsymbol{\Delta}^\dagger(\mathbf{H}^\dagger\mathbf{H})^{-1}\boldsymbol{\Delta}]
\end{aligned} \qquad (10.167)$$

This expression shows how the price paid for nulling the spatial interference is noise enhancement.

10.12.2 Linear MMSE interface

Here we choose the matrix $\mathbf{A}$ so as to minimize the mean-square value of the spatial interference plus noise. Define the mean-square error (MSE) as

$$\begin{aligned}
\varepsilon^2(\mathbf{A}) &\triangleq \mathbb{E}[\|\mathbf{A}\mathbf{Y} - \mathbf{X}\|^2] \\
&= \mathbb{E}[\mathrm{Tr}\,((\mathbf{A}\mathbf{H} - \mathbf{I}_t)\mathbf{X} + \mathbf{A}\mathbf{Z})((\mathbf{A}\mathbf{H} - \mathbf{I}_t)\mathbf{X} + \mathbf{A}\mathbf{Z})^\dagger]
\end{aligned} \qquad (10.168)$$

Using the simplifying assumption of iid zero-mean components of $\mathbf{x}$ (with second moment $\mathcal{E}$), we obtain the following expression:

$$\varepsilon^2(\mathbf{A}) = \mathrm{Tr}\,(\mathcal{E}(\mathbf{A}\mathbf{H} - \mathbf{I}_t)(\mathbf{A}\mathbf{H} - \mathbf{I}_t)^\dagger + N_0\mathbf{A}\mathbf{A}^\dagger) \qquad (10.169)$$

The variation of $\varepsilon^2(\mathbf{A})$ with respect to $\mathbf{A}$ is then given by

$$\delta(\varepsilon^2) = \mathrm{Tr}\,\Big\{\delta\mathbf{A}[\mathcal{E}\mathbf{H}(\mathbf{A}\mathbf{H} - \mathbf{I}_t)^\dagger + N_0\mathbf{A}^\dagger] + [\mathcal{E}(\mathbf{A}\mathbf{H} - \mathbf{I}_t)\mathbf{H}^\dagger + N_0\mathbf{A}]\delta\mathbf{A}^\dagger\Big\} \qquad (10.170)$$

The corresponding stationary point obtained by nulling this variation yields the MMSE solution:

$$\mathbf{A} = \mathbf{A}_{\mathrm{mmse}} \triangleq \mathbf{H}^{\dagger}(\mathbf{H}\mathbf{H}^{\dagger} + \delta_s\mathbf{I}_r)^{-1} = (\mathbf{H}^{\dagger}\mathbf{H} + \delta_s\mathbf{I}_t)^{-1}\mathbf{H}^{\dagger} \qquad (10.171)$$

where $\delta_s \triangleq N_0/\mathcal{E}$. From (10.162) we obtain

$$P(\mathbf{X} \to \widehat{\mathbf{X}}) = \mathbb{E}\left[Q\left(\frac{\|\boldsymbol{\Delta}\|^2 + 2(((\mathbf{H}^{\dagger}\mathbf{H} + \delta_s\mathbf{I}_t)^{-1}\mathbf{H}^{\dagger}\mathbf{H} - \mathbf{I}_t)\mathbf{X}, \boldsymbol{\Delta})}{\sqrt{2N_0\|\mathbf{H}(\mathbf{H}^{\dagger}\mathbf{H} + \delta_s\mathbf{I}_t)^{-1}\boldsymbol{\Delta}\|^2}}\right)\right]$$
$$(10.172)$$

Notice that, as $\delta_s \to 0$ (vanishingly small noise), the right-hand side of (10.172) tends to the PEP of the zero-forcing interface, as it should.

10.12.3 Asymptotics: Finite t and $r \to \infty$.

Here we consider the case $r \gg t$ by examining the asymptotic performance obtained when $r \to \infty$, while t remains constant. By the strong law of large numbers we can write, as $r \to \infty$,

$$\mathbf{H}^{\dagger}\mathbf{H} \to r\mathbf{I}_t \qquad \text{a.s.} \qquad (10.173)$$

and we have previously seen from (10.152) that with ML detection the pairwise error probability tends to that of a nonfading AWGN channel (no spatial interference). Using (10.173) in (10.167) and in (10.172), we see that, asymptotically, ZF and MMSE interfaces do not entail any loss of performance with respect to ML.

10.12.4 Asymptotics: $t, r \to \infty$ with $t/r \to \alpha > 0$.

Things change if both t and r grow to infinity while their ratio tends to a constant positive value α. In this case an SNR loss is expected, as we are going to illustrate for the ZF interface (see [10.9] for the MMSE case).

Theorem C.3.2 of Appendix C shows that, as $t, r \to \infty$ with $t/r \to \alpha$, the cumulative empirical eigenvalue distribution of $\mathbf{H}^{\dagger}\mathbf{H}/r$ converges to a function $F(\lambda; \alpha)$ whose derivative is given by:

$$\frac{\partial}{\partial\lambda}F(\lambda; \alpha) = f(\lambda; \alpha) \triangleq (1 - \alpha^{-1})_+ \delta(\lambda) + \alpha^{-1}\frac{\sqrt{(\lambda - \lambda_-)_+(\lambda_+ - \lambda)_+}}{2\pi\lambda}$$
$$(10.174)$$

where $\lambda_{\pm} \triangleq (\sqrt{\alpha} \pm 1)^2$. In particular, when $\alpha = 0$ or ∞, the pdf $f(\lambda; \alpha)$ tends to $\delta(\lambda - 1)$ or $\delta(\lambda)$, respectively.

The asymptotic PEP of the ML and ZF receivers can now be calculated by using Theorem C.3.3 of Appendix C, where the role of the matrix sequences $\mathbf{A}_n$ and $\mathbf{B}_n$

is played by $\mathbf{W} \triangleq \mathbf{H}^\dagger\mathbf{H}/r$ and $\mathbf{\Delta}\mathbf{\Delta}^\dagger$ as $r \to \infty$. Then for the ML receiver we have

$$
\begin{aligned}
\frac{\|\mathbf{H}\mathbf{\Delta}\|^2}{2N_0} &= \frac{1}{2N_0}\text{Tr}\,(\mathbf{H}^\dagger\mathbf{H}\mathbf{\Delta}\mathbf{\Delta}^\dagger) \\
&= \frac{rt}{2N_0}\tau(\mathbf{W}\mathbf{\Delta}\mathbf{\Delta}^\dagger) \\
&\to \frac{rt}{2N_0}\mathbb{E}[\tau(\mathbf{W})]\tau(\mathbf{\Delta}\mathbf{\Delta}^\dagger) \qquad \text{(a.s. as } t,r \to \infty, t/r \to \alpha)
\end{aligned}
$$

(10.175)

where $\tau(\mathbf{A}) \triangleq \text{Tr}\,(\mathbf{A})/n$ for an $n \times n$ matrix $\mathbf{A}$. Since

$$
\mathbb{E}[\tau(\mathbf{W})] \to \int_a^b \lambda f(\lambda;\alpha)\,d\lambda = 1 \tag{10.176}
$$

we obtain, a.s. as $t,r \to \infty, t/r \to \alpha$,

$$
\frac{\|\mathbf{H}\mathbf{\Delta}\|^2}{2N_0} \to \frac{r\|\mathbf{\Delta}\|^2}{2N_0} \tag{10.177}
$$

and hence

$$
\mathbb{P}(\mathbf{X} \to \widehat{\mathbf{X}}) \to Q\left(\sqrt{\frac{r\|\mathbf{\Delta}\|^2}{2N_0}}\right) \tag{10.178}
$$

For the ZF receiver we have, from (10.167),

$$
\sigma^2 = \frac{tN_0}{2r}\tau(\mathbf{W}^{-1}\mathbf{\Delta}\mathbf{\Delta}^\dagger) \to \frac{tN_0}{2r}\mathbb{E}[\tau(\mathbf{W}^{-1})]\tau(\mathbf{\Delta}\mathbf{\Delta}^\dagger) \tag{10.179}
$$

Since

$$
\mathbb{E}[\tau(\mathbf{W}^{-1})] \to \int_a^b \lambda^{-1} f(\lambda;\alpha)\,d\lambda = \frac{1}{1-\alpha} \tag{10.180}
$$

we obtain, a.s. as $t,r \to \infty, t/r \to \alpha$,

$$
\frac{\|\mathbf{\Delta}\|^4}{4\sigma^2} \to (1-\alpha)\frac{r\|\mathbf{\Delta}\|^2}{2N_0} \tag{10.181}
$$

and hence

$$
\mathbb{P}(\mathbf{X} \to \widehat{\mathbf{X}}) \to Q\left(\sqrt{(1-\alpha)\frac{r\|\mathbf{\Delta}\|^2}{2N_0}}\right) \tag{10.182}
$$

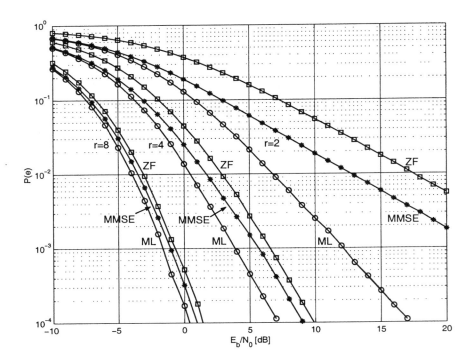

Figure 10.25: *Word error probability of the binary* $(8, 4, 4)$ *Reed-Muller code with binary PSK over a channel with* $t = 2$ *transmit antennas and* r *receive antennas with ML, MMSE, and ZF interfaces (computer simulation results).*

Thus, the asymptotic SNR loss with respect to the ML interface is equal to $(1 - \alpha)^{-1}$ for the ZF interface, which predicts that the choice $r = t$ with a large number of antennas yields a considerable loss in performance. From the above we may expect that these linear interfaces exhibit a PEP close to ML only for $r \gg t$; otherwise, the performance loss may be substantial. This is validated by Figure 10.25, which shows the error probability of a multiple-antenna system where the binary $(8, 4, 4)$ Reed-Muller code is used by splitting its code words evenly between two transmit antennas. The word-error probabilities shown are obtained through Monte Carlo simulation. Binary PSK is used, and the code rate is 1 bit per channel use. It is seen that for $r = 2$ both MMSE and ZF interface exhibit a considerable performance loss with respect to ML, while for $r = 8$ the losses are very moderate.

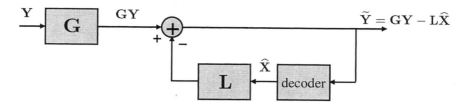

Figure 10.26: *General structure of a nonlinear interface.*

10.13 Nonlinear interfaces

The task of reducing the spatial interference affecting the received signal can be accomplished by first processing $\mathbf{Y}$ linearly and then subtracting from the result an estimate of the spatial interference obtained from preliminary decisions on the transmitted code word. The metric used for decoding is $\|\widetilde{\mathbf{Y}} - \mathbf{X}\|$, where $\widetilde{\mathbf{Y}}$ is the soft estimate of $\mathbf{X}$ given by

$$\widetilde{\mathbf{Y}} \triangleq \mathbf{GY} - \mathbf{L}\widehat{\mathbf{X}} \tag{10.183}$$

for a suitable choice of the two matrices $\mathbf{G}$ and $\mathbf{L}$ (Figure 10.26). The diagonal entries of the matrix $\mathbf{L}$ must be zero in order to have only spatial interference subtracted from $\mathbf{GY}$.

10.13.1 Vertical BLAST interface

One nonlinear interface is called vertical BLAST (this stands for Bell Laboratories Layered Space–Time Architecture). With V-BLAST, the data are divided into t substreams to be transmitted on different antennas. The receiver preprocesses linearly the received signal by forming the matrix $\mathbf{GY}$, which has t rows. Then it first decodes one of the substreams after reducing the spatial interference coming from the others. Next, the contribution of this substream is subtracted from the received signal, and the second substream is decoded after reducing the remaining spatial interference. This process is repeated t times.

Different implementations of the basic V-BLAST idea are possible, two of them being the zero-forcing *ZF V-BLAST* interface and the minimum-mean-square-error *MMSE V-BLAST* interface. These arise from the minimization of the mean-square error of the spatial interference without or with noise, respectively.

It should be observed that the performance of V-BLAST depends on the order in which the substreams are decoded (in the algorithm above, the actual numbering of the rows of $\mathbf{GY}$ is arbitrary), and on the data rate associated with each substream. Several strategies are possible here (see [10.10] and references therein, and [10.22, 10.65, 10.67]): the decoding order may be predefined, and the data rates may be the same; or an ordering may be chosen so as to maximize an SNR-related parameter, with equal data rates; or different data rates may be assigned to different substreams.

ZF V-BLAST When the presence of noise is disregarded, the MSE of the disturbance can be written as

$$\varepsilon^2(\mathbf{G}, \mathbf{L}) = \mathbb{E}[\|\widetilde{\mathbf{Y}} - \mathbf{X}\|^2] = \mathbb{E}[\|\mathbf{GHX} - \mathbf{L}\widehat{\mathbf{X}} - \mathbf{X}\|^2] \tag{10.184}$$

Under the approximations

$$\begin{aligned} \mathbb{E}[\mathbf{X}\widehat{\mathbf{X}}^\dagger] &\approx \mathbb{E}[\mathbf{XX}^\dagger] \\ \mathbb{E}[\widehat{\mathbf{X}}\widehat{\mathbf{X}}^\dagger] &\approx \mathbb{E}[\mathbf{XX}^\dagger] \end{aligned} \tag{10.185}$$

(which are justified by the assumption of having $\widehat{\mathbf{X}} \approx \mathbf{X}$ unless the error probability is high) we obtain

$$\begin{aligned} \varepsilon^2(\mathbf{G}, \mathbf{L}) &= \mathbb{E}[\|(\mathbf{GH} - \mathbf{L} - \mathbf{I}_t)\mathbf{X}\|^2] \\ &= \mathbb{E}[\mathrm{Tr}\{(\mathbf{GH} - \mathbf{L} - \mathbf{I}_t)^\dagger(\mathbf{GH} - \mathbf{L} - \mathbf{I}_t)\mathbf{XX}^\dagger\}] \\ &= N\mathcal{E}\mathbb{E}[\|\mathbf{GH} - \mathbf{L} - \mathbf{I}_t\|^2] \end{aligned} \tag{10.186}$$

since $\mathbb{E}[\mathbf{XX}^\dagger] = N\mathcal{E}\mathbf{I}_t$. From the QR decomposition of $\mathbf{H}$ (Section B.6.2 of Appendix B),

$$\underbrace{\mathbf{H}}_{r \times t} = \underbrace{\mathbf{Q}}_{r \times t} \underbrace{\mathbf{R}}_{t \times t}$$

(where $\mathbf{R}$ is an upper triangular matrix), we see that the MSE $\varepsilon^2(\mathbf{G}, \mathbf{L})$ vanishes by setting

$$\begin{cases} \mathbf{G} &= \mathrm{diag}^{-1}(\mathbf{R})\mathbf{Q}^\dagger \\ \mathbf{L} &= \mathrm{diag}^{-1}(\mathbf{R})\mathbf{R} - \mathbf{I}_t \end{cases} \tag{10.187}$$

The block diagram of Figure 10.26 illustrates that ZF V-BLAST corresponds to having a strictly upper triangular matrix $\mathbf{L}$. Explicitly, the steps of the ZF V-BLAST algorithm proceed as follows. Denoting by $(\mathbf{A})_i$ the ith row of matrix $\mathbf{A}$, by $(\mathbf{A})_{ij}$ its entry in ith row and jth column, and by $\Longrightarrow$ the result of decoding, we have

$$\begin{cases} (\widetilde{\mathbf{Y}})_t &= (\mathbf{GY})_t & \Longrightarrow (\widehat{\mathbf{X}})_t \\ (\widetilde{\mathbf{Y}})_{t-1} &= (\mathbf{GY})_{t-1} - (\mathbf{L})_{t-1,t}(\widehat{\mathbf{X}})_t & \Longrightarrow (\widehat{\mathbf{X}})_{t-1} \\ (\widetilde{\mathbf{Y}})_{t-2} &= (\mathbf{GY})_{t-2} - (\mathbf{L})_{t-2,t}(\widehat{\mathbf{X}})_t - (\mathbf{L})_{t-2,t-1}(\widehat{\mathbf{X}})_{t-1} & \Longrightarrow (\widehat{\mathbf{X}})_{t-2} \\ &\vdots \\ (\widetilde{\mathbf{Y}})_1 &= (\mathbf{GY})_1 - (\mathbf{L})_{1,t}(\widehat{\mathbf{X}})_t - \cdots - (\mathbf{L})_{1,2}(\widehat{\mathbf{X}})_2 & \Longrightarrow (\widehat{\mathbf{X}})_1 \end{cases}$$

The soft estimate of $\mathbf{X}$ can be written as

$$
\begin{aligned}
\widetilde{\mathbf{Y}} &= \mathrm{diag}^{-1}(\mathbf{R})\mathbf{Q}^{\dagger}\mathbf{Y} - [\mathrm{diag}^{-1}(\mathbf{R})\mathbf{R} - \mathbf{I}_t]\widehat{\mathbf{X}} \\
&= \underbrace{\mathbf{X}}_{①} + \underbrace{[\mathrm{diag}^{-1}(\mathbf{R})\mathbf{R} - \mathbf{I}_t]\boldsymbol{\Delta}}_{②} + \underbrace{\mathrm{diag}^{-1}(\mathbf{R})\mathbf{Q}^{\dagger}\mathbf{Z}}_{③}
\end{aligned}
\tag{10.188}
$$

The three terms in the last expression are: ① the useful term (which is free of spatial interference, thus justifying the name *zero forcing* associated with this interface); ② the interference due to past wrong decisions; and ③ colored noise.

MMSE V-BLAST This minimizes the MSE of the disturbance $\widetilde{\mathbf{Y}} - \mathbf{X}$, taking into account the presence of noise. Again, under the approximations (10.185), we can write the MSE as

$$
\begin{aligned}
\varepsilon^2(\mathbf{G}, \mathbf{L}) &= \mathbb{E}[\|\mathbf{G}\mathbf{Y} - \mathbf{L}\mathbf{X} - \mathbf{X}\|^2] \\
&= \mathbb{E}[\|(\mathbf{G}\mathbf{H} - \mathbf{L} - \mathbf{I}_t)\mathbf{X} + \mathbf{G}\mathbf{Z}\|^2] \\
&= N\mathcal{E}\Big[\|\mathbf{G}\mathbf{H} - \mathbf{L} - \mathbf{I}_t\|^2 + \delta_s\|\mathbf{G}\|^2\Big]
\end{aligned}
\tag{10.189}
$$

where $\delta_s \triangleq N_0/\mathcal{E}$. The minimum MSE can be found in two steps:

i) Minimizing $\varepsilon^2(\mathbf{G}, \mathbf{L})$ over the set of matrices $\mathbf{G} \in \mathbb{C}^{t \times r}$ leads to

$$
\mathbf{G}_{\mathrm{mmse}} = (\mathbf{L} + \mathbf{I}_t)(\mathbf{H}^{\dagger}\mathbf{H} + \delta_s\mathbf{I}_t)^{-1}\mathbf{H}^{\dagger} .
\tag{10.190}
$$

The corresponding minimum MSE is

$$
\varepsilon^2_{\mathrm{mmse}}(\mathbf{L}) = NN_0\mathrm{Tr}\Big[(\mathbf{L} + \mathbf{I}_t)(\mathbf{H}^{\dagger}\mathbf{H} + \delta_s\mathbf{I}_t)^{-1}(\mathbf{L} + \mathbf{I}_t)^{\dagger}\Big]
\tag{10.191}
$$

ii) Next, $\varepsilon^2_{\mathrm{mmse}}(\mathbf{L})$ is minimized over the set of $t \times t$ strictly upper triangular matrices (i.e., such that $[\mathbf{L}]_{ij} = 0$ whenever $i \geq j$). This can be done by using the Cholesky factorization $\mathbf{H}^{\dagger}\mathbf{H} + \delta_s\mathbf{I}_t = \mathbf{S}^{\dagger}\mathbf{S}$, where $\mathbf{S}$ is an upper triangular matrix (Section B.6.1 of Appendix B). After using basic multiplication properties of triangular matrices, we obtain the following result:

$$
\begin{aligned}
\varepsilon^2_{\mathrm{mmse}}(\mathbf{L}) &= NN_0\mathrm{Tr}\Big[(\mathbf{L} + \mathbf{I}_t)(\mathbf{H}^{\dagger}\mathbf{H} + \delta_s\mathbf{I}_t)^{-1}(\mathbf{L} + \mathbf{I}_t)^{\dagger}\Big] \\
&= NN_0\|(\mathbf{L} + \mathbf{I}_t)\mathbf{S}^{-1}\|^2 \\
&\geq NN_0\|\mathrm{diag}((\mathbf{L} + \mathbf{I}_t)\mathbf{S}^{-1})\|^2 \\
&= NN_0\|\mathrm{diag}(\mathbf{S}^{-1})\|^2 = NN_0\sum_{i=1}^{t}|[\mathbf{S}]_{i,i}|^{-2}
\end{aligned}
\tag{10.192}
$$

The minimum is attained by setting $\mathbf{L} = \text{diag}^{-1}(\mathbf{S})\mathbf{S} - \mathbf{I}_t$. Thus, $\varepsilon^2(\mathbf{G}, \mathbf{L})$ is minimized by setting

$$\begin{cases} \mathbf{G} = \mathbf{G}_{\text{mmse}} \triangleq \text{diag}^{-1}(\mathbf{S})\mathbf{S}^{-\dagger}\mathbf{H}^\dagger \\ \mathbf{L} = \mathbf{L}_{\text{mmse}} \triangleq \text{diag}^{-1}(\mathbf{S})\mathbf{S} - \mathbf{I}_t \end{cases} \tag{10.193}$$

and

$$\varepsilon^2_{\text{mmse}} \triangleq \varepsilon^2(\mathbf{G}_{\text{mmse}}, \mathbf{L}_{\text{mmse}}) = NN_0 \sum_{i=1}^{t} |[\mathbf{S}]_{i,i}|^{-2} \tag{10.194}$$

As a result, the soft estimate $\widetilde{\mathbf{Y}}$ can be written as

$$\begin{aligned} \widetilde{\mathbf{Y}} &= \text{diag}^{-1}(\mathbf{S})\mathbf{S}^{-\dagger}\mathbf{H}^\dagger\mathbf{Y} - (\text{diag}^{-1}(\mathbf{S})\mathbf{S} - \mathbf{I}_t)\widehat{\mathbf{X}} \\ &= \underbrace{(\mathbf{I}_t - \text{diag}^{-1}(\mathbf{S})\mathbf{S}^{-\dagger})\mathbf{X}}_{\textcircled{1}} + \underbrace{(\text{diag}^{-1}(\mathbf{S})\mathbf{S} - \mathbf{I}_t)\boldsymbol{\Delta}}_{\textcircled{2}} \\ &\quad + \underbrace{\text{diag}^{-1}(\mathbf{S})\mathbf{S}^{-\dagger}\mathbf{H}^\dagger\mathbf{Z}}_{\textcircled{3}} \end{aligned} \tag{10.195}$$

where the three terms in the last expression are: $\textcircled{1}$ the (biased) useful term; $\textcircled{2}$ the interference due to past wrong decisions; and $\textcircled{3}$ colored noise.

10.13.2   Diagonal BLAST interface

Consider the transmission scheme of Figure 10.27, referred to as Diagonal BLAST (D-BLAST). Here, a, b, c, . . ., denote different data substreams. As discussed in Section 10.14 *infra*, this scheme differs from V-BLAST because each symbol in a data substream is transmitted by a different antenna and hence achieves a larger diversity. To obtain this, the information stream is demultiplexed into t substreams, which are transmitted by t antennas through a *diagonal* interleaving scheme. The interleaver is designed so that the symbols of a given substream are cyclically sent over all the t antennas in order to guarantee the necessary diversity order. Diagonals are written from top to bottom, and the letters in each rectangle denote the corresponding code symbol index, i.e., indicate the sequence in which diagonals are filled. Each rectangle in Figure 10.27 may actually contain an arbitrary number $\kappa \geq 1$ of coded symbols. Each column of t symbols of the diagonal interleaver array is transmitted in parallel, from the t antennas.

To illustrate the operation of D-BLAST, consider a simple case with two transmit antennas. The transmitted matrix has the form

$$\mathbf{X} = \begin{bmatrix} x_{11} & x_{12} & x_{13} & \cdots \\ 0 & x_{21} & x_{22} & \cdots \end{bmatrix} \tag{10.196}$$

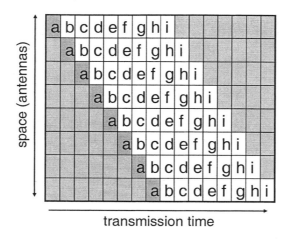

Figure 10.27: *An example of diagonal interleaving with* $t = 8$.

where x_{ij} is the signal transmitted by the ith antenna in the jth substream. The receiver first detects x_{11}, which is not affected by spatial interference. Then, it detects x_{21}; this is affected by the spatial interference caused by x_{12}, which can be reduced or nulled, by using, for example, a zero-forcing filter. Next, the estimates of x_{11} and x_{21} are sent to the decoder of the first substream. Once this first substream has been decoded, its contribution is subtracted out before decoding the second substream, and so forth. Notice that D-BLAST entails a rate loss due to the overhead symbols necessary to start the decoding process (these are shaded in Figure 10.27).

10.13.3 Threaded space–time architecture

To avoid the rate loss implied by D-BLAST, the latter architecture can be generalized by wrapping substreams around, as shown in Figure 10.28. This figure shows a simple special case of *threaded layering*, whereby the symbols are distributed in the code word matrix so as to achieve full spatial span t (which guarantees the right spatial diversity order) and full temporal span N (which guarantees the right temporal diversity order in the case of fast fading) [10.19].

10.13.4 Iterative interface

An alternative to BLAST consists of performing an iterative spatial interference cancellation. Referring again to the block diagram of Figure 10.26, at iteration k,

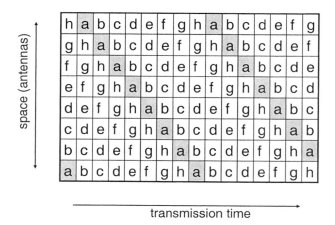

Figure 10.28: *An example of threading with $t = 8$ (each letter represents a layer).*

$k = 0, 1, \ldots,$ an estimate of the spatial interference is generated in the form

$$\mathbf{W}^{(k)} = (\mathbf{GH} - \text{diag}\,(\mathbf{GH}))\widehat{\mathbf{X}}^{(k)} \tag{10.197}$$

Here $\widehat{\mathbf{X}}^{(k)}$ is the decoded word at iteration k, computed by minimizing the metric $\|\widetilde{\mathbf{Y}}^{(k)} - \mathbf{X}\|^2$, where

$$\begin{aligned}
\widetilde{\mathbf{Y}}^{(k+1)} &= \widetilde{\mathbf{Y}} - \mathbf{W}^{(k)} \\
&= \widetilde{\mathbf{Y}} - (\mathbf{GH} - \text{diag}\,(\mathbf{GH}))\widehat{\mathbf{X}}^{(k)}
\end{aligned} \tag{10.198}$$

and for $k = 0$ we define $\widehat{\mathbf{X}}^{(0)} \triangleq \mathbf{0}$. It can be easily seen that, if decoding is perfect (that is, if $\widehat{\mathbf{X}}^{(k)} = \mathbf{X}$ for some k), then

$$\widetilde{\mathbf{Y}}^{(k)} = \text{diag}\,(\mathbf{GH})\,\mathbf{X} + \mathbf{GZ} \tag{10.199}$$

which shows that the spatial interference is completely removed.

10.14   The fundamental trade-off

As we briefly mentioned in Section 10.1.1, the use of multiple antennas provides at the same time a *rate gain* and a *diversity gain*. The former is due to the fact that multiple, independent transmission paths generate a multiplicity of independent "spatial" channels that can simultaneously be used for transmission. The latter is obtained by exploiting the independent fading gains that affect the same signal and that can be averaged through to increase the reliability of its detection. Here we examine how these two performance measures are related by fundamental limits that reflect the ubiquitous trade-off between rate and transmission quality of a transmission system.

We focus our attention on the nonergodic fading channel of Section 10.6, with channel state information available at the receiver only, and to a high-SNR situation. The latter restriction refers to a system whose performance is not power limited. We have seen (Section 4.2.2 and Observation 10.3.2) that, as the SNR $\zeta \to \infty$, the capacity $C(\zeta)$ of an ergodic Rayleigh fading channel with SNR ζ grows as $m \log \zeta$, with $m \triangleq \min\{t, r\}$. Recalling the high-SNR expression of the capacity of the single-antenna ergodic Rayleigh fading channel, which is $\log \zeta$, the result above can be interpreted by saying that the maximum number of independent parallel channels (or, in a different parlance, the number of degrees of freedom) created by t transmit and r receive antennas equals m, which is the maximum rate gain we can achieve. Consider next the number of independently faded paths: in our model this is equal to tr, which is indeed the maximum achievable diversity gain with maximum-likelihood detection (Observation 10.9.1).

We discuss here the fact that, while both gains can be achieved by MIMO systems, higher rate gains come at the expenses of diversity gains. We start our discussion by defining precisely what we mean by rate gain and diversity gain in the present context. In a situation where different data rates are involved, *a sequence of codes with increasing rate*, rather than a single code, must be considered. For a fair comparison among codes with different rates, the rate gain is defined by the ratio between the actual code rate $\rho(\zeta)$ and the capacity of the scalar channel at that SNR:

$$\mu \triangleq \lim_{\zeta \to \infty} \frac{\rho(\zeta)}{C(\zeta)} \tag{10.200}$$

This indicates how far the system is operating from the capacity limit. Notice that the capacity increases with the SNR ζ, so to approach capacity the code rate $\rho(\zeta)$ must also increase with ζ; if a single code were used, the rate gain would vanish, because, as ζ increases, the ratio (10.200) would tend to zero. As for the diversity gain δ, this is defined as the exponent of ζ^{-1} in the expression of the average error probability of the system: formally,

$$\delta \triangleq -\lim_{\zeta \to \infty} \frac{\log P(e)}{\log \zeta} \tag{10.201}$$

The main point here is that the maximum values of rate gain and diversity gain *cannot be achieved simultaneously*; μ and δ are connected by a trade-off curve that

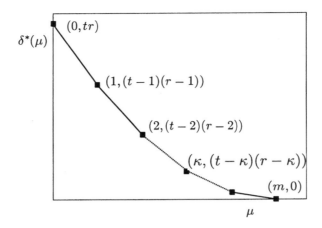

Figure 10.29: *Diversity–rate trade-off for multiple-antenna systems with t transmit and r receive antennas.*

we are going to introduce and discuss. This curve plots, as a function of the rate gain μ, the maximum achievable diversity gain, denoted $\delta^*(\mu)$.

The trade-off curve, in the special but important case of a code with length $N \geq t + r - 1,$[6] is given by the piecewise-linear function connecting the points $(\kappa, \delta^*(\kappa))$, $\kappa \in \{0, 1, \ldots, m\}$, where

$$\delta^*(\kappa) \triangleq (t - \kappa)(r - \kappa) \qquad (10.202)$$

as shown in Figure 10.29. We see that the maximum values that μ and δ can achieve are m and tr, respectively, as discussed before. Equation (10.202) also shows that the maximum diversity gain can only be achieved for zero rate gain, and the maximum rate gain can only be achieved for zero diversity gain. More generally, (10.202) shows that, out of the total number of t transmit and r receive antennas, κ transmit and κ receive antennas are allocated to increase the rate, and the remaining $t - \kappa$ and $r - \kappa$ create diversity.

Proof

Under the above assumptions, and asymptotically as $\zeta \to \infty$, the outage probability (see Section 10.6) corresponding to an information rate $\rho = \mu \log \zeta$ can be written

---

[6]See [10.67] for lower values of N. Here, it suffices to observe that no more diversity gain can be obtained if the block length of the code exceeds $t + r - 1$, which consequently expresses the infinite-block-length performance.

as

$$P_{\text{out}}(\mu, \zeta) \triangleq \mathbb{P}\left(\sum_{i=1}^{m} \log(1 + \lambda_i \zeta) \leq \mu \log \zeta \right) \tag{10.203}$$

where λ_i is the ith ordered eigenvalue of the matrix $\mathbf{HH}^\dagger$. If $\mu \geq m$ the outage probability is always 1 (since $\log \zeta$ dominates asymptotically the other terms as $\zeta \to \infty$), so we restrict ourselves to the case $\mu < m$. The joint pdf of the λ_i's is given by (C.28) of Appendix C. Defining the new variables $\alpha_i \triangleq -\log \lambda_i / \log \zeta$, we can write the outage probability (10.203) as follows:

$$P_{\text{out}}(\mu, \zeta) = \frac{(\ln \zeta)^m}{\Gamma_m(m)\Gamma_m(n)} \cdot$$

$$\cdot \int_{\boldsymbol{\alpha} \in \mathbb{R}^m, \alpha_1 \geq \dots \geq \alpha_m} \left[\sum_{i=1}^{m} \log(1 + \zeta^{1-\alpha_i}) \leq \mu \log \zeta \right]$$

$$\cdot \prod_{i=1}^{m} e^{-\zeta^{-\alpha_i}} \zeta^{-(n-m+1)\alpha_i} \prod_{i<j} (\zeta^{-\alpha_i} - \zeta^{-\alpha_j})^2 d\boldsymbol{\alpha} \tag{10.204}$$

Since $\zeta \to \infty$, several simplifications can be used:

- The Iverson function in the integral tends to the following limit:

$$\left[\sum_{i=1}^{m} \log(1 + \zeta^{1-\alpha_i}) \leq \mu \log \zeta \right] \to \left[\sum_{i=1}^{m} (1 - \alpha_i)_+ \leq \mu \right] \tag{10.205}$$

- Since $\exp(-\zeta^{-\alpha_i}) \to 0$ for $\alpha_i < 0$ and $\exp(-\zeta^{-\alpha_i}) \to 1$ for $\alpha_i > 0$, the integration domain where the integrand is not asymptotically small reduces to $\mathbb{R}_+^m$.

- $(\zeta^{-\alpha_i} - \zeta^{-\alpha_j})^2 \to \zeta^{-2\alpha_j}$, since $\alpha_i > \alpha_j$ for $i < j$ except for a set of measure zero.

Collecting the above observations, we obtain, as $\zeta \to \infty$,

$$P_{\text{out}}(\mu, \zeta) \to \frac{(\ln \zeta)^m}{\Gamma_m(m)\Gamma_m(n)} \int_{\alpha_1 \geq \dots \geq \alpha_m \geq 0} \left[\sum_{i=1}^{m} (1 - \alpha_i)_+ \leq \mu \right]$$

$$\cdot \exp\left(-\ln \zeta \sum_{i=1}^{m} (n - m + 2i - 1)\alpha_i \right) d\boldsymbol{\alpha} \tag{10.206}$$

Using Laplace's method of asymptotic multidimensional integral approximation (see, e.g., [10.12]), it can be shown that

$$P_{\text{out}}(\mu, \zeta) \to \zeta^{-d_{\text{out}}(\mu)} \tag{10.207}$$

where

$$d_{\text{out}}(\mu) \triangleq \min_{\alpha_1 \geq \dots \geq \alpha_m \geq 0, \sum_{i=1}^{m}(1-\alpha_i)_+ \leq \mu} \sum_{i=1}^{m} (n - m + 2i - 1)\alpha_i \tag{10.208}$$

The minimization above is a linear programming problem with nonlinear constraints, equivalent to the computation of

$$\bar{\alpha}^* \triangleq \operatorname*{arg\,max}_{\bar{\alpha}_1 \leq \ldots \leq \bar{\alpha}_m \leq 1, \ \sum_i (\bar{\alpha}_i)_+ \leq \mu} \sum_{i=1}^{m} (n - m + 2i - 1)\bar{\alpha}_i \qquad (10.209)$$

In order to compute (10.209), one sets $\bar{\alpha}_i = 1$ for $i = m, m-1, \ldots, m - \lfloor \mu \rfloor + 1$. Next, one has to set $\alpha_{m-\lfloor \mu \rfloor} = \mu - \lfloor \mu \rfloor < 1$. This corresponds to setting

$$\alpha_i = \begin{cases} 1 & i = 1, \ldots, m - \lfloor \mu \rfloor - 1 \\ 1 - (\mu - \lfloor \mu \rfloor) & i = m - \lfloor \mu \rfloor \\ 0 & i = m - \lfloor \mu \rfloor + 1 \ldots, m \end{cases} \qquad (10.210)$$

As a result, we can write the following expression for the diversity:

$$d_{\text{out}} = (n - \lfloor \mu \rfloor)(m - \lfloor \mu \rfloor) - (\mu - \lfloor \mu \rfloor)(n + m - 1 - 2\lfloor \mu \rfloor) \qquad (10.211)$$

which, for integer μ, yields

$$d_{\text{out}} = (n - \mu)(m - \mu) \qquad (10.212)$$

In this case, $\bar{\alpha}_i$ represents an indicator of the usage of the ith equivalent channel: $\bar{\alpha}_i = 0$ means that the ith channel is not used, and vice versa for $\bar{\alpha}_i = 1$. In fact, if $\bar{\alpha}_i = 0$ and hence $\alpha_i = 1$, the ith eigenvalue $\lambda_i = \zeta^{-\alpha_i} \to 0$ as $\zeta \to \infty$. That implies a rate loss due to the inability of using the ith channel. Meanwhile, the diversity $d_{\text{out}}(\mu)$ is increased by $(n - m + 2i - 1)$ units as shown by (10.206). $\square$

This diversity–rate trade-off curve can be used to compare different schemes and to interpret their behavior, as shown in the examples that follow. In particular, we shall see how orthogonal schemes are attractive when high diversity gain is sought, while BLAST interfaces favor rate gain.

2×2 schemes

Consider two transmit and two receive antennas, and a block length chosen to comply with the condition of validity of (10.202), viz., $N \geq t + r - 1$. The maximum diversity gain is $tr = 4$, achieved if each transmitted signal passes through all four propagation paths. The maximum rate gain is $t = r = 2$. The optimal trade-off curve for this system is shown by the continuous line of Figure 10.30.

A simple scheme that achieves maximum diversity is a repetition code:

$$\mathbf{X} = \begin{bmatrix} x_1 & 0 \\ 0 & x_1 \end{bmatrix} \qquad (10.213)$$

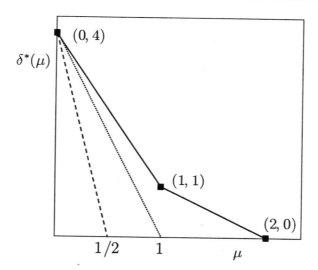

Figure 10.30: *Diversity–rate trade-off for* 2×2 *systems. Continuous line: Optimal trade-off. Dotted line: Alamouti code. Dashed line: Repetition code.*

where x_1 is a signal from a suitable constellation (we may think of this scheme as an inner code concatenated with an outer code that generates x_1). Figure 10.30 shows the trade-off curve for this "repetition" system. Since it takes two channel uses to transmit one symbol, the maximum rate gain is $1/2$. When maximum diversity is achieved, the rate gain is 0. In fact, if a data rate $\mu \log \zeta$ must be supported, the size of the constellation from which x_1 is drawn must increase, and consequently the minimum distance decreases, as does the achievable diversity gain.

The Alamouti code can also be used on this channel. Here

$$\mathbf{X} = \left[\begin{array}{cc} x_1 & -x_2^* \\ x_2 & x_1^* \end{array} \right] \tag{10.214}$$

This achieves full diversity gain. Two symbols are transmitted every two channel uses, and hence the maximum rate gain is 1. Its trade-off curve is shown in Figure 10.30. Notice that, although both the repetition and Alamouti code achieve the optimum diversity at $\mu = 0$, their behavior is markedly different when the diversity–rate trade-off is taken into consideration.

Orthogonal designs

Consider first the special case $t = 2$, with $\mathbf{X}$ given again by (10.214). The optimal trade-off can be computed, and yields

$$\delta^*(\mu) = tr(1 - \mu)_+ \qquad (10.215)$$

More generally, since orthogonal designs with full rate (that is, $\mu = 1$) do not exist for $t > 2$, one can observe that their maximum rate gain is strictly less than 1. Hence, although they achieve maximum diversity at $\mu = 0$, they are strictly suboptimum in terms of the diversity–rate trade-off.

Zero-forcing vertical BLAST

Consider now zero-forcing vertical BLAST (ZF V-BLAST), with m transmit and receive antennas and independent substreams transmitted by each antenna. Its performance, as discussed in Section 10.13.1, depends on the order of detection of the substreams and on the data rates of the substreams. For all versions of V-BLAST, the trade-off curve is suboptimal, especially for low rate gains: in fact, every transmitted substream experiences only m independent fading gains, and, even with no spatial interference between substreams, the trade-off curve cannot exceed $\delta(\kappa) = m - \kappa$.

Zero-forcing diagonal BLAST

This system, which has coding over signals transmitted on different antennas, promises a higher diversity gain. Here, if the rate loss caused by the overhead symbols is disregarded, the trade-off curve connects the points $(m - \kappa, \kappa(\kappa + 1)/2)$, $\kappa = 0, \ldots, m$. Observe that the maximum diversity gain is now $m(m + 1)/2$, better than for V-BLAST but still short of the theoretical maximum m^2. It is recognized [10.2, 10.67] that this performance loss is caused by the zero-forcing step. If MMSE filtering is used instead of ZF, then D-BLAST achieves the optimum trade-off curve (apart from the rate loss mentioned before). This behavior can be justified by observing that D-BLAST achieves the optimum mutual information of the MIMO channel for any realization of channel $\mathbf{H}$ [10.42, Sec. 12.4.1]

10.15 Bibliographical notes

In this chapter we have focused on narrowband channels only. For treatments of MIMO broadband fading channels, and in particular of the impact of frequency selectivity on capacity and on receiver structures, see, e.g., [10.2, 10.13]. The rich-scattering MIMO channel model was introduced and extensively studied in [10.20, 10.21]. The separately correlated channel model is studied in [10.40, 10.51]. The keyhole model is discussed in [10.17, 10.24]. Rician MIMO channels are examined in [10.53]. In [10.60], the authors show how the properties of the physical MIMO channel are reflected into the random-matrix model.

Capacity calculations closely follow [10.58]. Equation (10.28) was obtained by refining a result in [10.50], which in turn was derived by elaborating on (10.45), derived in [10.58]). An alternative expression for the asymptotic capacity variance in the form of an integral is obtained in [10.3]). The limits on communication over a MIMO channel without CSI were derived by Marzetta and Hochwald [10.30, 10.37]. Nonergodic channels and their outage probability were originally examined in [10.20, 10.21, 10.23, 10.58]. The asymptotic normality of the instantaneous mutual information $C(\mathbf{H})$ was obtained independently by several authors under slightly different technical assumptions [10.25, 10.31, 10.41, 10.48, 10.52] (see also [10.5]).

Delay diversity was proposed in [10.49, 10.64]. Space–time trellis codes were introduced in [10.57]. To avoid the rate loss of orthogonal designs, algebraic codes can be designed that, for any number of transmit and receive antennas, achieve maximum diversity, such as Alamouti codes, while the rate is t symbols per channel use (see [10.36] and references therein). Linear codes, called *linear dispersion* codes, were introduced in [10.28], while a more general treatment can be found in [10.35]. Differential space–time coding has been advocated for noncoherent channels: see, e.g., [10.32, 10.33].

Deeeeper discussions of iterative interfaces can be found in [10.7, 10.8, 10.45–10.47] and references therein.

Our discussion of the fundamental trade-off follows [10.67].

10.16 Problems

1. Consider the deterministic MIMO channel. Express its capacity under the constraint that equal energies are transmitted over all channels associated with nonzero singular values of $\mathbf{H}$. Using Jensen's inequality, prove that, among all channels with the same "total power gain" $\|\mathbf{H}\|^2$, the one whose singular values are all equal has the largest constrained capacity.

2. Consider an ergodic fading MIMO channel with $\mathbb{E}|h_{ij}|^2 = 1$ and CSI at the receiver only. Show that, at low SNR, its capacity exceeds that of the

AWGN channel by a factor that depends on r, but neither on t nor on the fading correlation.

3. Prove (10.139).

4. Show that the Alamouti code with $r = 2$ is also an orthogonal design.

5. Recast V-BLAST as a special case of linear space–time codes by computing the matrices $\mathbf{A}_\ell, \mathbf{B}_\ell, \ell = 1, \ldots, L$.

6. Consider a MIMO system with t antennas all transmitting the same symbol $x \in \mathbb{C}$. Assume that the channel matrix $\mathbf{H}$ has independent and identically distributed entries, and that its realization (the CSI) is known at the transmitter and at the receiver. The transmitted symbol is weighted by vector $(\mathbf{gH})^\dagger / a$, where $\mathbf{g} \in \mathbb{C}^r$ and $a \triangleq |\mathbf{gH}|$ is a normalization factor, so that the received signal has the form

$$ \mathbf{y} = \frac{x}{a} \mathbf{H}(\mathbf{gH})^\dagger + \mathbf{z} $$

The receiver estimates x by forming $\hat{x} = \mathbf{gy}$. Compute the vector $\mathbf{g}$ that maximizes the SNR of estimate $\hat{x}$.

7. Prove that, for an uncorrelated keyhole channel, the constrained capacity C^* defined in Section 10.4 satisfies

$$ C^* \leq \log(1 + r\zeta) $$

(This result shows that this channel, regardless of the number of antennas, yields no rate gain).

8. Consider a MIMO system with $r = 1$, and the following transmission scheme. The receiver broadcasts a known probe symbol x_0. Transmitter i, $1 \leq i \leq t$, receives $h_i x_0 + z_i$, z_i a noise term (assume for simplicity that $\mathbb{E}|z_i|^2 = N_0$ for all i), and sends its information symbol x in the form $(h_i^* x_0^* + z_i^*)x$. Assuming binary PSK signaling and an independent Rayleigh fading channel, compute numerically the error probability of this scheme as a function of t and of the ratio between the transmitted energy per bit and the energy of the probe signal.

References

[10.1] S. M. Alamouti, "A simple transmit diversity technique for wireless communications," *IEEE J. Select. Areas Commun.*, Vol. 16, No. 8, pp. 1451–1458, October 1998.

[10.2] S. L. Ariyavisitakul, "Turbo space–time processing to improve wireless channel capacity," *IEEE Trans. Commun.*, Vol. 48, No. 8, pp. 1347–1359, Aug. 2000.

[10.3] Z. D. Bai and J. W. Silverstein, "CLT for linear spectral statistics of large dimensional sample covariance matrices," *Ann. Probabil.*, Vol. 32, No. 1A, pp. 553–605, 2004.

[10.4] S. Benedetto and E. Biglieri, *Principles of Digital Transmission with Wireless Applications*. New York: Kluwer/Plenum, 1999.

[10.5] E. Biglieri and G. Taricco, "Transmission and reception with multiple antennas: Theoretical foundations," *Foundations and Trends in Communications and Information Theory*, Vol. 1, No. 2, 2004.

[10.6] E. Biglieri, G. Caire, and G. Taricco, "Limiting performance of block-fading channels with multiple antennas," *IEEE Trans. Inform. Theory*, Vol. 47, No. 4, pp. 1273–1289, May 2001.

[10.7] E. Biglieri, A. Nordio, and G. Taricco, "Suboptimum receiver interfaces and space-time codes," *IEEE Trans. Sig. Proc.*, Vol. 51, No. 11, pp. 2720–2728, November 2003.

[10.8] E. Biglieri, A. Nordio, and G. Taricco, "Doubly-iterative decoding of space–time turbo codes with a large number of antennas," *Proc. 2004 IEEE Int. Conf. Communications (ICC 2004)*, Paris, France, June 20–24, 2004.

[10.9] E. Biglieri, G. Taricco, and A. Tulino, "Performance of space–time codes for a large number of antennas," *IEEE Trans. Inform. Theory*, Vol. 48, No. 7, pp. 1794–1803, July 2002.

[10.10] E. Biglieri, G. Taricco, and A. Tulino, "Decoding space-time codes with BLAST architectures," *IEEE Trans. Sig. Proc.*, Vol. 50, No. 10, pp. 2547–2552, October 2002.

[10.11] M. Bilodeau and D. Brenner, *Theory of Multivariate Statistics*. New York: Springer, 1999.

[10.12] N. Bleistein and R.A. Handelsman, *Asymptotic Expansions of Integrals*. Dover, 1986

[10.13] H. Boelcskei, D. Gesbert, and A. J. Paulraj, "On the capacity of OFDM-based spatial multiplexing systems," *IEEE Trans. Commun.*, Vol. 50, No. 2, pp. 225–234, February 2002.

[10.14] J. K. Cavers and P. Ho, "Analysis of the Error Performance of Trellis-Coded Modulations in Rayleigh Fading Channels," *IEEE Trans. Commun.*, Vol. 40, No. 1, pp. 74–83, January 1992.

[10.15] J.-J. Chang, D.-J. Hwang, and M.-C. Lin, "Some extended results on the search for good convolutional codes," *IEEE Trans. Inform. Theory*, Vol. 43, No. 5, pp. 1682–1697, September 1997.

[10.16] N. Chiurtu, B. Rimoldi, and E. Telatar, "Dense multiple antenna systems," *Proc. IEEE Information Theory Workshop ITW2001*, Cairns, Australia, pp. 108–109, September 2–7, 2001.

[10.17] D. Chizhik, G.J. Foschini, M.J. Gans, and R.A. Valenzuela, "Keyholes, correlations, and capacities of multielement transmit and receive antennas," *IEEE Trans. Wireless Commun.*, Vol. 1, No. 2, pp. 361–368, April 2002

[10.18] C.-N. Chuah, D. N. C. Tse, J. M. Kahn, and R. A. Valenzuela, "Capacity bounds via duality with applications to multiple-antenna systems on flat-fading channels," *IEEE Trans. Inform. Theory*, Vol. 48, No. 3, pp. 637–650, March 2002.

[10.19] H. El-Gamal and A. R. Hammons, Jr., "A new approach to layered space–time coding and signal processing," *IEEE Trans. Inform. Theory*, Vol. 47, No. 6, pp. 2321–2334, September 2001.

[10.20] G. J. Foschini, "Layered space-time architecture for wireless communication in a fading environment when using multi-element antennas," *Bell Labs Tech. J.*, Vol. 1, No. 2, pp. 41–59, Autumn 1996.

[10.21] G. J. Foschini and M. J. Gans, "On limits of wireless communications in a fading environment when using multiple antennas," *Wireless Personal Commun.*, Vol. 6, No. 3, pp. 311–335, March 1998.

[10.22] G. J. Foschini, G. D. Golden, R. A. Valenzuela, and P. W. Wolniansky, "Simplified processing for high spectral efficiency wireless communication employing multi-element arrays," *IEEE J. Select. Areas Commun.*, Vol. 17, No. 11, pp. 1841–1852, November 1999.

[10.23] G. J. Foschini and R. A. Valenzuela, "Initial estimation of communication efficiency of indoor wireless channels," *Wireless Networks*, Vol. 3, No. 2, pp. 141–154, 1997.

[10.24] D. Gesbert, H. Bölcskei, D. Gore, and A. Paulraj, "Outdoor MIMO wireless channels: Models and performance prediction," *IEEE Trans. Commun.*, Vol. 50, No. 12, pp. 1926–1934, December 2002.

[10.25] V. L. Girko, "A refinement of the central limit theorem for random determinants," *Theory Probab. Appl.*, Vol. 42, No. 1, pp. 121–129, 1997.

[10.26] A. Goldsmith, S. A. Jafar, N. Jindal, and S. Vishwanath, "Capacity limits of MIMO channels," *IEEE J. Select. Areas Commun.*, Vol. 21, No. 5, pp. 684–702, June 2003.

[10.27] A. R. Hammons, Jr., and H. El Gamal, "On the theory of space–time codes for PSK modulation," *IEEE Trans. Inform. Theory*, Vol. 46, No. 2, pp. 524–542, March 2000.

[10.28] B. Hassibi and B. M. Hochwald, "High-rate codes that are linear in space and time," *IEEE Trans. Inform. Theory*, Vol. 48, No. 7, pp. 1804–1824, July 2002.

[10.29] B. Hassibi, B. M. Hochwald, A. Shokrollahi, and W. Sweldens, "Representation theory for high-rate multiple-antenna code design," *IEEE Trans. Inform. Theory*, Vol. 47, No. 6, pp. 2335–2367, September 2001.

[10.30] B. Hochwald and T. Marzetta, "Unitary space-time modulation for multiple-antenna communication in Rayleigh flat-fading," *IEEE Trans. Inform. Theory*, Vol. 46, No. 2, pp. 543–564, March 2000.

[10.31] B. M. Hochwald, T. L. Marzetta, and V. Tarokh, "Multi-antenna channel-hardening and its implications for rate feedback and scheduling," *IEEE Trans. Inform. Theory*, Vol. 50, No. 9, pp. 1893–1909, September 2004.

[10.32] B. Hochwald and W. Sweldens, "Differential unitary space–time modulation," *IEEE Trans. Commun.*, Vol. 48, No. 12, pp. 2041–2052, December 2000.

[10.33] B. L. Hughes, "Differential space-time modulation," *IEEE Trans. Inform. Theory*, Vol. 46, No. 11, pp. 2567–2578, November 2000.

[10.34] A. Lapidoth and S. M. Moser, "Capacity bounds via duality with applications to multiple-antenna systems on flat-fading channels," *IEEE Trans. Inform. Theory*, Vol. 49, No. 10, pp. 2426–2467, October 2003.

[10.35] E.G. Larsson and P. Stoica, *Space–Time Block Coding for Wireless Communications*. Cambridge, UK: Cambridge University Press, 2003.

[10.36] X. Ma and G. B. Giannakis, "Full-diversity full-rate complex-field space–time coding," *IEEE Trans. Sig. Proc.*, Vol. 51, No. 11, pp. 2917–2930, November 2003.

[10.37] T. L. Marzetta and B. M. Hochwald, "Capacity of a mobile multiple-antenna communication link in Rayleigh flat fading," *IEEE Trans. Inform. Theory*, Vol. 45, No. 1, pp. 139–157, January 1999.

[10.38] T. L. Marzetta, B. M. Hochwald, and B. Hassibi, "New approach to single-user multiple-antenna wireless communication," *Proc. 2000 Conference on Information Sciences and Systems*, Princeton University, pp. WA4-16–WA4-21, March 15–17, 2000.

[10.39] A. F. Molish, M. Z. Win, and J. H. Winters, "Capacity of MIMO systems with antenna selection," *Proc. IEEE International Conference on Communications (ICC 2001)*, Helsinki, Finland, June 11–14, 2001.

[10.40] A. L. Moustakas, H. U. Baranger, L. Balents, A. M. Sengupta, and S.H. Simon, "Communication through a diffusive medium: Coherence and capacity," *Science*, Vol. 287, pp. 287–290, January 2000.

[10.41] A. L. Moustakas, S. H. Simon, A. M. Sengupta, "MIMO capacity through correlated channels in the presence of correlated interferers and noise: A (not so) large N analysis," *IEEE Trans. Inform. Theory* Vol. 49, No. 10, pp. 2545–2561, October 2003.

[10.42] A. Paulraj, R. Nabar, and D. Gore, *Introduction to Space-Time Wireless Communications*. Cambridge, UK: Cambridge University Press, 2003.

[10.43] P. B. Rapajic and D. Popescu, "Information capacity of a random signature multiple-input multiple-output channel," *IEEE Trans. Commun.*, Vol. 48, No. 8, pp. 1245–1248, August 2000.

[10.44] M. Schwartz, W. R. Bennett, and S. Stein, *Communications Systems and Techniques*. New York: McGraw-Hill, 1966.

[10.45] M. Sellathurai and S. Haykin, "Turbo-BLAST for wireless communications: Theory and experiments," *IEEE Trans. Sig. Proc.*, Vol. 50, No. 10, pp. 2538–2546, October 2002.

[10.46] M. Sellathurai and S. Haykin, "Turbo-BLAST: Performance evaluation in correlated Rayleigh-fading environment," *IEEE J. Select. Areas Commun.*, Vol. 21, No. 3, pp. 340–349, April 2003.

[10.47] M. Sellathurai and S. Haykin, "T-BLAST for wireless communications: First experimental results," *IEEE Trans. Vehicular Technology*, Vol. 52, No. 3, pp. 530–535, May 2003.

[10.48] A. M. Sengupta and P. P. Mitra, "Capacity of multivariate channels with multiplicative noise: Random matrix techniques and large-N expansions for full transfer matrices," *LANL arXiv:physics*, 31 October 2000.

[10.49] N. Seshadri and J. H. Winters, "Two signaling schemes for improving the error performance of frequency-division duplex (FDD) transmission systems using transmitter antenna diversity," *Proc. 43rd IEEE Vehicular Technology Conference (VTC 1993)*, pp. 508–511, May 18–20, 1993.

[10.50] H. Shin and J. H. Lee, "Capacity of multiple-antenna fading channels: Spatial fading correlation, double scattering, and keyhole," *IEEE Trans. Inform. Theory*, Vol. 49, No. 10, pp. 2636–2647, October 2003.

[10.51] D. Shiu, G. Foschini, M. J. Gans, and J. M. Kahn, "Fading correlation and its effect on the capacity of multielement antenna systems", *IEEE Trans. Commun.*, Vol. 48, No. 3, pp. 502–513, March 2000.

[10.52] P. J. Smith and M. Shafi, "On a Gaussian approximation to the capacity of wireless MIMO systems," *Proc. IEEE International Conference on Communications (ICC 2002)*, pp. 406–410, New York, April 28–May 2, 2002.

[10.53] P. Soma, D. S. Baum, V. Erceg, R. Krishnamoorty, and A. J. Paulraj, "Analysis and modeling of multiple-input multiple-output (MIMO) radio channel based on outdoor measurements conducted at 2.5 GHz for fixed BWA applications," in *Proc. IEEE ICC 2002*, Vol. 1, pp. 272–276, New York, NY, May 28–April 2, 2002.

[10.54] G. Szegö, *Orthogonal Polynomials.* Providence, RI: American Mathematical Society, 1939.

[10.55] G. Taricco and E. Biglieri, "Space–time decoding with imperfect channel estimation," *IEEE Trans. Wireless Commun.*, in press, 2005.

[10.56] V. Tarokh and T. K. Y. Lo, "Principal ratio combining for fixed wireless applications when transmitter diversity is employed," *IEEE Commun. Lett.*, Vol. 2, No. 8, pp. 223–225, August 1998.

[10.57] V. Tarokh, N. Seshadri, and A. R. Calderbank, "Space-time codes for high data rate wireless communication: Performance criterion and code construction," *IEEE Trans. Inform. Theory* Vol. 44, No. 2, pp. 744 – 765, March 1998.

[10.58] E. Telatar, "Capacity of multi-antenna Gaussian channels," *Eur. Trans. Telecomm.*, Vol. 10, No. 6, pp. 585–595, November–December 1999.

[10.59] D. Tse and V. Hanly, "Multi-access fading channels—Part I: Polymatroid structure, optimal resource allocation and throughput capacities," *IEEE Trans. Inform. Theory,* Vol. 44, No. 7, pp. 2796–2815, November 1998.

[10.60] D. Tse and P. Viswanath, *Fundamentals of Wireless Communication.* Cambridge, UK: Cambridge University Press, 2005, to be published.

[10.61] S. Verdú, *Multiuser Detection.* Cambridge, UK: Cambridge University Press, 1998.

[10.62] S. Verdú and S. Shamai (Shitz), "Spectral efficiency of CDMA with random spreading," *IEEE Trans. Inform. Theory*, Vol. 45, No. 2, pp. 622–640, March 1999.

[10.63] H. Wang and X.-G. Xia, "Upper bounds of rates of complex orthogonal space–time block codes," IEEE Trans. Inform. Theory, Vol. 49, No. 10, pp. 2788–2796, October 2003.

[10.64] A. Wittneben, "Basestation modulation diversity for digital SIMULCAST," *Proc. 41st Vehicular Technology Conference (VTC 1991)*, pp. 848–853, May 19–22, 1991.

[10.65] D. Wuebben, R. Boehnke, J. Rinas, V. Kuehn, and K. D. Kammeyer, "Efficient algorithm for decoding layered space-time codes," *Electron. Lett.*, Vol. 37, No. 22, pp. 1348–1349, 25 October 2001.

[10.66] L. Zheng and D. N. C. Tse, "Communication on the Grassman manifold: A geometric approach to the noncoherent multiple-antenna channel," *IEEE Trans. Inform. Theory*, Vol. 48, No. 2, pp. 359–383, February 2002.

[10.67] L. Zheng and D. N. C. Tse, "Diversity and multiplexing: A fundamental tradeoff in multiple antenna channels," *IEEE Trans. Inform. Theory*, Vol. 49, No. 5, pp. 1073–1096, May 2003.

A

There's Ada,

Facts from information theory

In this appendix we review the basic definitions and provide some facts from information theory that are needed in the rest of the book. In particular, we compute the capacity of MIMO channels.

A.1   Basic definitions

We start with the definition of quantity of information carried by a signal x, chosen from an constellation $\mathcal{X}$ and transmitted with probability $p(x)$. The information content of signal x, denoted by $I(x)$, increases with the uncertainty of its transmission:

$$I(x) \triangleq \log \frac{1}{p(x)} \tag{A.1}$$

where log, here and throughout this book, denotes logarithm to base 2. This information is measured in *bits*. One bit of information is conveyed, for example, by the transmission of one out of two equally likely signals.

Denote by $\mathbb{E}_x$ the expectation taken with respect to the probability measure $p(x)$ on $\mathcal{X}$, that is, $\mathbb{E}_x[f(x)] \triangleq \sum_x p(x)f(x)$ for any function $f(x)$. Then the average information content of x is

$$H(x) \triangleq \mathbb{E}_x[I(x)] = \mathbb{E}_x \log \frac{1}{p(x)} \tag{A.2}$$

This is called the *entropy* of $\mathcal{X}$, and is measured in bit/signal.[1]

Example A.1

If the source constellation consists of M equally likely signals, then $p(x) = 1/M$, and we have

$$H(x) = \sum_{i=1}^{M} \frac{1}{M} \log M = \log M \qquad \text{bit/signal}$$

When $\mathcal{X}$ consists of two signals with probabilities p and $1 - p$, the entropy of x is

$$H(x) = p \log \frac{1}{p} + (1 - p) \log \frac{1}{1 - p} \triangleq H(p) \tag{A.3}$$

It can be seen that the maximum of the function $H(p)$ defined in (A.3) occurs for $p = 0.5$, that is, when the two signals are equally likely. More generally, it can be proved that, if x is chosen from a finite constellation $\mathcal{X}$, then its entropy is maximum when $p(x)$ is uniform, i.e., $p(x) = 1/|\mathcal{X}|$. □

---

[1]The notation here is not felicitous, as $H(x)$ is not a function of x, but rather of $\mathcal{X}$ and of the probability measure defined on $\mathcal{X}$. However, it has the advantage of simplicity, and should not be confusing.

A communication channel is the physical medium used to connect a source of information with its user. The channel accepts signals x belonging to the input constellation $\mathcal{X}$, and outputs signals y belonging to the output constellation $\mathcal{Y}$ (unless specified otherwise, we assume that both $\mathcal{X}$ and $\mathcal{Y}$ are discrete). A channel is characterized by the two constellations $\mathcal{X}$ and $\mathcal{Y}$ and by the conditional probabilities $p(y \mid x)$ of receiving the signal y given that the signal x has been transmitted. ($p(y \mid x)$ may be a probability or a probability density function, according to the structure of the input and output constellations.) A channel is called *stationary memoryless* if

$$p(y_1, \ldots, y_n \mid x_1, \ldots, x_n) = \prod_{i=1}^{n} p(y_i \mid x_i) \tag{A.4}$$

where $x_1, \ldots, x_n$ and $y_1, \ldots, y_n$ represent n consecutive transmitted and received signals, respectively.

Given the input and output channel constellations $\mathcal{X}$ and $\mathcal{Y}$, and their probabilistic dependence specified by the *channel transition function* $p(y \mid x)$, we can define five entropies, viz.,

(a) The *input* entropy $\mathrm{H}(x)$,

$$\mathrm{H}(x) \triangleq \mathbb{E}_x \log \frac{1}{p(x)} \quad \text{bit/signal} \tag{A.5}$$

which measures the average information content of the input constellation, that is, the information we want to transfer through the channel.

(b) The *output* entropy $\mathrm{H}(y)$,

$$\mathrm{H}(y) \triangleq \mathbb{E}_y \log \frac{1}{p(y)} \quad \text{bit/signal} \tag{A.6}$$

which measures the average information content of the signal observed at channel output.

(c) The *joint* entropy $\mathrm{H}(x, y)$,

$$\mathrm{H}(x, y) \triangleq \mathbb{E}_{x,y} \log \frac{1}{p(x, y)} \quad \text{bit/signal pair} \tag{A.7}$$

which measures the average information content of a pair of input and output signals. This is the average uncertainty of the communication system formed by the input constellation, the channel, and the output constellation as a whole.

(d) The *conditional* entropy $H(y \mid x)$,

$$H(y \mid x) \triangleq \mathbb{E}_{x,y} \log \frac{1}{p(y \mid x)} \quad \text{bit/signal} \qquad (A.8)$$

which measures the average information quantity needed to specify the output signal y when the input signal x is known. In other words, this measures the average residual uncertainty on signal y when signal x is known.

(e) The *conditional* entropy $H(x \mid y)$,

$$H(x \mid y) \triangleq \mathbb{E}_{x,y} \log \frac{1}{p(x \mid y)} \quad \text{bit/signal} \qquad (A.9)$$

which measures the average information quantity needed to specify the input (or transmitted) signal x when the output (or received) signal y is known. Equivalently, this is the average residual uncertainty on the transmitted signal x when the received signal y is observed at the output of the channel: thus, this conditional entropy (often called *channel equivocation*) represents the average amount of information that has been lost on the channel. (A limiting case is the noiseless channel, for which the channel output equals its input, so we have $H(x \mid x) = 0$. Another limiting case is the infinitely noisy channel, for which input and output are statistically independent, and hence $H(x \mid y) = H(x)$.)

Using these definitions, the following equations and inequalities can be proved:

$$H(x, y) = H(y, x) = H(x) + H(y \mid x) = H(y) + H(x \mid y) \qquad (A.10)$$

$$H(x \mid y) \leq H(x) \qquad (A.11)$$

$$H(y \mid x) \leq H(y) \qquad (A.12)$$

These are summarized in Figure A.1. Inequalities (A.11) and (A.12) become equalities if $x \perp\!\!\!\perp y$, that is, $p(x, y) = p(x)p(y)$.

As mentioned before, the conditional pdf $p(y \mid x)$, $x \in \mathcal{X}$, $y \in \mathcal{Y}$, characterizes a stationary memoryless channel. If $\mathcal{X}$ and $\mathcal{Y}$ are finite constellations, the values of $p(y \mid x)$ can be arranged in a matrix $\mathbf{P}$ whose entries are $(\mathbf{P})_{x,y} = p(y \mid x)$. An important channel example occurs when $\mathbf{P}$ is such that at the same time all its rows are permutations of the first one and all its columns are permutations of the first one. In this case the channel is said to be *symmetric*. A central property of the symmetric channel is that $H(\mathcal{Y} \mid \mathcal{X})$ is independent of the input probabilities $p(x)$ and hence depends only on the channel matrix $\mathbf{P}$.

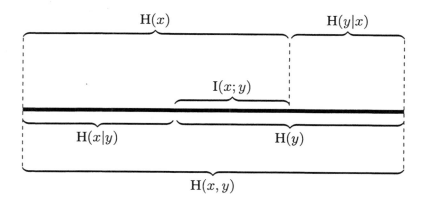

Figure A.1: *Relationships among different entropies and mutual information.*

Example A.2 (Binary symmetric channel)

An important special case of a symmetric channel occurs when $|\mathcal{X}| = |\mathcal{Y}| = 2$, and

$$\mathbf{P} = \left[\begin{array}{cc} 1-p & p \\ p & 1-p \end{array} \right]$$

This is called the *Binary Symmetric Channel* (BSC).   ⬜

A.2 Mutual information and channel capacity

A part of the information $H(x)$ transmitted over a noisy channel is lost. This part is measured by the channel equivocation $H(x \mid y)$. Thus, it seems natural to define an average information flow through the channel, called the *mutual information* between x and y and denoted $I(x; y)$, as

$$I(x; y) \triangleq H(x) - H(x \mid y) \qquad \text{bit/signal} \qquad (A.13)$$

Using (A.10) (see also Figure A.1), the following alternative forms can be derived:

$$I(x; y) = H(y) - H(y \mid x) = H(x) + H(y) - H(x, y) \qquad (A.14)$$

Comparing (A.13) with the first equality of (A.14), it is apparent that $I(x; y) = I(y; x)$.

Example A.3

Consider again the two limiting cases of a noiseless channel and an infinitely noisy channel. In the first case, we observe $y = x$, so there is no uncertainty left on x: we have $H(x \mid y) = 0$, and hence from (A.13) we have $I(x; y) = H(x)$: the mutual information is exactly the entropy of the source, as nothing is lost over the channel. In the second case, y is independent of x: no information on x can be gathered by observing y, and hence $H(x \mid y) = H(x)$ (after observing y, the uncertainty on x is the same as without any observation). We have $I(x; y) = 0$ (all information is lost over the channel). □

Example A.4

Let us consider again the BSC and see how $I(x; y)$ depends on the probability distribution of the input signals. Direct calculation gives

$$I(x; y) = H(y) - H(p) \tag{A.15}$$

where the function $H(p)$ was defined in (A.3). The maximum value of $I(x; y)$, irrespective of the value of p, is obtained when the input signals are equally likely. Thus, equally likely input signals yield the maximum information flow through a BSC. This is given by

$$\max_{p(x)} I(x; y) = 1 - H(p) = 1 + p \log p + (1 - p) \log(1 - p) \tag{A.16}$$

where the maximum is taken with respect to all possible probability distributions of the input signals. If the channel is not symmetric (for example, transmitted 0s are more affected by noise than the 1s), then the more "robust" input signals should be transmitted more often for maximum information transfer. □

In general, the maximum value of $I(\mathcal{X}; \mathcal{Y})$ for a given memoryless channel is called its *capacity* and is denoted by C.

Example A.5

In the BSC, capacity is maximum when p is equal to 0 or equal to 1, since both these situations lead to a noiseless channel. For $p = 0.5$, the capacity is zero, since the output signals turn out to be independent from the input signals, and no information is transferred through the channel. □

A.2.1   Channel depending on a parameter

Suppose that the channel depends on a random parameter H independent of x and y. The following equality can be proved:

$$I(x; y, H) = I(x; H) + I(x; y \mid H) \tag{A.17}$$

A.3   Measure of information in the continuous case

Assume now that x is a continuous RV taking values in $\mathcal{X}$ with pdf $p(x)$. The entropy of $\mathcal{X}$ can still be defined formally through (A.2); however, some differences arise, the main one being that $H(x)$ may be arbitrarily large, positive, or negative. As we did for discrete $\mathcal{X}$ and $\mathcal{Y}$, we can define, for two continuous random variables x and y having a joint probability density function $p(x, y)$, the joint entropy $H(x, y)$ and the conditional entropies $H(x \mid y)$ and $H(y \mid x)$. If both $H(x)$ and $H(y)$ are finite, the relationships represented pictorially in Figure A.1 still hold. As in the discrete case, the inequalities (A.11) and (A.12) become equalities if x and y are statistically independent.

In the discrete case, the entropy of x is a maximum if all $x \in \mathcal{X}$ are equally likely. In the continuous case, the following theorems hold.

Theorem A.3.1 *Let x be a real, zero-mean continuous RV with pdf $p(x)$. If x has finite variance σ_x^2, then $H(x)$ satisfies the inequality*

$$H(x) \le \frac{1}{2} \log(2\pi e \sigma_x^2) \tag{A.18}$$

with equality if and only if $x \sim \mathcal{N}(0, \sigma_x^2)$.

A case more general than the above is obtained by considering, instead of a scalar random variable x, a complex random vector $\mathbf{x}$:

Theorem A.3.2 *Let $\mathbf{x}$ be a zero-mean complex random vector with covariance matrix $\mathbf{R_x}$. Then $H(\mathbf{x})$ satisfies the inequality*

$$H(\mathbf{x}) \le \log \det(\pi e \mathbf{R_x}) \tag{A.19}$$

with equality if and only if $\mathbf{x} \sim \mathcal{N}_c(\mathbf{0}, \mathbf{R_x})$.

Proof

Let $\mathbf{x}_G \sim \mathcal{N}_c(\mathbf{0}, \mathbf{R}_{\mathbf{x}_G})$, and calculate its entropy by using the pdf (C.24):

$$
\begin{aligned}
H(\mathbf{x}_G) &= \mathbb{E}\left[\log \det\left(\pi \mathbf{R}_{\mathbf{x}_G}\right) + \mathbf{x}_G^\dagger \mathbf{R}_{\mathbf{x}_G}^{-1} \mathbf{x}_G \log e\right] \\
&= \log \det\left(\pi \mathbf{R}_{\mathbf{x}_G}\right) + \mathbb{E}\left[\mathbf{x}_G^\dagger \mathbf{R}_{\mathbf{x}_G}^{-1} \mathbf{x}_G\right] \log e \\
&= \log \det\left(\pi \mathbf{R}_{\mathbf{x}_G}\right) + \mathrm{Tr}\left(\mathbf{R}_{\mathbf{x}_G}^{-1} \mathbb{E}[\mathbf{x}_G \mathbf{x}_G^\dagger]\right) \log e \\
&= \log \det\left(\pi e \mathbf{R}_{\mathbf{x}_G}\right) \qquad\qquad\qquad\qquad\qquad \text{(A.20)}
\end{aligned}
$$

Let $p(\mathbf{x})$ and $p_G(\mathbf{x})$ be the pdfs of $\mathbf{x}$ and $\mathbf{x}_G$, respectively. The theorem follows by

$$
\begin{aligned}
H(\mathbf{x}) - H(\mathbf{x}_G) &= \int \left(p_G(\mathbf{x}) \log p_G(\mathbf{x}) - p(\mathbf{x}) \log p(\mathbf{x})\right) d\mathbf{x} \\
&= \int p(\mathbf{x}) \left(\log p_G(\mathbf{x}) - \log p(\mathbf{x})\right) d\mathbf{x} \\
&= \int p(\mathbf{x}) \log \frac{p_G(\mathbf{x})}{p(\mathbf{x})} d\mathbf{x} \\
&\leq \int p(\mathbf{x}) \left(\frac{p_G(\mathbf{x})}{p(\mathbf{x})} - 1\right) d\mathbf{x} \\
&= 0 \qquad\qquad\qquad\qquad\qquad\qquad\qquad\qquad \text{(A.21)}
\end{aligned}
$$

where we used the equality $\mathbb{E}\left[\log p_G(\mathbf{x})\right] = \mathbb{E}\left[\log p_G(\mathbf{x}_G)\right]$ and $\ln u \leq u - 1$ for $u > 0$, with equality if and only if $u = 1$.[2] Notice that equality holds if and only if $p(\mathbf{x}) = p_G(\mathbf{x})$, i.e., if and only if $\mathbf{x}$ is circularly symmetric Gaussian. $\qquad \square$

A.4   Shannon theorem on channel capacity

The celebrated Shannon theorem on channel capacity states that the information capacity of a channel is also the maximum rate at which information can be transmitted through it. Specifically, if the channel capacity is C then a sequence of rate-C codes with block length n exists such that as $n \to \infty$ their word-error probability tends to zero. Conversely, if a sequence of rate-ρ codes with block length n has word-error probabilities tending to 0 as $n \to \infty$, then necessarily $\rho \leq C$.

Quantitatively, we can say that if we transmit data over a noisy channel, then there exists a code with block length n and rate ρ bit/dimension for which the error

---

[2]In fact, the function $f(u) \triangleq \ln u - u + 1$ has derivative $f'(u) = u^{-1} - 1$, which is positive for $0 < u < 1$ and negative for $u > 1$. Hence, it has a maximum at $u = 1$, i.e., $f(u) \leq f(1) = 0$.

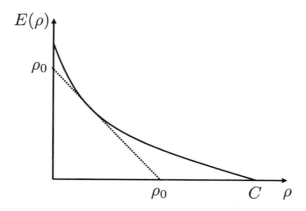

Figure A.2: *Reliability function and cutoff rate of a channel.*

probability is bounded above by

$$P(e) \leq 2^{-nE(\rho)} \tag{A.22}$$

where $E(\rho)$, the *channel reliability function*, is a convex $\cup$, decreasing, nonnegative function of ρ for $0 \leq \rho \leq C$, and C is the channel capacity. We notice that $E(\rho)$ can be taken as a measure of the channel quality when a rate-ρ code is to be used on it: in fact, the larger the value of $E(\rho)$, the smaller the upper bound to error probability for any given block length n. From (A.22) we can also see that by increasing the block length n the error probability can be decreased at will, provided that the transmission occurs at a rate strictly less than C. Thus, computing C we obtain a range of rates for which reliable transmission is possible, and hence an indication of the quality of the channel when coding is to be used on it.

Another parameter for the comparison of coding channels is offered by the so-called *cutoff rate* of the channel. This is obtained by lower-bounding the reliability function $E(\rho)$ with a straight line with slope $-45°$ and tangent to $E(\rho)$ (see Figure A.2).

The intercept of this line over the abscissa and the ordinate axis is ρ_0. Thus, we have

$$P(e) \leq 2^{-n(\rho_0-\rho)}, \qquad \rho \leq \rho_0 \tag{A.23}$$

It is seen that while C yields a range of rates where reliable transmission is possible, ρ_0 yields both a range of rates and an exponent to error probability, i.e., an indication of how fast $P(e)$ tends to zero when n is increased. Moreover, it is generally easier to compute than the capacity. For a long time it was widely believed that ρ_0 was also the rate beyond which reliable transmission would become very complex, so ρ_0 was considered as a *practical* bound to the transmission rate. However, the discovery of classes of codes that admit a practical decoding algorithm and yet have performance close to capacity (LDPC codes, turbo codes) has somewhat diminished the importance of ρ_0 as a performance parameter.

A.5   Capacity of the Gaussian MIMO channel

Let the channel input–output relationship be

$$\mathbf{y} = \mathbf{H}\mathbf{x} + \mathbf{z} \tag{A.24}$$

where $\mathbf{H}$ is a constant $r \times t$ matrix whose realization is called the *channel-state information* (CSI), $\mathbf{x}$ is a t-vector, and $\mathbf{y}$ and $\mathbf{z}$ are r-vectors. Assume $\mathbf{x}\perp\!\!\!\perp\mathbf{z}$ and $\mathbf{z} \sim \mathcal{N}_c(\mathbf{0}, N_0\mathbf{I}_r)$. From the mutual information

$$I(\mathbf{x}; \mathbf{y}) = H(\mathbf{y}) - H(\mathbf{y} \mid \mathbf{x}) = H(\mathbf{y}) - H(\mathbf{z}) \tag{A.25}$$

we seek the channel capacity under the constraint

$$\mathrm{Tr}\,(\mathbf{R_x}) \leq t\mathcal{E} \tag{A.26}$$

For a given $\mathbf{R_x}$, the covariance matrix of $\mathbf{y}$ is $\mathbf{R_y} = \mathbf{H}\mathbf{R_x}\mathbf{H}^\dagger + N_0\mathbf{I}_r$, and $H(\mathbf{y})$ is maximum for $\mathbf{y} \sim \mathcal{N}_c(\mathbf{0}, \mathbf{R_y})$ (Theorem A.3.2). Moreover, the maximum mutual information is given by

$$I(\mathbf{R_x}) = \log \det\,(\mathbf{I}_r + N_0^{-1}\mathbf{R_x}\mathbf{H}^\dagger\mathbf{H}) \tag{A.27}$$

The channel capacity can be calculated according to different assumptions:

(a) The receiver has perfect CSI and the transmitter has no CSI.

(b) The receiver and the transmitter have perfect CSI.

With assumption (a), the transmitter divides the available power evenly among the transmit antennas, and the capacity is

$$C_{\mathrm{rx}} = \log \det\,\left(\mathbf{I}_r + \frac{\zeta}{t}\mathbf{H}^\dagger\mathbf{H}\right) \tag{A.28}$$

where $\zeta \triangleq t\mathcal{E}/N_0$. With assumption (b), the capacity can be written as

$$C_{\mathrm{tx/rx}} = \max_{\mathbf{R_x}\geq 0,\,\mathrm{Tr}[\mathbf{R_x}]\leq t\mathcal{E}} \log \det\,(\mathbf{I}_r + N_0^{-1}\mathbf{R_x}\mathbf{H}^\dagger\mathbf{H}) \tag{A.29}$$

From Hadamard's inequality (B.23) and the orthogonal diagonalization

$$N_0^{-1}\mathbf{H}^\dagger\mathbf{H} = \mathbf{U}\mathbf{D}\mathbf{U}^\dagger$$

(where $\mathbf{U}$ is unitary and $\mathbf{D}$ diagonal), we have

$$
\begin{aligned}
\log\det\left(\mathbf{I}_r + N_0^{-1}\mathbf{R_x}\mathbf{H}^\dagger\mathbf{H}\right) &= \log\det\left(\mathbf{I}_r + \widetilde{\mathbf{R}}\mathbf{D}\right) \\
&\leq \sum_{(\mathbf{D})_{i,i}>0} \log(1 + (\widetilde{\mathbf{R}})_{i,i}(\mathbf{D})_{i,i}) \quad \text{(A.30)}
\end{aligned}
$$

where $\widetilde{\mathbf{R}} \triangleq \mathbf{U}^\dagger\mathbf{R_x}\mathbf{U}$, with equality if and only if $\widetilde{\mathbf{R}}$ is diagonal. Since constraint (A.26) translates into $\mathrm{Tr}\,(\widetilde{\mathbf{R}}) \leq t\mathcal{E}$, the maximization problem admits the *water-filling* solution [A.1]:

$$(\widetilde{\mathbf{R}})_{i,i} = \left(\mu - (\mathbf{D})_{i,i}^{-1}\right)_+ \quad \text{(A.31)}$$

with μ obtained as the solution of

$$\sum_{(\mathbf{D})_{i,i}>0}\left(\mu - (\mathbf{D})_{i,i}^{-1}\right)_+ = t\mathcal{E} \quad \text{(A.32)}$$

The channel input pdf achieving capacity is circularly Gaussian:

$$\mathbf{x} \sim \mathcal{N}_c(\mathbf{0}, \mathbf{U}\mathrm{diag}((\mu - (\mathbf{D})_{i,i}^{-1})_+)\mathbf{U}^\dagger)$$

A.5.1 Ergodic capacity

Consider the separately correlated fading channel described in Section 10.2.1, with channel matrix given by (10.11). Assume that the receiver has perfect CSI but the transmitter has no CSI. Here we calculate the average capacity under the power constraint $\mathbb{E}[\|\mathbf{x}\|^2] \leq t\mathcal{E} = \zeta N_0$ in the case $\mathbf{T} = \mathbf{I}_t$. This derivation follows the guidelines of [A.6] for the case of $\mathbf{R} = \mathbf{I}_r$ and [A.4] in the more general setting of $\mathbf{R} \neq \mathbf{I}_r$, although we restrict ourselves to consideration of iid, $\sim \mathcal{N}_c(0,1)$ entries of $\mathbf{H}_u$.

From the expression of the mutual information (A.25), capacity is given by

$$C = \max_{\mathbf{R_x}\geq 0,\mathrm{Tr}[\mathbf{R_x}]\leq t\mathcal{E}} \mathbb{E}[\log\det\left(\mathbf{I}_r + N_0^{-1}\mathbf{H}\mathbf{R_x}\mathbf{H}^\dagger\right)] \quad \text{(A.33)}$$

Using the orthogonal diagonalization $\mathbf{R_x} = \mathbf{U}\mathbf{D}\mathbf{U}^\dagger$ (with matrix $\mathbf{U}$ unitary and $\mathbf{D}$ diagonal), we notice that capacity can also be obtained as follows:

$$C = \max_{\mathbf{D}\geq 0,\mathrm{Tr}[\mathbf{D}]\leq\zeta} \mathbb{E}[\log\det\left(\mathbf{I}_r + \mathbf{H}\mathbf{D}\mathbf{H}^\dagger\right)] \quad \text{(A.34)}$$

where the maximum is sought over the set of diagonal matrices with nonnegative diagonal entries and trace upperbounded by ζ. The equivalence derives from the fact that $\mathbf{H}_u$ and $\mathbf{H}_u\mathbf{U}$ (and hence $\mathbf{H}$ and $\mathbf{HU}$) have the same joint pdf. Let us write

$$\Psi(\mathbf{D}) \triangleq \mathbb{E}\left[\log \det\left(\mathbf{I}_r + \mathbf{HDH}^\dagger\right)\right] \tag{A.35}$$

Since the log-det function is concave [A.1, Th. 16.8.1], we have, for every vector (α_i) such that $\alpha_i \geq 0$ and $\sum_i \alpha_i = 1$,

$$
\begin{aligned}
\Psi\left(\sum_i \alpha_i \mathbf{D}_i\right) &= \mathbb{E}\left[\log \det\left(\sum_i \alpha_i(\mathbf{I}_r + \mathbf{HD}_i\mathbf{H}^\dagger)\right)\right] \\
&\geq \sum_i \alpha_i \mathbb{E}\left[\log \det\left(\mathbf{I}_r + \mathbf{HD}_i\mathbf{H}^\dagger\right)\right] \\
&= \sum_i \alpha_i \Psi(\mathbf{D}_i)
\end{aligned}
\tag{A.36}
$$

Now, let $\mathbf{P}_i$ denote the ith permutation matrix $(i = 1, \dots, t!)$. For a given matrix $\mathbf{D}$ such that $\mathrm{Tr}\,\mathbf{D} = \zeta$, define $\mathbf{D}_i \triangleq \mathbf{P}_i\mathbf{DP}_i'$, i.e., the diagonal matrix obtained by applying the ith permutation on the diagonal elements. We notice that

(a) $\Psi(\mathbf{D}_i) = \Psi(\mathbf{D})$, since

$$\mathbb{E}\left[\log \det\left(\mathbf{I}_r + \mathbf{HP}_i\mathbf{DP}_i'\mathbf{H}^\dagger\right)\right] = \mathbb{E}\left[\log \det\left(\mathbf{I}_r + \mathbf{HDH}^\dagger\right)\right]$$

as $\mathbf{H}$ and $\mathbf{HP}_i$ have the same joint pdf.

(b) $\sum_i \mathbf{D}_i/t! = (\rho/t)\mathbf{I}_t$, since every diagonal entry of $\mathbf{D}$ appears the same number of times at each position of the matrix sum.

Hence, we have

$$\Psi((\zeta/t)\mathbf{I}_t) \geq \frac{1}{t!}\sum_i \Psi(\mathbf{D}_i) = \Psi(\mathbf{D}) \tag{A.37}$$

which proves that the maximum $\Psi(\mathbf{D})$, i.e., capacity, is attained for $\mathbf{D} = (\zeta/t)\mathbf{I}_t$, i.e., uniform power allocation.

References

[A.1] T. M. Cover and J. A. Thomas, *Elements of Information Theory*. New York: Wiley, 1991.

[A.2] R. G. Gallager, *Information Theory and Reliable Communication*. New York: Wiley, 1968.

[A.3] R. J. McEliece, *The Theory of Information and Coding*, 2nd Edition. Cambridge, UK: Cambridge University Press, 2002.

[A.4] W. Rhee and J.M. Cioffi, "On the capacity of multiuser wireless channels with multiple antennas," *IEEE Trans. Inform. Theory*, Vol. 49, No. 10, pp. 2580–2595, October 2003.

[A.5] Р. Л. Стратонович, Теория Информации. Москва: Советское Радио, 1975.

[A.6] E. Telatar, "Capacity of multi-antenna Gaussian channels," *Eur. Trans. Telecomm.*, Vol. 10, No. 6, pp. 585–595, November–December 1999.

[A.7] A. J. Viterbi and J. K. Omura, *Principles of Digital Communication and Coding*. New York: McGraw-Hill, 1979.

B

Bett,

Facts from matrix theory

In this appendix we collect some useful definitions and properties about matrices. As we assume that the reader has some familiarity with this topic, results are stated without proof.

B.1   Basic matrix operations

A real (complex) $m \times n$ matrix is a rectangular array of mn real (complex) numbers, arranged in m rows and n columns and indexed as follows:

$$
\begin{bmatrix}
a_{11} & a_{12} & \cdots & a_{1n} \\
a_{21} & a_{22} & \cdots & a_{2n} \\
& & \cdots & \\
a_{m1} & a_{m2} & \cdots & a_{mn}
\end{bmatrix}
\tag{B.1}
$$

We write $\mathbf{A} = (a_{ij})$ as shorthand for the matrix (B.1), and sometimes we use the notation $\mathbf{A} \in \mathbb{R}^{m \times n}$ or $\mathbf{A} \in \mathbb{C}^{m \times n}$ to indicate a real or complex matrix, respectively. If $m = n$, $\mathbf{A}$ is called a *square matrix*; if $n = 1$, $\mathbf{A}$ is called a *column vector*, and if $m = 1$, $\mathbf{A}$ is called a *row vector*. We denote column vectors using boldface lowercase letters, such as $\mathbf{x}$, $\mathbf{y}$, $\ldots$.

Standard operations for matrices are the following:

(a) Multiplication of $\mathbf{A}$ by the real or complex number c. The result, denoted by $c\mathbf{A}$, is the $m \times n$ matrix with entries ca_{ij}.

(b) Sum of two $m \times n$ matrices $\mathbf{A} = (a_{ij})$ and $\mathbf{B} = (b_{ij})$. The result is the $m \times n$ matrix whose entries are $a_{ij} + b_{ij}$. The sum is commutative, i.e., $\mathbf{A} + \mathbf{B} = \mathbf{B} + \mathbf{A}$.

(c) Product of the $m \times k$ matrix $\mathbf{A}$ by the $k \times n$ matrix $\mathbf{B}$. The result is the $m \times n$ matrix $\mathbf{C}$ with entries

$$
c_{ij} \triangleq \sum_{\ell=1}^{k} a_{i\ell} b_{\ell j} , \qquad i = 1, \ldots, m , \quad j = 1, \ldots, n
$$

When $\mathbf{AB} = \mathbf{BA}$, the two matrices $\mathbf{A}$ and $\mathbf{B}$ are said to *commute*. The matrix product is not commutative (i.e., in general $\mathbf{AB} \neq \mathbf{BA}$), but it is associative (i.e., $\mathbf{A}(\mathbf{BC}) = (\mathbf{AB})\mathbf{C}$) and distributive with respect to the sum (i.e., $\mathbf{A}(\mathbf{B} + \mathbf{C}) = \mathbf{AB} + \mathbf{AC}$ and $(\mathbf{A} + \mathbf{B})\mathbf{C} = \mathbf{AC} + \mathbf{BC}$).

The notation $\mathbf{A}^k$ is used to denote the kth power of a square matrix $\mathbf{A}$ (i.e., the product of $\mathbf{A}$ by itself performed $k - 1$ times). If we define the *identity matrix* $\mathbf{I} \triangleq (\delta_{ij})$ as the square matrix all of whose elements are 0 unless $i = j$, in which case they are equal to 1, multiplication of any square matrix $\mathbf{A}$ by $\mathbf{I}$ gives $\mathbf{A}$ itself, and we can set

$$
\mathbf{A}^0 = \mathbf{I}
$$

(d) The *transpose* of the $m \times n$ matrix $\mathbf{A}$ with entries a_{ij} is the $n \times m$ matrix with entries a_{ji}, which we denote by $\mathbf{A}'$. If $\mathbf{A}$ is a complex matrix, its *conjugate* $\mathbf{A}^*$ is the matrix with elements a_{ij}^*, and its conjugate (or Hermitian) transpose $\mathbf{A}^\dagger \triangleq (\mathbf{A}')^*$ has entries a_{ji}^*. The following properties hold:

$$(\mathbf{AB})' = \mathbf{B}'\mathbf{A}' \qquad (\mathbf{AB})^\dagger = \mathbf{B}^\dagger \mathbf{A}^\dagger$$

(e) Given a square matrix $\mathbf{A}$, there may exist a matrix, which we denote by $\mathbf{A}^{-1}$, such that $\mathbf{AA}^{-1} = \mathbf{A}^{-1}\mathbf{A} = \mathbf{I}$. If $\mathbf{A}^{-1}$ exists, it is called the *inverse* of $\mathbf{A}$, and $\mathbf{A}$ is said to be *nonsingular*. We have

$$(\mathbf{AB})^{-1} = \mathbf{B}^{-1}\mathbf{A}^{-1} \qquad (\mathbf{A}')^{-1} = (\mathbf{A}^{-1})' \qquad (\mathbf{A}^\dagger)^{-1} = (\mathbf{A}^{-1})^\dagger$$

B.2 Some numbers associated with a matrix

(a) Trace. Given a square $n \times n$ matrix $\mathbf{A}$, its *trace* (or *spur*) is the sum of the elements of the *main diagonal* of $\mathbf{A}$:

$$\mathrm{Tr}\,(\mathbf{A}) \triangleq \sum_{i=1}^{n} a_{ii} \tag{B.2}$$

The trace operation is linear; that is, for any two complex numbers α, β, and two square $n \times n$ matrices $\mathbf{A}$, $\mathbf{B}$, we have

$$\mathrm{Tr}\,(\alpha\mathbf{A} + \beta\mathbf{B}) = \alpha\,\mathrm{Tr}\,(\mathbf{A}) + \beta\,\mathrm{Tr}\,(\mathbf{B}) \tag{B.3}$$

In general, $\mathrm{Tr}\,(\mathbf{AB}) = \mathrm{Tr}\,(\mathbf{BA})$ even if $\mathbf{AB} \neq \mathbf{BA}$. In particular, the following properties hold:

$$\mathrm{Tr}\,(\mathbf{A}^{-1}\mathbf{BA}) = \mathrm{Tr}\,(\mathbf{B}) \tag{B.4}$$

$$\mathrm{Tr}\,(\mathbf{ABC}) = \mathrm{Tr}\,(\mathbf{CAB}) = \mathrm{Tr}\,(\mathbf{BCA}) \tag{B.5}$$

and

$$\mathbf{x}'\mathbf{y} = \mathrm{Tr}\,(\mathbf{xy}') \tag{B.6}$$

(b) Determinant. Given an $n \times n$ square matrix $\mathbf{A}$, its *determinant* is the number defined as the sum of the products of the elements in any row of $\mathbf{A}$ with their respective *cofactors* γ_{ij}:

$$\det \mathbf{A} \triangleq \sum_{j=1}^{n} a_{ij}\gamma_{ij}, \qquad \text{for any } i = 1, 2, \ldots, n \tag{B.7}$$

The cofactor of a_{ij} is defined as $\gamma_{ij} \triangleq (-1)^{i+j} m_{ij}$, where the *minor* m_{ij} is the determinant of the $(n-1) \times (n-1)$ submatrix obtained from $\mathbf{A}$ by removing its ith row and jth column. The determinant has the following properties:

$$\det \mathbf{A} \quad = \quad 0 \text{ if one row of } \mathbf{A} \text{ is zero}$$
$$\text{or } \mathbf{A} \text{ has two equal rows} \qquad (B.8)$$
$$\det \mathbf{A}' \quad = \quad \det \mathbf{A} \qquad (B.9)$$
$$\det \mathbf{A}^{\dagger} \quad = \quad (\det \mathbf{A})^* \qquad (B.10)$$
$$\det (\mathbf{A}^{-1}) \quad = \quad (\det \mathbf{A})^{-1} \qquad (B.11)$$
$$\det (\mathbf{AB}) \quad = \quad \det \mathbf{A} \cdot \det \mathbf{B} \qquad (B.12)$$
$$\det(c\mathbf{A}) \quad = \quad c^n \cdot \det \mathbf{A} \quad \text{for any number } c \qquad (B.13)$$

A matrix is nonsingular if and only if its determinant is nonzero.

(c) Rank. The *rank* of an $m \times n$ matrix $\mathbf{A}$ is the maximum number of linearly independent columns or rows. The rank has the following properties:

$$\text{rank}(\mathbf{A}) \leq \min(m, n) \qquad (B.14)$$

$$\text{rank}(\mathbf{A} + \mathbf{B}) \leq \text{rank}(\mathbf{A}) + \text{rank}(\mathbf{B}) \qquad (B.15)$$

$$\text{rank}(\mathbf{AB}) \leq \min(\text{rank}(\mathbf{A}), \text{rank}(\mathbf{B})) \qquad (B.16)$$

(d) Eigenvalues. Given an $n \times n$ square matrix $\mathbf{A}$ and a column vector $\mathbf{u}$ with n entries, consider the set of n linear equations

$$\mathbf{Au} = \lambda \mathbf{u} \qquad (B.17)$$

where λ is a constant and the entries of $\mathbf{u}$ are unknown. There are only n values of λ (not necessarily distinct) such that (B.17) has a nonzero solution. These numbers are the *eigenvalues* of $\mathbf{A}$, and the corresponding vectors $\mathbf{u}$ are the *eigenvectors* associated with them. Note that if $\mathbf{u}$ is an eigenvector associated with the eigenvalue λ then, for any complex number c, $c\mathbf{u}$ is also an eigenvector.

The polynomial $a(\lambda) \triangleq \det(\lambda \mathbf{I} - \mathbf{A})$ in the indeterminate λ is called the *characteristic polynomial* of $\mathbf{A}$. The equation

$$\det (\lambda \mathbf{I} - \mathbf{A}) = 0 \qquad (B.18)$$

is the *characteristic equation* of $\mathbf{A}$, and its roots are the eigenvalues of $\mathbf{A}$. The Cayley–Hamilton theorem states that every square $n \times n$ matrix $\mathbf{A}$ satisfies its characteristic equation. That is, if the characteristic polynomial of $\mathbf{A}$ is $a(\lambda) = \lambda^n + \alpha_1 \lambda^{n-1} + \cdots + \alpha_n$, then

$$a(\mathbf{A}) \triangleq \mathbf{A}^n + \alpha_1 \mathbf{A}^{n-1} + \cdots + \alpha_n \mathbf{I} = \mathbf{0} \tag{B.19}$$

where $\mathbf{0}$ is the null matrix (i.e., the matrix all of whose elements are zero). The monic polynomial $\mu(\lambda)$ of lowest degree such that $\mu(\mathbf{A}) = \mathbf{0}$ is called the *minimal polynomial* of $\mathbf{A}$.

The following properties hold:

(i) If $f(x)$ is a polynomial in the indeterminate x, and $\mathbf{u}$ is an eigenvector of $\mathbf{A}$ associated with the eigenvalue λ, then

$$f(\mathbf{A})\mathbf{u} = f(\lambda)\mathbf{u} \tag{B.20}$$

That is, $f(\lambda)$ is an eigenvalue of $f(\mathbf{A})$ and $\mathbf{u}$ is the corresponding eigenvector.

(ii) The product and the sum of the eigenvalues $\lambda_1, \ldots, \lambda_n$ of the $n \times n$ matrix $\mathbf{A}$ satisfy

$$\det(\mathbf{A}) = \prod_{i=1}^{n} \lambda_i \tag{B.21}$$

and

$$\mathrm{Tr}\,(\mathbf{A}) = \sum_{i=1}^{n} \lambda_i \tag{B.22}$$

(iii) From (B.21), it is immediately seen that $\mathbf{A}$ is nonsingular if and only if none of its eigenvalues is zero.

(iv) The eigenvalues of $\mathbf{A}^m$, m a positive integer, are λ_i^m.

(v) If $\mathbf{A}$ is nonsingular, then the eigenvalues of $\mathbf{A}^{-1}$ are λ^{-1}, and the eigenvectors are the eigenvectors of $\mathbf{A}$.

(vi) The eigenvalues of $\mathbf{A} + \sigma^2 \mathbf{I}$ are $\lambda_i + \sigma^2$, and the eigenvectors are the eigenvectors of $\mathbf{A}$.

(vii) The two matrices $\mathbf{A}\mathbf{A}^\dagger$ and $\mathbf{A}^\dagger\mathbf{A}$ share the same set of nonzero eigenvalues.

B.3   Gauss–Jordan elimination

The Gauss–Jordan algorithm transforms a given matrix into row echelon form by using elementary row operations. A matrix is said to be in row echelon form if the following conditions are satisfied:

- If a row does not consist entirely of zeros, then the first nonzero entry in the row is a 1.

- The rows all of whose entries are zero are grouped at the bottom of the matrix.

- In any two successive nonzero rows, the leading 1 in the lower row occurs farther to the right than the leading 1 in the upper row.

B.4   Some classes of matrices

Let $\mathbf{A}$ be an $n \times n$ square matrix.

(a) $\mathbf{A}$ is called *symmetric* if $\mathbf{A}' = \mathbf{A}$.

(b) $\mathbf{A}$ is called *Hermitian* if $\mathbf{A}^\dagger = \mathbf{A}$.

(c) $\mathbf{A}$ is called *orthogonal* if $\mathbf{A}^{-1} = \mathbf{A}'$.

(d) $\mathbf{A}$ is called *unitary* if $\mathbf{A}^{-1} = \mathbf{A}^\dagger$.

(e) $\mathbf{A}$ is called *diagonal* if its entries a_{ij} are zero unless $i = j$. A useful notation for a square diagonal matrix is

$$\mathbf{A} = \operatorname{diag}(a_{11}, a_{22}, \ldots, a_{nn})$$

This definition also holds for nonsquare matrices.

(f) $\mathbf{A}$ is called *scalar* if $\mathbf{A} = c\mathbf{I}$ for some constant c; that is, $\mathbf{A}$ is diagonal with equal entries on the main diagonal.

(g) A symmetric real matrix $\mathbf{A}$ is called *positive (nonnegative) definite* if all its eigenvalues are positive (nonnegative). Equivalently, $\mathbf{A}$ is positive (nonnegative) definite if and only if for any nonzero column vector $\mathbf{x}$ the quadratic form $\mathbf{x}^\dagger \mathbf{A} \mathbf{x}$ is positive (nonnegative). If $\mathbf{A}$ is nonnegative definite, then the number of positive eigenvalues equals rank $(\mathbf{A})$, and the remaining eigenvalues are zero. If $\mathbf{A}$ is positive definite (we write $\mathbf{A} > 0$), then all its eigenvalues are positive. The *Hadamard inequality* states that, for every nonnegative definite matrix $\mathbf{A}$,

$$\det \mathbf{A} \leq \prod_i (\mathbf{A})_{ii} \tag{B.23}$$

with equality if and only if $\mathbf{A}$ is diagonal.

Example B.1

Let $\mathbf{A}$ be Hermitian. Then the quadratic form $f \triangleq \mathbf{x}^\dagger \mathbf{A} \mathbf{x}$ is real. In fact

$$f^* = (\mathbf{x}^\dagger \mathbf{A} \mathbf{x})^* = \mathbf{x}' \mathbf{A}^* \mathbf{x}^* = (\mathbf{A}^* \mathbf{x}^*)' \mathbf{x} = \mathbf{x}^\dagger \mathbf{A}^\dagger \mathbf{x} \tag{B.24}$$

Since $\mathbf{A}^\dagger = \mathbf{A}$, this is equal to $\mathbf{x}^\dagger \mathbf{A} \mathbf{x} = f$, which shows that f is real. $\qquad\square$

Example B.2

Consider the random column vector $\mathbf{x} = [x_1, x_2, \ldots, x_N]'$ and its *correlation matrix*

$$\mathbf{R} \triangleq \mathrm{E}\left[\mathbf{x}\mathbf{x}^\dagger\right] \tag{B.25}$$

It is easily seen that $\mathbf{R}$ is Hermitian. Also, $\mathbf{R}$ is nonnegative definite; in fact, for any nonzero deterministic column vector $\mathbf{a}$,

$$\mathbf{a}^\dagger \mathbf{R} \mathbf{a} = \mathbf{a}^\dagger \mathrm{E}\left[\mathbf{x}\mathbf{x}^\dagger\right]\mathbf{a} = \mathrm{E}\left[\mathbf{a}^\dagger \mathbf{x}\mathbf{x}^\dagger \mathbf{a}\right] = \mathrm{E}\left[|\mathbf{a}^\dagger \mathbf{x}|^2\right] \geq 0 \tag{B.26}$$

with equality only if $\mathbf{a}^\dagger \mathbf{x} = 0$ almost surely; that is, the components of $\mathbf{x}$ are *linearly dependent*. $\qquad\square$

B.5   Scalar product and Frobenius norms

Given two $m \times n$ matrices $\mathbf{A}$ and $\mathbf{B}$, their scalar product is defined as

$$(\mathbf{A}, \mathbf{B}) \triangleq \sum_{i=1}^{m} \sum_{j=1}^{n} a_{ij} b_{ij}^* \tag{B.27}$$

The scalar products has the properties

(a) $(\mathbf{A}, \mathbf{B})^* = (\mathbf{B}, \mathbf{A})$

(b) $(\mathbf{A}, \mathbf{B}) = \text{Tr}\,(\mathbf{A}\mathbf{B}^\dagger) = \text{Tr}\,(\mathbf{B}^\dagger\mathbf{A})$

(c) The scalar product of the $n \times 1$ vector $\mathbf{x}$ by itself,

$$\mathbf{A} = \mathbf{x}\mathbf{x}^\dagger$$

is a Hermitian matrix.

(d) Two vectors $\mathbf{x}$, $\mathbf{y}$ such that $(\mathbf{x}, \mathbf{y}) = 0$ are called *orthogonal*. If in addition $\|\mathbf{x}\| = \|\mathbf{y}\| = 1$, they are called *orthonormal*.

We define the *Frobenius* (or *Euclidean*) norm of $\mathbf{A}$ as

$$\|\mathbf{A}\| \triangleq \sqrt{(\mathbf{A}, \mathbf{A})} = \sqrt{\text{Tr}\,(\mathbf{A}\mathbf{A}^\dagger)} = \sqrt{\text{Tr}\,(\mathbf{A}^\dagger\mathbf{A})} = \sqrt{\sum_{i=1}^{m}\sum_{j=1}^{n} |a_{ij}|^2} \quad \text{(B.28)}$$

The Frobenius norm has the properties

(a) $\mathbf{A} = \mathbf{0}$ if and only if $\|\mathbf{A}\| = 0$

(b) $\|\mathbf{A} + \mathbf{B}\| \le \|\mathbf{A}\| + \|\mathbf{B}\|$ (triangle inequality)

(c) $|(\mathbf{A}, \mathbf{B})| \le \|\mathbf{A}\| \cdot \|\mathbf{B}\|$

(d) $\|\mathbf{A}\| = \|\mathbf{A}^\dagger\|$

(e) $\|\mathbf{A}\mathbf{B}\| \le \|\mathbf{A}\| \cdot \|\mathbf{B}\|$

(f) $\|\mathbf{A}\|^2 = \text{Tr}\,(\mathbf{A}\mathbf{A}^\dagger) = \text{Tr}\,(\mathbf{A}^\dagger\mathbf{A})$

B.6 Matrix decompositions

B.6.1 Cholesky factorization

Given a Hermitian $n \times n$ matrix $\mathbf{A}$, its *Cholesky factorization* is

$$\mathbf{A} = \mathbf{L}\mathbf{L}^\dagger \quad \text{(B.29)}$$

where $\mathbf{L}$ is lower triangular (that is, $(\mathbf{L})_{ij} = 0$ unless $j > i$) and has nonnegative diagonal entries.

B.6.2 QR decomposition

An $m \times n$ matrix $\mathbf{A}$, $m \geq n$, can be decomposed in the form

$$\mathbf{A} = \mathbf{Q}\mathbf{R} \tag{B.30}$$

where $\mathbf{R}$ is an upper triangular $n \times n$ matrix, and $\mathbf{Q}$ is $m \times n$ with orthonormal columns: $\mathbf{Q}^\dagger\mathbf{Q} = \mathbf{I}_n$. When $\mathbf{A}$ has rank n, all diagonal entries of $\mathbf{R}$ are positive.

Notice that $\mathbf{Q}$ is not necessarily unitary. To make it a unitary $m \times m$ matrix we can append to it an additional $m - n$ orthonormal columns $\mathbf{q}_{n+1}, \cdots, \mathbf{q}_m$. Correspondingly, we append rows of zeros to $\mathbf{R}$ so that

$$\mathbf{A} = \begin{bmatrix} \mathbf{Q} & \mathbf{q}_{n+1} & \cdots & \mathbf{q}_m \end{bmatrix} \begin{bmatrix} \mathbf{R} \\ \mathbf{0} \end{bmatrix} \tag{B.31}$$

B.6.3 Spectral decomposition

If we define the $n \times n$ matrix $\mathbf{U}$ whose columns are the orthonormalized eigenvectors of the square matrix $\mathbf{A}$, and the $n \times n$ diagonal matrix of its eigenvalues $\mathrm{diag}\,(\lambda_1, \cdots, \lambda_n)$, then we have the decomposition

$$\mathbf{A} = \mathbf{U}\mathbf{\Lambda}\mathbf{U}^\dagger \tag{B.32}$$

Notice that a square $\mathbf{U}$ may not exist: for example, the matrix

$$\mathbf{A} = \begin{bmatrix} 1 & 1 \\ 0 & 1 \end{bmatrix}$$

has only one nonzero eigenvector, $\mathbf{u} = (1 \quad 0)'$. In this case the singular-value decomposition (see below) can be used instead of the spectral decomposition. A simple sufficient condition for the spectral decomposition to exist is that $\mathbf{A}$ be positive definite.

B.6.4 Singular-value decomposition

Singular-value decomposition (SVD) of the $m \times n$ matrix $\mathbf{A}$ yields

$$\mathbf{A} = \mathbf{U}\begin{bmatrix} \mathbf{\Sigma} & \mathbf{0} \end{bmatrix}\mathbf{V}^\dagger \tag{B.33}$$

where $\mathbf{U}$ is a unitary $m \times m$ matrix, $\mathbf{V}$ is a unitary $n \times n$ matrix, and $\mathbf{\Sigma}$ is $m \times n$ diagonal, with nonnegative entries $\sigma_1 \geq \sigma_2 \geq \cdots \geq 0$, called the *singular values* of $\mathbf{A}$. The singular values are the square roots of the eigenvalues of $\mathbf{A}\mathbf{A}^\dagger$, and the number of their nonzero values equals the rank of the matrix $\mathbf{A}$. The columns of $\mathbf{U}$ are the eigenvectors of $\mathbf{A}\mathbf{A}^\dagger$, and the columns of $\mathbf{V}$ are the eigenvectors of $\mathbf{A}^\dagger\mathbf{A}$.

The SVD of a matrix expresses the fact that any linear transformation can be decomposed into a rotation, a scaling operation, and another rotation.

B.7 Pseudoinverse

Given an $m \times n$ matrix $\mathbf{A}$ whose SVD is (B.33), its *(Moore–Penrose) pseudoinverse* $\mathbf{A}^+$ is defined as

$$
\mathbf{A}^+ \triangleq \begin{cases} \mathbf{V} \begin{bmatrix} \mathbf{\Sigma}^+ \\ \mathbf{0} \end{bmatrix} \mathbf{U}^\dagger & \text{for } m \leq n \\ \mathbf{V} \begin{bmatrix} \mathbf{\Sigma}^+ & \mathbf{0} \end{bmatrix} \mathbf{U}^\dagger & \text{for } n \leq m \end{cases}
\tag{B.34}
$$

where

$$
\mathbf{\Sigma}^+ = \text{diag}\,(\sigma_1^{-1}, \sigma_2^{-1}, \ldots, \sigma_p^{-1}, 0, \ldots, 0)
$$

and p is the number of nonzero singular values of $\mathbf{A}$.

The pseudoinverse has the following properties:

$$
\mathbf{A}\mathbf{A}^+\mathbf{A} = \mathbf{A} \qquad \mathbf{A}^+\mathbf{A}\mathbf{A}^+ = \mathbf{A}^+ \qquad (\mathbf{A}\mathbf{A}^+)^\dagger = \mathbf{A}\mathbf{A}^+ \qquad (\mathbf{A}^+\mathbf{A})^\dagger = \mathbf{A}^+\mathbf{A}
\tag{B.35}
$$

Moreover, if the $m \times n$ matrix $\mathbf{A}$ has full rank (i.e., $\text{rank}\,(\mathbf{A}) = \min(m, n)$), then

$$
\mathbf{A}^+ = \begin{cases} \mathbf{A}^\dagger(\mathbf{A}\mathbf{A}^\dagger)^{-1} & \text{if } m \leq n \\ \mathbf{A}^{-1} & \text{if } m = n \\ (\mathbf{A}^\dagger\mathbf{A})^{-1}\mathbf{A}^\dagger & \text{if } m \geq n \end{cases}
\tag{B.36}
$$

$\mathbf{A}^+$ is the unique solution to the approximation problem

$$
\min \|\mathbf{A}\mathbf{X} - \mathbf{I}_m\|
\tag{B.37}
$$

where the minimum is taken over all complex $n \times m$ matrices $\mathbf{X}$, and $\mathbf{I}_m$ denotes the $m \times m$ identity matrix.

References

[B.1] R. E. Bellman, *Matrix Analysis*. New York: McGraw-Hill, 1968.

[B.2] F. R. Gantmacher, *The Theory of Matrices*, Vols. I and II. New York: Chelsea, 1959.

[B.3] G. H. Golub and C. F. Van Loan, *Matrix Computations,* 2nd edition. Baltimore, MD: Johns Hopkins University Press, 1989.

[B.4] R. Horn and C. Johnson, *Matrix Analysis*. New York: Cambridge University Press, 1985.

C

Celia,

Random variables, vectors, and matrices

In this appendix we collect some facts about complex random variables, random vectors, and random matrices. In particular, we list some results on the probability distribution of the eigenvalues of large random matrices.

C.1   Complex random variables

A complex random variable (RV) has the form

$$Z = X + jY \tag{C.1}$$

where X and Y are real RVs with mean values μ_X and μ_Y and variances $\mathbb{V}[X]$, $\mathbb{V}[Y]$, respectively; Z can also be viewed as a two-dimensional vector RV (see *infra*). The mean value and the variance of Z are defined as $\mu_Z \triangleq \mathbb{E}[Z]$ and $\mathbb{V}[Z] \triangleq \mathbb{E}[|Z - \mu_Z|^2]$, respectively. We also have, directly from the definitions,

$$\mathbb{E}[Z] = \mathbb{E}[X] + j\mathbb{E}[Y] \tag{C.2}$$

and

$$\mathbb{V}[Z] = \mathbb{E}[X^2] + \mathbb{E}[Y^2] - \mu_X^2 - \mu_Y^2 \tag{C.3}$$

so that[1]

$$\mathbb{V}[Z] = \mathbb{V}[X] + \mathbb{V}[Y] \tag{C.4}$$

The *cross-covariance* of the two complex RVs W and Z is defined as

$$R_{WZ} \triangleq \mathbb{E}[(W - \mu_W)(Z - \mu_Z)^*] \tag{C.5}$$

As a special case of this definition, we have the *covariance* of Z:

$$R_Z \triangleq R_{ZZ} = \mathbb{V}[Z] \tag{C.6}$$

Moreover, we have

$$R_{WZ} = R_{ZW}^* \tag{C.7}$$

If X is real Gaussian with mean μ and variance σ^2, i.e., its pdf is

$$p(x) = \frac{1}{\sqrt{2\pi\sigma^2}} e^{-(x-\mu)^2/2\sigma^2} \tag{C.8}$$

We write $X \sim \mathcal{N}(\mu, \sigma^2)$, a notation stressing the fact that the pdf of X is completely specified by μ and σ^2. If the real and imaginary parts of the complex RV Z are independent with the same variance $\sigma^2/2$, and $\mu \triangleq \mathbb{E}(Z) \in \mathbb{C}$, then we say that Z is *circularly symmetric*, and we write $Z \sim \mathcal{N}_c(\mu, \sigma^2)$. Its pdf is the product of its real and imaginary part:

$$p(z) = \frac{1}{\pi\sigma^2} e^{-|z-\mu|^2/\sigma^2} \tag{C.9}$$

---

[1]To avoid any confusion, observe that if $Z \triangleq X + Y$, then (C.4) holds only for uncorrelated X and Y, while if $Z \triangleq X + jY$ then it holds in general.

C.2   Random vectors

A random vector is a column vector $\mathbf{x} = (X_1, X_2, \ldots, X_n)'$ whose components are random variables.

C.2.1   Real random vectors

A real random vector is a column vector $\mathbf{x} = (X_1, X_2, \ldots, X_n)'$ whose components are real random variables. Its mean value is defined as

$$\boldsymbol{\mu}_{\mathbf{x}} \triangleq \big(\mathbb{E}[X_1], \mathbb{E}[X_2], \ldots, \mathbb{E}[X_n]\big)' \tag{C.10}$$

The expectation of the squared norm of $\mathbf{x}$,

$$\mathbb{E}\left[\|\mathbf{x}\|^2\right] = \mathbb{E}\left[\mathbf{x}'\mathbf{x}\right] = \sum_{i=1}^{n} \mathbb{E}\left[X_i^2\right] \tag{C.11}$$

is often referred to as the *average energy* of $\mathbf{x}$. The *covariance matrix* of $\mathbf{x}$ is defined as the (nonnegative-definite) $n \times n$ matrix

$$\mathbf{R}_{\mathbf{x}} \triangleq \mathbb{E}\left[(\mathbf{x} - \boldsymbol{\mu}_{\mathbf{x}})(\mathbf{x} - \boldsymbol{\mu}_{\mathbf{x}})'\right] = \mathbb{E}\left[\mathbf{x}\mathbf{x}'\right] - \boldsymbol{\mu}_{\mathbf{x}}\boldsymbol{\mu}_{\mathbf{x}}' \tag{C.12}$$

Notice that the diagonal elements of $\mathbf{R}_{\mathbf{x}}$ are the variances of the components of $\mathbf{x}$. The $n \times n$ matrix

$$\mathbf{C}_{\mathbf{x}} \triangleq \mathbb{E}\left[\mathbf{x}\mathbf{x}'\right] \tag{C.13}$$

is called the *correlation matrix* of $\mathbf{x}$. We observe that the trace of $\mathbf{C}_{\mathbf{x}}$ equals the average energy of $\mathbf{x}$:

$$\mathbb{E}[\|\mathbf{x}\|^2] = \mathbb{E}[\mathbf{x}'\mathbf{x}] = \mathbb{E}\left[\mathrm{Tr}\left(\mathbf{x}\mathbf{x}'\right)\right] = \mathrm{Tr}\,\mathbf{C}_{\mathbf{x}} \tag{C.14}$$

The *cross-covariance matrix* of the two random vectors $\mathbf{x}$ and $\mathbf{y}$, with dimensions n and m, respectively, is defined as the $n \times m$ matrix

$$\mathbf{R}_{\mathbf{x},\mathbf{y}} \triangleq \mathbb{E}\left[(\mathbf{x} - \boldsymbol{\mu}_{\mathbf{x}})(\mathbf{y} - \boldsymbol{\mu}_{\mathbf{y}})'\right] = \mathbb{E}\left[\mathbf{x}\mathbf{y}'\right] - \boldsymbol{\mu}_{\mathbf{x}}\boldsymbol{\mu}_{\mathbf{y}}' \tag{C.15}$$

Real Gaussian random vectors

A real random vector $\mathbf{x} = (X_1, \ldots, X_n)'$ is called *Gaussian* if its components are jointly Gaussian, that is, if their joint pdf is

$$
\begin{aligned}
p(\mathbf{x}) &= \frac{1}{(2\pi)^{n/2}\det\left(\mathbf{R}_{\mathbf{x}}\right)^{1/2}} \exp\left\{-\frac{1}{2}(\mathbf{x} - \boldsymbol{\mu}_{\mathbf{x}})'\mathbf{R}_{\mathbf{x}}^{-1}(\mathbf{x} - \boldsymbol{\mu}_{\mathbf{x}})\right\} \\
&= \frac{1}{(2\pi)^{n/2}\det\left(\mathbf{R}_{\mathbf{x}}\right)^{1/2}} \exp\left\{-\frac{1}{2}\mathrm{Tr}\left[\mathbf{R}_{\mathbf{x}}^{-1}(\mathbf{x} - \boldsymbol{\mu}_{\mathbf{x}})(\mathbf{x} - \boldsymbol{\mu}_{\mathbf{x}})'\right]\right\}
\end{aligned} \tag{C.16}
$$

where $\mathbf{R_x}$ is a nonnegative definite $n \times n$ matrix, the covariance matrix of $\mathbf{x}$. We write $\mathbf{x} \sim \mathcal{N}(\boldsymbol{\mu_x}, \mathbf{R_x})$, which stresses the fact that the pdf of a real Gaussian random vector is completely specified by its mean value and its covariance matrix.

C.2.2   Complex random vectors

A complex random vector is a column vector $\mathbf{z} = (Z_1, Z_2, \ldots, Z_n)'$ whose components are complex random variables. The *covariance matrix* of $\mathbf{z}$ is defined as the nonegative-definite $n \times n$ matrix

$$\mathbf{R_z} \triangleq \mathbb{E}[(\mathbf{z} - \boldsymbol{\mu_z})(\mathbf{z} - \boldsymbol{\mu_z})^\dagger] = \mathbb{E}[\mathbf{zz}^\dagger] - \boldsymbol{\mu_z}\boldsymbol{\mu_z}^\dagger \tag{C.17}$$

The diagonal entries of $\mathbf{R_z}$ are the variances of the entries of $\mathbf{z}$. If $\mathbf{z} = \mathbf{x} + j\mathbf{y}$,

$$\mathbf{R_z} = (\mathbf{R_x} + \mathbf{R_y}) + j(\mathbf{R_{yx}} - \mathbf{R_{xy}}) \tag{C.18}$$

Thus, knowledge of $\mathbf{R_z}$ does not yield knowledge of $\mathbf{R_x}$, $\mathbf{R_y}$, $\mathbf{R_{yx}}$, and $\mathbf{R_{xy}}$, i.e., of the complete second-order statistics of $\mathbf{z}$. The latter is completely specified if, in addition to $\mathbf{R_z}$, the *pseudocovariance matrix*

$$\begin{aligned} \widetilde{\mathbf{R}}_{\mathbf{z}} &\triangleq \mathbb{E}[(\mathbf{z} - \boldsymbol{\mu_z})(\mathbf{z} - \boldsymbol{\mu_z})'] \\ &= (\mathbf{R_x} - \mathbf{R_y}) + j(\mathbf{R_{yx}} + \mathbf{R_{xy}}) \end{aligned} \tag{C.19}$$

is also known [C.5]. We have the following relations:

$$\begin{cases} \mathbf{R_x} &= \frac{1}{2}\Re(\mathbf{R_z} + \widetilde{\mathbf{R}}_{\mathbf{z}}) \\ \mathbf{R_y} &= \frac{1}{2}\Re(\mathbf{R_z} - \widetilde{\mathbf{R}}_{\mathbf{z}}) \\ \mathbf{R_{xy}} &= \frac{1}{2}\Im(\mathbf{R_z} - \widetilde{\mathbf{R}}_{\mathbf{z}}) \\ \mathbf{R_{yx}} &= \frac{1}{2}\Im(\mathbf{R_z} + \widetilde{\mathbf{R}}_{\mathbf{z}}) \end{cases} \tag{C.20}$$

Proper complex random vectors, endowed with the additional property

$$\widetilde{\mathbf{R}}_{\mathbf{z}} = 0 \tag{C.21}$$

(see [C.5] for a justification of the term), are completely specified by $\mathbf{R_z}$ as far as their second-order statistics are concerned.

Similar properties can be derived for the $n \times n$ matrix $\mathbf{C_z} \triangleq \mathbb{E}[\mathbf{zz}^\dagger]$, which is called the *correlation matrix* of $\mathbf{z}$.

Complex Gaussian random vectors

A complex random vector $\mathbf{z} = \mathbf{x} + j\mathbf{y} \in \mathbb{C}^n$ is called *Gaussian* if its real part $\mathbf{x}$ and imaginary part $\mathbf{y}$ are jointly Gaussian, or, equivalently, if the real random vector

$$\check{\mathbf{z}} \triangleq \begin{bmatrix} \mathbf{x} \\ \mathbf{y} \end{bmatrix} \in \mathbb{R}^{2n}$$

is Gaussian.

Unlike real Gaussian random vectors, their complex counterparts are *not* completely specified by their mean values and covariance matrices (the pseudocovariance matrices are also needed). In fact, to specify the pdf of $\mathbf{z}$, and hence of $\check{\mathbf{z}}$, we need, in addition to $\mathbb{E}[\mathbf{z}]$, the covariance matrix of $\check{\mathbf{z}}$:

$$\mathbf{R}_{\check{\mathbf{z}}} = \begin{bmatrix} \mathbf{R}_{\mathbf{x}} & \mathbf{R}_{\mathbf{xy}} \\ \mathbf{R}'_{\mathbf{xy}} & \mathbf{R}_{\mathbf{y}} \end{bmatrix} \tag{C.22}$$

which is completely specified by $\mathbf{R}_{\mathbf{z}}$ and $\widetilde{\mathbf{R}}_{\mathbf{z}}$. In order to be able to uniquely determine $\mathbf{R}_{\mathbf{x}}$, $\mathbf{R}_{\mathbf{y}}$, and $\mathbf{R}_{\mathbf{xy}}$ from $\mathbf{R}_{\mathbf{z}}$, we need to restrict our attention to the subclass of *proper* Gaussian random vectors, also called *circularly symmetric*. The covariance matrix of $\check{\mathbf{z}}$ can be written as follows:

$$\mathbf{R}_{\check{\mathbf{z}}} = \frac{1}{2} \begin{bmatrix} \Re\mathbf{R}_{\mathbf{z}} & -\Im\mathbf{R}_{\mathbf{z}} \\ \Im\mathbf{R}_{\mathbf{z}} & \Re\mathbf{R}_{\mathbf{z}} \end{bmatrix} \tag{C.23}$$

Hence, a circularly symmetric complex Gaussian random vector is characterized by $\mu_{\mathbf{z}}$ and $\mathbf{R}_{\mathbf{z}}$. We write[2] $\mathbf{z} \sim \mathcal{N}_c(\mu_{\mathbf{z}}, \mathbf{R}_{\mathbf{z}})$. The probability density function of $\mathbf{z}$ is given by

$$\begin{aligned} p(\mathbf{z}) &\triangleq p(\check{\mathbf{z}}) \\ &= \det(2\pi\mathbf{R}_{\check{\mathbf{z}}})^{-1/2} \exp\left\{ -(\check{\mathbf{z}} - \mu_{\check{\mathbf{z}}})^{\dagger}\mathbf{R}_{\check{\mathbf{z}}}^{-1}(\check{\mathbf{z}} - \mu_{\check{\mathbf{z}}}) \right\} \\ &= \det(\pi\mathbf{R}_{\mathbf{z}})^{-1} \exp\left\{ -(\mathbf{z} - \mu_{\mathbf{z}})^{\dagger}\mathbf{R}_{\mathbf{z}}^{-1}(\mathbf{z} - \mu_{\mathbf{z}}) \right\} \end{aligned} \tag{C.24}$$

The following theorem [C.5] describes important properties of circularly symmetric Gaussian random vectors.

---

[2]This notation is meant to avoid confusion with the real case and as a reminder that, in our context, circular symmetry is a property of *Gaussian* complex random vectors.

Theorem C.2.1 *If* $\mathbf{z} \sim \mathcal{N}_c(\boldsymbol{\mu}_{\mathbf{z}}, \mathbf{R}_{\mathbf{z}})$, *then every affine transformation*

$$\mathbf{y} = \mathbf{A}\mathbf{z} + \mathbf{b}$$

yields a circularly symmetric Gaussian RV $\mathbf{y} \sim \mathcal{N}_c(\mathbf{A}\boldsymbol{\mu}_{\mathbf{z}} + \mathbf{b}, \mathbf{A}\mathbf{R}_{\mathbf{z}}\mathbf{A}^\dagger)$. *If* $\mathbf{z}_1$ *and* $\mathbf{z}_2$ *are independent, circularly symmetric Gaussian RVs, the linear combination* $a_1\mathbf{z}_1 + a_2\mathbf{z}_2$, *where* $a_1 \neq 0$ *and* a_2 *are complex numbers, is circularly symmetric Gaussian.*

C.3   Random matrices

A random matrix is a matrix whose entries are random variables. Consequently, a random matrix is described by assigning the joint probability density function (pdf) of its entries, which is especially easy when these are independent. For example, an $m \times n$ matrix $\mathbf{A}$, $m \leq n$, whose elements are independent identically distributed $\mathcal{N}(0, 1)$ real RVs has pdf [C.2]

$$(2\pi)^{-mn/2}\mathrm{etr}\left(\mathbf{A}\mathbf{A}'/2\right) \tag{C.25}$$

An $m \times n$ matrix $\mathbf{B}$ with iid complex Gaussian $\mathcal{N}_c(0, 1)$ entries has pdf [C.2]

$$(\pi)^{-mn}\mathrm{etr}\left(\mathbf{B}\mathbf{B}^\dagger\right) \tag{C.26}$$

We have the following theorem [C.4]:

Theorem C.3.1 *If* $\mathbf{A}$ *is a given* $m \times m$ *Hermitian matrix such that* $\mathbf{I}_m + \mathbf{A} > 0$ *and* $\mathbf{B}$ *is an* $m \times n$ *matrix whose entries are iid as* $\mathcal{N}_c(0, 1)$, *then*

$$\mathbb{E}[\mathrm{etr}\left(-\mathbf{A}\mathbf{B}\mathbf{B}^\dagger\right)] = \det\left(\mathbf{I}_m + \mathbf{A}\right)^{-n} \tag{C.27}$$

The eigenvalues of a random matrix are also random variables and hence can be described by their joint probability density function. An important special case occurs is that of a *complex Wishart matrix*, that is, a random complex Hermitian square $m \times m$ matrix $\mathbf{W} \triangleq \mathbf{B}\mathbf{B}^\dagger$, with $\mathbf{B}$ as in (C.26). The pdf of the ordered eigenvalues $\boldsymbol{\lambda} = (\lambda_1, \ldots, \lambda_m)$, $0 \leq \lambda_1 \leq \ldots \leq \lambda_m$, of $\mathbf{W}$ is given by [C.2, C.8]

$$\frac{1}{\Gamma_m(n)\Gamma_m(m)} \exp\left(-\sum_{i=1}^{m} \lambda_i\right) \prod_{i=1}^{m} \lambda_i^{n-m} \prod_{j=i+1}^{m} (\lambda_i - \lambda_j)^2 \tag{C.28}$$

where $\Gamma_m(a) \triangleq \prod_{i=0}^{m-1} \Gamma(a - i)$. The joint pdf of the *unordered* eigenvalues is obtained from (C.28) by dividing it by $m!$.

It is interesting to observe the limiting distribution of the eigenvalues of a Wishart matrix as its dimensions grow to infinity. To do this, we define the *empirical distribution* of the eigenvalues of an $n \times n$ random matrix $\mathbf{A}$ as the function $F(\lambda)$ that yields the fraction of eigenvalues of $\mathbf{A}$ not exceeding λ. Formally,

$$F(\lambda) \triangleq \frac{1}{n} |\{\lambda_i(\mathbf{A}) : \lambda_i(\mathbf{A}) < \lambda\}| \qquad (C.29)$$

The empirical distribution is generally a random process. However, under certain mild technical conditions [C.7], as $n \to \infty$ the empirical distribution converges to a nonrandom cumulative distribution function. For a Wishart matrix we have the following theorem, a classic in random-matrix theory [C.2]:

Theorem C.3.2 *Consider the sequence of $n \times m$ matrices $\mathbf{A}_n$, with iid entries having variances $1/n$; moreover, let $m = m(n)$, with $\lim_{n \to \infty} m(n)/n = c > 0$ and finite. Next, let $\mathbf{B}_n \triangleq \mathbf{A}_n \mathbf{A}_n^\dagger$. As $n \to \infty$, the empirical eigenvalue distribution of $\mathbf{B}_n$ tends to the probability density function*

$$p(\lambda) = [1 - c]_+ \, \delta(\lambda) + \frac{1}{2\pi\lambda} \sqrt{(\lambda - \lambda_-)_+ [(\lambda_+ - \lambda)_+]} \qquad (C.30)$$

with $\lambda_\pm \triangleq (\sqrt{c} \pm 1)^2$.

The theorem that follows [C.1] describes an important asymptotic property of a class of matrices. This is a special case of a general theory described in [C.3].

Theorem C.3.3 *Let $(\mathbf{H}_n(s))_{s \in \mathbb{S}}$ be an independent family of $n \times n$ matrices whose entries are iid complex Gaussian random variables with independent, equally distributed real and imaginary parts. Let $\mathbf{A}_n(s) \triangleq f(\mathbf{H}_n(s)^\dagger \mathbf{H}_n(s))$ where f is a real continuous function on $\mathbb{R}$. Let $(\mathbf{B}_n(t))_{t \in \mathcal{T}}$ be a family of deterministic matrices with eigenvalues $\lambda_1(n, t), \ldots, \lambda_n(n, t)$ such that for all $t \in \mathcal{T}$*

$$\sup_n \max_i \lambda_i(n, t) < \infty$$

and $(\mathbf{B}_n(t), \mathbf{B}_n^\dagger(t))_{t \in \mathcal{T}}$ has a limit distribution. Then $\mathbf{A}_n(s)$ converges in distribution almost surely to a compactly supported probability measure on $\mathbb{R}$ for each $s \in \mathbb{S}$ and, almost surely as $n \to \infty$,

$$\frac{1}{n} \mathrm{Tr}\,(\mathbf{A}_n \mathbf{B}_n) \to \frac{1}{n} \mathbb{E}\,[\mathrm{Tr}\,(\mathbf{A}_n)] \cdot \frac{1}{n} \mathbb{E}\,[\mathrm{Tr}\,(\mathbf{B}_n)] \qquad (C.31)$$

References

[C.1] E. Biglieri, G. Taricco, and A. Tulino, "Performance of space–time codes for a large number of antennas," *IEEE Trans. Inform. Theory,* Vol. 48, No. 7, pp. 1794–1803, July 2002.

[C.2] A. Edelman, *Eigenvalues and Condition Numbers of Random matrices.* PhD Thesis, Department of Mathematics, Massachusetts Institute of Technology, Cambridge, MA, 1989.

[C.3] F. Hiai and D. Petz, *The Semicircle Law, Free Random Variables and Entropy.* Providence, RI: American Mathematical Society, 2000.

[C.4] A.T. James, "Distribution of matrix variates and latent roots derived from normal samples," *Ann. Math. Statistics*, Vol. 35, pp. 475–501, 1964.

[C.5] F. D. Neeser and J. L. Massey, "Proper complex random processes with applications to information theory," *IEEE Trans. Inform. Theory*, Vol. 39, No. 4, pp. 1293–1302, July 1993.

[C.6] B. Picinbono, *Random Signals and Systems.* Englewood Cliffs, NJ: Prentice-Hall, 1993.

[C.7] J. W. Silverstein, "Strong convergence of the empirical distribution of eigenvalues of large dimensional random matrices," *J. Multivariate Anal.*, Vol. 55, pp. 331–339, 1995.

[C.8] E. Telatar, "Capacity of multi-antenna Gaussian channels," *Eur. Trans. Telecomm.*, Vol. 10, No. 6, pp. 585–595, November–December 1999.

D

Delia,

Computation of error probabilities

Here we provide some useful formulas for the calculation of error probabilities. We first give a closed-form expression for the expectation of a function of a chi-square-distributed random variable. Next, we describe a technique for the evaluation of pairwise error probabilities. Based on numerical integration, it allows the computation of pairwise error probabilities within any degree of accuracy.

D.1   Calculation of an expectation involving the Q function

Define the random variable

$$X = \sum_{i=1}^{n} X_k \tag{D.1}$$

where $X_i \triangleq A\alpha_i^2$, A a constant and α_i, $i = 1, \ldots, n$, a set of independent, identically Rayleigh-distributed random variables with common mean value $\overline{X} \triangleq \mathbb{E}\,\alpha_i^2$. The RV X is chi-square-distributed with $2n$ degrees of freedom, i.e., its probability density function is

$$p_X(x) = \frac{1}{(n-1)!\overline{X}^n} x^{n-1} e^{-x/\overline{X}} \tag{D.2}$$

We have the following result [D.4, p. 781]:

$$\mathbb{E}\,Q(\sqrt{X}) = \left(\frac{1-\mu}{2}\right)^n \sum_{k=0}^{n-1} \binom{n-1+k}{k} \left(\frac{1+\mu}{2}\right)^k \tag{D.3}$$

where

$$\mu \triangleq \sqrt{\frac{\overline{X}}{2+\overline{X}}} \tag{D.4}$$

Moreover, for large enough $\overline{X}$, we have

$$\frac{1}{2}(1+\mu) \approx 1, \qquad \frac{1}{2}(1-\mu) \approx \frac{1}{2\overline{X}}$$

and

$$\sum_{k=0}^{n-1} \binom{n-1+k}{k} = \binom{2n-1}{n}$$

so that

$$\mathbb{E}\,Q(\sqrt{X}) \approx \binom{2n-1}{n} \left(\frac{1}{2\overline{X}}\right)^n \tag{D.5}$$

D.2   Numerical calculation of error probabilities

Consider the evaluation of the probability $P \triangleq \mathbb{P}(\nu > x)$, where ν and x are independent random variables whose moment-generating functions (MGFs)

$$\Phi_\nu(s) \triangleq \mathbb{E}[\exp(-s\nu)] \quad \text{and} \quad \Phi_x(s) \triangleq \mathbb{E}[\exp(-sx)]$$

are known. Defining $\Delta \triangleq x - \nu$, we have $P = \mathbb{P}(\Delta < 0)$. We describe a method for computing the value of P based on numerical integration. Assume that the MGF of Δ, which, due to the independence of ν and x, can be written as

$$\Phi_\Delta(s) \triangleq \mathbb{E}[\exp(-s\Delta)] = \Phi_x(s)\,\Phi_\nu(-s) \tag{D.6}$$

is analytically known. Using the Laplace inversion formula, we obtain

$$\mathbb{P}(\Delta < 0) = \frac{1}{2\pi j} \int_{c-j\infty}^{c+j\infty} \frac{\Phi_\Delta(s)}{s}\,ds \tag{D.7}$$

where we assume that c is in the region of convergence (ROC) of $\Phi_\Delta(s)$. This is given by the intersection of the ROC of $\Phi_x(s)$ and the ROC of $\Phi_\nu(-s)$. Integral (D.7) can be computed exactly by using the method of residues [D.3, D.8]. This method works well when the integrand exhibits simple poles, but it becomes long and intricate when multiple poles or essential singularities are present. Here we describe a general approach based on numerical calculation of the integral.

Expand the real and imaginary parts in (D.7). We have the following result:

$$\mathbb{P}(\Delta < 0) = \frac{1}{2\pi} \int_{-\infty}^{\infty} \frac{\Phi_\Delta(c+j\omega)}{c+j\omega}\,d\omega$$

$$= \frac{1}{2\pi} \int_{-\infty}^{\infty} \frac{c\Re\{\Phi_\Delta(c+j\omega)\} + \omega\Im\{\Phi_\Delta(c+j\omega)\}}{c^2 + \omega^2}\,d\omega$$

The change of variables $\omega = c\sqrt{1-x^2}/x$ yields

$$\frac{1}{2\pi} \int_{-1}^{1} \left\{ \Re\left[\Phi_\Delta\left(c + jc\frac{\sqrt{1-x^2}}{x} \right) \right] \right.$$

$$\left. + \frac{\sqrt{1-x^2}}{x} \Im\left[\Phi_\Delta\left(c + jc\frac{\sqrt{1-x^2}}{x} \right) \right] \right\} \frac{dx}{\sqrt{1-x^2}} \tag{D.8}$$

and this integral can be approximated numerically by using a Gauss–Chebyshev numerical quadrature rule with L nodes [D.6]. We have

$$\mathbb{P}(\Delta < 0)$$

$$= \frac{1}{2L} \sum_{k=1}^{L} \left\{ \Re[\Phi_\Delta(c(1+j\tau_k))] + \tau_k\,\Im[\Phi_\Delta(c(1+j\tau_k))] \right\} + E_L \tag{D.9}$$

where $\tau_k \triangleq \tan((k-1/2)\pi/L)$, and $E_L \to 0$ as $L \to \infty$. In numerical calculations, a rule-of-thumb choice yields $L = 64$.

Example D.1.

As a special case of the above, consider the calculation of the expectation

$$P \triangleq \mathbb{E}[Q(\sqrt{\xi})] \qquad \text{(D.10)}$$

where $Q(y)$ is again the Gaussian tail function, i.e., $Q(y) \triangleq \mathbb{P}(\nu > y)$ with $\nu \sim \mathcal{N}(0,1)$, and ξ is a nonnegative random variable. Defining $\Delta \triangleq \xi - \nu^2$, we have $P = (1/2)\mathbb{P}[\Delta < 0]$. Thus,

$$\Phi_\Delta(s) = \Phi_\xi(s)\Phi_{\nu^2}(-s) = \Phi_\xi(s)(1 - 2s)^{-1/2}$$

Here the ROC of $\Phi_\Delta(s)$ includes the complex region defined by $\{0 < \Re(s) < 1/2\}$. Therefore, we can safely assume $0 < c < 1/2$: a good choice is $c = 1/4$, corresponding to an integration line in the middle of the minimal ROC of $\Phi_\Delta(s)$. The latter integral can be evaluated numerically by using (D.9). □

D.3 Application: MIMO channel

Here we apply the general technique outlined above to calculate pairwise error probabilities for MIMO channels affected by fading.

D.3.1 Independent-fading channel with coding

The channel equation can be written as

$$\mathbf{y}_i = \mathbf{H}_i\mathbf{x}_i + \mathbf{z}_i \qquad\qquad i = 1, \ldots, N \qquad \text{(D.11)}$$

where N is the code block length, $\mathbf{H}_i \in \mathbb{C}^{rt}$ is the ith channel gain matrix, $\mathbf{x}_i \in \mathbb{C}^t$ is the ith transmitted symbol vector (each entry transmitted from a different antenna), $\mathbf{y}_i \in \mathbb{C}^r$ is the ith received sample vector (each entry received from a different antenna), and $\mathbf{z}_i \in \mathbb{C}^r$ is the ith received noise sample vector (each entry received from a different antenna). We assume that the channel gain matrices $\mathbf{H}_i$ are elementwise independent and independent of each other with $[\mathbf{H}_i]_{jk} \sim \mathcal{N}_c(0, 1)$. Also, the noise samples are independent with $[\mathbf{z}]_i \sim \mathcal{N}_c(0, N_0)$.

It is straightforward to obtain the PEP associated with the two code words $\mathbf{X} = (\mathbf{x}_1, \ldots, \mathbf{x}_N)$ and $\widehat{\mathbf{X}} = (\widehat{\mathbf{x}}_1, \ldots, \widehat{\mathbf{x}}_N)$ as follows:

$$
\begin{aligned}
P(\mathbf{X} &\to \widehat{\mathbf{X}}) \\
&= \mathbb{P}\left(\sum_{i=1}^{N}\{\|\mathbf{y}_i - \mathbf{H}_i\widehat{\mathbf{x}}_i\|^2 - \|\mathbf{y}_i - \mathbf{H}_i\mathbf{x}_i\|^2\} < 0\right) \\
&= \mathbb{P}\left(\sum_{i=1}^{N}\{\|\mathbf{H}_i(\mathbf{x}_i - \widehat{\mathbf{x}}_i) + \mathbf{z}_i\|^2 - \|\mathbf{z}_i\|^2\} < 0\right) \\
&= \mathbb{E}\left[Q\left(\sqrt{\frac{1}{2N_0}\sum_{i=1}^{N}\|\mathbf{H}_i(\mathbf{x}_i - \widehat{\mathbf{x}}_i)\|^2}\right)\right]
\end{aligned}
\tag{D.12}
$$

Setting

$$
\xi \triangleq \frac{1}{2N_0}\sum_{i=1}^{N}\|\mathbf{H}_i(\mathbf{x}_i - \widehat{\mathbf{x}}_i)\|^2
\tag{D.13}
$$

a straightforward computation yields

$$
\Phi_\xi(s) = \prod_{i=1}^{N}[1 + s\|\mathbf{x}_i - \widehat{\mathbf{x}}_i\|^2/(2N_0)]^{-r}
\tag{D.14}
$$

and the result of Example D.1 applies.

D.3.2 Block-fading channel with coding

Here we assume that the channel gain matrices $\mathbf{H}_i$ are independent of the time index i and are equal to $\mathbf{H}$: under this assumption the channel equation is

$$
\mathbf{Y} = \mathbf{H}\mathbf{X} + \mathbf{Z}
\tag{D.15}
$$

where $\mathbf{H} \in \mathbb{C}^{rt}$, $\mathbf{X} = (\mathbf{x}_1, \ldots, \mathbf{x}_N) \in \mathbb{C}^{tN}$, $\mathbf{Y} \in \mathbb{C}^{rN}$, and $\mathbf{Z} \in \mathbb{C}^{rN}$. We assume iid entries $[\mathbf{H}]_{ij} \sim \mathcal{N}_c(0, 1)$ and i.i.d. $[\mathbf{Z}]_{ij} \sim \mathcal{N}_c(0, N_0)$. We obtain

$$
\mathbb{P}(\mathbf{X} \to \widehat{\mathbf{X}}) = \mathbb{E}\left[Q\left(\frac{\|\mathbf{H}\boldsymbol{\Delta}\|}{\sqrt{2N_0}}\right)\right]
\tag{D.16}
$$

where $\boldsymbol{\Delta} \triangleq \mathbf{X} - \widehat{\mathbf{X}}$.

Setting

$$
\xi \triangleq \frac{\|\mathbf{H}\boldsymbol{\Delta}\|^2}{2N_0}
\tag{D.17}
$$

we can evaluate the PEP by resorting to (D.10). Apply Theorem C.3.1. First, notice that ξ can be written in the form

$$
\begin{aligned}
\xi &= \frac{1}{2N_0} \sum_{i=1}^{r} \mathbf{h}_i \mathbf{\Delta}\mathbf{\Delta}^\dagger \mathbf{h}_i^\dagger \\
&= \frac{1}{2N_0} [\mathbf{h}_1, \dots, \mathbf{h}_r][\mathbf{I}_r \otimes (\mathbf{\Delta}\mathbf{\Delta}^\dagger)][\mathbf{h}_1, \dots, \mathbf{h}_r]^\dagger \qquad \text{(D.18)}
\end{aligned}
$$

where $\mathbf{h}_i$ denotes the ith row of matrix $\mathbf{H}$. Setting $\mathbf{z} = [\mathbf{h}_1, \dots, \mathbf{h}_r]^\dagger$, we have $\boldsymbol{\mu} = \mathbf{0}$ and $\boldsymbol{\Sigma} = \mathbb{E}[\mathbf{z}\mathbf{z}^\dagger] = \mathbf{I}_{rt}$. Finally, setting $\mathbf{A} = [\mathbf{I}_r \otimes (\mathbf{\Delta}\mathbf{\Delta}^\dagger)]/(2N_0)$ in (C.27), we obtain

$$
\begin{aligned}
\Phi_\xi(s) \triangleq \mathbb{E}[\exp(-s\xi)] &= \mathbb{E}[\exp(-s\mathbf{z}^\dagger \mathbf{A}\mathbf{z})] \\
&= \det(\mathbf{I} + s\boldsymbol{\Sigma}\mathbf{A})^{-1} \\
&= \det\left[\mathbf{I}_t + s\mathbf{\Delta}\mathbf{\Delta}^\dagger/2N_0\right]^{-r} \qquad \text{(D.19)}
\end{aligned}
$$

and the result of Example D.1 applies.

References

[D.1] A. Annamalai, C. Tellambura, and V. K. Bhargava, "Efficient computation of MRC diversity performance in Nakagami fading channel with arbitrary parameters," *Electron. Lett.*, Vol. 34, No. 12, pp. 1189–1190, 11 June 1998.

[D.2] E. Biglieri, G. Caire, G. Taricco, and J. Ventura, "Simple method for evaluating error probabilities," *Electron. Lett.*, vol. 32, no. 3, pp. 191–192, February 1996.

[D.3] J. K. Cavers and P. Ho, "Analysis of the error performance of trellis-coded modulations in Rayleigh fading channels," *IEEE Trans. Commun.*, Vol. 40, No. 1, pp. 74–83, January 1992.

[D.4] J. G. Proakis, *Digital Communications*, 3rd edition. New York: Mc-Graw-Hill, 1995.

[D.5] M. K. Simon and M.-S. Alouini, *Digital Communications over Fading Channels*. New York: Wiley, 2000.

[D.6] G. Szegö, *Orthogonal Polynomials*. Providence, RI: American Mathematical Society, 1939.

[D.7] C. Tellambura, "Evaluation of the exact union bound for trellis-coded modulations over fading channels," *IEEE Trans. Commun.*, Vol. 44, No. 12, pp. 1693–1699, December 1996.

[D.8] M. Uysal and N. C. Georghiades, "Error performance analysis of space–time codes over Rayleigh fading channels," *J. Commun. Networks*, Vol. 2, No. 4, pp. 344–350, December 2000.

Notations and Acronyms

- ☞ a.s., Almost surely
- ☞ ACS, Add-compare-select
- ☞ APP, A posteriori probability
- ☞ AWGN, Additive white Gaussian noise
- ☞ BER, Bit error rate
- ☞ BICM, Bit-interleaved coded modulation
- ☞ BSC, Binary Symmetric Channel
- ☞ cdf, Cumulative distribution function
- ☞ CSI, Channel state information
- ☞ FER, Frame-error rate
- ☞ GSM, Global system for mobile communications (a digital cellular telephony standard)
- ☞ GU, Geometrically uniform
- ☞ iid, Independent and identically distributed
- ☞ IS-136, An American digital cellular telephony standard
- ☞ LDPC, Low-density parity-check
- ☞ LLR, Log-likelihood ratio
- ☞ ln, Natural logarithm
- ☞ log, Logarithm in base 2
- ☞ MD, Maximum-distance
- ☞ ML, Maximum-likelihood
- ☞ MGF, Moment-generating function
- ☞ MIMO, Multiple-input, multiple-output
- ☞ MMSE, Minimum-mean-square error
- ☞ MPEG, A standard algorithm for coding of moving pictures and associated audio
- ☞ MRC, Maximum-ratio combining
- ☞ MSE, Mean-square error
- ☞ pdf, Probability density function
- ☞ PEP, Pairwise error probability
- ☞ PN, Pseudonoise
- ☞ ROC, Region of convergence

- ☞ RV, Random variable
- ☞ RX, Reception
- ☞ SIMO, Single-input, multiple-output
- ☞ SISO, Soft-input, soft-output
- ☞ SNR, Signal-to-noise ratio
- ☞ SPA, Sum–product algorithm
- ☞ TCM, Trellis-coded modulation
- ☞ TX, Transmission
- ☞ UMTS, A third-generation digital cellular telecommunication standard
- ☞ VA, Viterbi algorithm
- ☞ $\sum_{\sim x_i} f(x_1, \ldots, x_n)$, Sum with respect of all variables except x_i
- ☞ a^*, Conjugate of complex number a
- ☞ $(a)_+ \triangleq \max(0, a)$, Equal to a if $a > 0$, equal to 0 otherwise.
- ☞ $\mathbf{A}^+$, (Moore-Penrose) pseudoinverse of matrix $\mathbf{A}$
- ☞ $\mathbf{A}'$, Transpose of matrix $\mathbf{A}$
- ☞ $\mathbf{A}^\dagger$, Conjugate (or Hermitian) transpose of matrix $\mathbf{A}$
- ☞ $\|\mathbf{A}\|$, Frobenius norm of matrix $\mathbf{A}$
- ☞ $\mathbb{C}$, The set of complex numbers
- ☞ d_E, Euclidean distance
- ☞ d_H, Hamming distance
- ☞ $\deg g(D)$, Degree of polynomial $g(D)$
- ☞ δ_{ij}, Kronecker symbol ($\delta_{ij} = 1$ if $i = j$, $= 0$ otherwise)
- ☞ $\mathbb{E}[X]$, Expectation of the random variable X
- ☞ $\mathrm{etr}(\cdot) \triangleq \exp(\mathrm{Tr}(\cdot))$
- ☞ $\mathbb{F}_2$, The binary field $\{0, 1\}$ equipped with modulo-2 sum and product.
- ☞ γ, Asymptotic power efficiency of a signal constellation
- ☞ η, Asymptotic coding gain
- ☞ $\mathbf{I}_n$, The $n \times n$ identity matrix

☞ $\Im$, Imaginary part

☞ log, Logarithm to base 2

☞ $Q(x) \triangleq (2\pi)^{-1/2} \int_x^\infty \exp(-z^2/2)\,dz$, The Gaussian tail function

☞ R_b, Transmission rate, in bit/s

☞ $\Re$, Real part

☞ $\mathbb{R}$, The set of real numbers

☞ $\mathbb{R}_+$, The set of nonnegative real numbers

☞ $\mathbb{V}[X]$, Variance of the random variable X

☞ W, Shannon bandwidth

☞ w_H, Hamming weight

☞ $\triangleq$, Equal by definition

☞ $X \sim \mathcal{N}(\mu, \sigma^2)$, X is a real Gaussian RV with mean μ and variance σ^2

☞ $X \sim \mathcal{N}_c(\mu, \sigma^2)$, X is a circularly distributed complex Gaussian RV with mean μ and $\mathbb{E}[|X|^2] = \sigma^2$

☞ $X \perp\!\!\!\perp Y$, The RVs X and Y are statistically independent

☞ $a \propto b$, a is proportional to b

☞ $\Gamma(x) \triangleq \int_0^\infty u^{x-1}e^{-u}\,du$, The Gamma function

☞ ρ, Transmission rate, in bit/dimension

☞ $\mathrm{Tr}\,(\mathbf{A})$, Trace of matrix $\mathbf{A}$

☞ $\mathrm{vec}(\mathbf{A})$, The column vector obtained by stacking the columns of $\mathbf{A}$ on top of each other

☞ $[\mathcal{A}]$, Equal to 1 if proposition $\mathcal{A}$ is true, equal to 0 if it is false

☞ $\mathcal{A} \setminus a$, The set $\mathcal{A}$ without its element a

Index